Transforms in
Signals and Systems

MODERN APPLICATIONS OF MATHEMATICS

Consulting Editors   **Glyn James**
Coventry University

Richard Clements
University of Bristol

OTHER TITLES IN THE SERIES

Engineering Mathematics: A Modern Foundation for Electronic, Electrical and Control Engineers   *A. Croft, R. Davison and M. Hargreaves*

Modern Engineering Mathematics   *Glyn James*

Transforms in Signals and Systems

Peter Kraniauskas

ADDISON-WESLEY PUBLISHING COMPANY

Wokingham, England • Reading, Massachusetts • Menlo Park, California • New York
Don Mills, Ontario • Amsterdam • Bonn • Sydney • Singapore
Tokyo • Madrid • San Juan • Milan • Paris • Mexico City • Seoul • Taipei

Cover designed by Hybert Design and Type, Maidenhead
and printed by The Riverside Printing Co. (Reading) Ltd.
Illustrations by Chartwell Ltd.
Typeset by Keytec Typesetting Ltd, Bridport, Dorset.
Transferred to digital print on demand, 2002
Printed and bound by Antony Rowe Ltd, Eastbourne

British Library Cataloguing in Publication Data

Kraniauskas, Peter
    Transforms in signals and systems.
    1. Systems. Signals. Processing
    I. Title
    003

    ISBN 0-201-19694-8

Library of Congress Cataloging-in-Publication Data

Kraniauskas, Peter.
    Transforms in signals and systems / Peter Kraniauskas.
            p.   cm.
    Includes bibliographical references and index.
    ISBN 0-201-19694-8 :
    1. Signal processing—Mathematics. 2. Transformations
    (Mathematics) 3. System analysis. I. Title.
TK5102.5.K665   1991
621.382′2—dc20   91-11701
                                                                           CIP

Contents

Chapter 7
Discrete-time Laplace Transform and *z*-transform

Chapter 8
System Description   354

Chapter 9
System Solution   382

PART 3

System Synthesis   405

Chapter 10
Ideal and Realizable Continuous-time Systems   407

Chapter 11
Discrete-time Systems and System Simulation   491

Preface

Discrete-time systems and digital signal processing are appearing in all branches of science and engineering. Despite predictions that they would totally displace continuous-time systems these are still flourishing. For the foreseeable future the two approaches will remain complementary, while boundaries between areas of application continue to shift. This text attempts to reconcile the two system classes and to unify their treatment.

The book is aimed primarily at the undergraduate engineering student, who requires a sound grounding in signals and systems in preparation for subsequent courses in communication systems, control systems and other advanced applications. It is also suitable as an introductory signals and systems course at postgraduate level and the material has been presented very successfully in high-tech industry as an introduction to discrete-time systems. It is assumed that the reader is familiar with calculus and differential equations at undergraduate level and had exposure to a course on electrical networks or mechanical systems.

The formulation of many aspects of continuous-time systems has matured enough for the accumulated knowledge to be consolidated on sound analytic foundations. In the 1930s, electrical networks had such foundations laid when the duality of time domain and frequency domain (spectral) representations was formulated in terms of the Laplace transform and its subset, the Fourier transform. Similar foundations exist for mechanical vibrations.

Discrete-time systems, being more recent, had less time to mature. They too have their foundations set on time domain and frequency domain concepts, now related by the z-transform and its subset, the discrete-time Fourier transform. But they developed via numerical analysis, statistics and other more esoteric realms of mathematics, a very different path from continuous-time systems.

Separate developments led to separate notation, terminology and expressions, which obscure the unity of underlying concepts and make it difficult to distinguish fundamental concepts that are valid for all systems from those specific to a class or a particular application.

Recent books on signals and systems recognize the need for unifying the hitherto fragmented presentation. This book makes it a priority. It exploits the key roles played by the **family of time domain to frequency domain transforms** in representing continuous-time and discrete-time signals and systems, in solving such systems and, most importantly, in providing a framework for linking the two system classes.

To this end the transforms are derived and related as a united family headed by the all-embracing Laplace transform, the others becoming specialized subsets. Comprehensive graphical aids are developed to convey the underlying mathematical concepts and to link the transforms. These take various forms, as graphical interpretations of the transforms, conceptual diagrams and functional interrelation charts. The graphical aids are also extended to derive a coherent set of solution methods for systems of both classes.

The book is divided into three parts covering the analysis of continuous-time signals and systems, the analysis of discrete-time signals and systems, and the synthesis of continuous-time and of discrete-time systems. It is structured to progressively build up and culminate in the synthesis of a discrete-time system by simulating a previously synthesized continuous-time system.

A preliminary chapter employs the Fourier series to introduce the concept of linking time domain and frequency domain representations by transforms, at the same time introducing functions of complex variables and developing the basic graphics for transform representations.

Part 1 is devoted to the analysis of continuous-time signals and systems. The first half covers signal analysis, in the sense of relating their time and frequency representations by the Fourier transform in Chapter 2 and by the more general Laplace transform in Chapter 3. These transforms are employed in the second half of Part 1 to analyse the corresponding systems. In Chapter 4 systems are characterized in both domains in terms of responses to elementary signals, and these responses are used in Chapter 5 to solve systems for arbitrary input signals.

Part 2 extends the concepts and results of Part 1 to the analysis of discrete-time signals and systems. In the first half, Chapter 6 examines the discrete forms of the Fourier transform and Chapter 7 derives the discrete-time Laplace transform and the z-transform. The second half presents the description and solution of discrete-time systems, echoing the second half of Part 1.

The two chapters of **Part 3** present and unify key aspects of system synthesis. Chapter 10 first examines some ideal systems and limitations of realizable systems. It then introduces the methodology of classical filter approximations and frequency mapping concepts involved in the design of continuous-time systems. Chapter 11 deals with the synthesis of discrete-time systems and with system simulation. It particularly stresses the overall time domain and frequency domain considerations involved in synthesizing a discrete-time system by simulating a related continuous-time system, as expressed in the chart of Figure 11.27.

Although the book is organized as a self-contained one-year course, it can be subdivided in two ways into shorter courses. It can be split into a first course covering the continuous-time material (Chapter 1 + Part 1 + Chapter 10), followed by a subsequent course covering the discrete-time material (Part 2 + Chapter 11). Alternatively, the transforms alone (Chapters 1, 2, 3, 6 and 7) can be presented in an engineering mathematics course, leaving the rest for a subsequent signals and systems course.

The text does not contain computer programs, since well documented software packages are commercially available for the analysis and design of signal and systems, and engineering departments often develop their own. Depending on the course objectives, the instructor can guide students on whether to use a standard package or write their own. In the latter case they should be encouraged to make full use of complex number instructions.

Acknowledgements

A book that gathers and attempts to unify and consolidate the fundamentals of a whole discipline, does not emerge from a vacuum. Of the many texts consulted, those found most influential in shaping this book are listed in the short bibliography.

Many friends and former colleagues have contributed valuable advice. I am particularly indebted to Professor Joe Hammond from the Institute of Sound and Vibration Research and to Dr David Thomas from the Department of Electronics and Computer Science, both at Southampton University, who gave generously of their time in frequent discussions.

My thanks are due to Dr Chris Ash and Dr Edward Stansfield from Racal Research Ltd for lengthy consultations and advice that make this text more relevant to high-tech industry. Also to the hundreds of engineers in the electronics industry, who attended training courses based on the material of this book, for their incisive questions, criticisms and suggestions that helped to shape it.

I am also grateful to the staff of Addison-Wesley, in particular Sarah Mallen, Tim Pitts, Stephen Bishop and Stephen Troth. Also to Professor Glyn James from Coventry University and Dr Richard Clements from Bristol University for early editorial reviews.

I would also like to thank a group of friends and ex-colleagues, Graham Assinder, Mick Balme, Neil Barnett, Alan Bond, Dave Croxford, Dave Edwards, Allen Mornington-West, who gave of their time at evening gatherings to discuss new ideas and to review portions of early manuscripts.

And, not least, I would like to thank Val, my wife, for her continuous help, encouragement and patience over the years in which this book was taking shape.

CHAPTER 1

Introduction to Signals, Systems and Transforms

This chapter introduces the functions, signals and systems considered in the book, thereby setting scope and nomenclature. The Fourier series is then used as a vehicle for introducing complex number notation and to develop a basis for the graphical representation of all subsequent transforms.

1.1 Signals and systems view of physical phenomena

The scientific method endeavours to explain complex physical phenomena in terms of cause and effect. This involves constructing a simplified physical or mathematical model, called a **system**. Stimulated by a suitable **excitation** or **input signal**, representing the cause, the system modifies it to produce a **response** or **output signal**, representing the effect of the physical phenomenon.

Phenomena and the associated signals are diverse, but the discipline of **signals and systems** brings together and unifies, at a more abstract level, concepts from network analysis, electric circuit theory, dynamic systems, optimal systems and numerical analysis, among many others. Such cross-fertilization brings better understanding of individual disciplines and gives deeper insight into concepts that are fundamental to all.

In most cases signals are temporal, in that they are expressed as functions of the form $f(t)$, where t represents real time. Spatial signals, such as the luminous intensity $i(x)$ along a line of an optical image, are

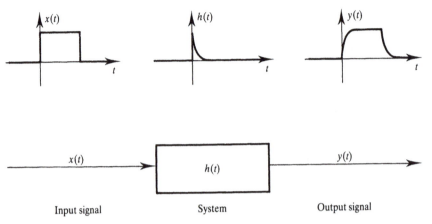

Figure 1.1   Typical system with input and output signals.

expressed in terms of linear position x. Mathematically the two types lead to formally similar representations. We restrict presentation to temporal systems, such as that indicated by the function $h(t)$ in Figure 1.1, which modifies an input signal $x(t)$ to give an output signal $y(t)$.

1.1.1   Time and frequency representations of signals

A signal can be viewed from two alternative standpoints. The most intuitive is the **time domain** approach, typically represented by the trace of an oscilloscope, whose vertical beam deflection represents the signal's amplitude $f(t)$ displayed against the horizontal deflection, which represents the independent time variable t.

The second representation is in the **frequency domain**, typified by the trace of a spectrum analyser, where the signal $F(\omega)$ is displayed against the independent frequency variable ω.

A given signal can be fully described in either of these two domains, both representations being equivalent. The equivalence is typically expressed by the Fourier transform $\mathcal{F}$ of the signal, which permits a free conversion from one form to the other, as symbolized by the expression

$$f(t) \quad \overset{\mathcal{F}}{\longleftrightarrow} \quad F(\omega)$$

A more general frequency representation is found in the Laplace domain, as expressed by the Laplace transform $F(s)$ of the function $f(t)$

$$f(t) \quad \overset{\mathcal{L}}{\longleftrightarrow} \quad F(s)$$

where s represents the generalized frequency variable and $\mathscr{L}$ the Laplace transform operator.

Depending on the process, one domain tends to offer simpler operations and can be more easily visualized than the other. Often a knowledge of both domains is necessary to understand the process fully. Transforms provide the tools to alternate between the domains. Spatial signals have corresponding spatial frequencies.

1.1.2 Signal classes

One time domain classification divides signals according to whether they are represented by an **aperiodic** function $f(t)$ or a **periodic** function $\widetilde{f}(t)$. For analytical purposes a process of short duration, such as a decaying transient, can be interpreted as a signal extending to infinite time, with negligible amplitude beyond a certain time value. In contrast, a signal that repeats itself with period T_0 over a long time interval, can often be considered, without significant error, to remain periodic to infinity.

Another classification refers to continuity or discreteness of the time variable. A **continuous-time** signal has a value $f(t)$, which can be zero, associated with every instant t. In contrast, a **discrete-time** signal only exists at discrete points $t = nT$ of real time, where T is the interval between adjacent sampling points, and has values associated with those points only. These two classifications are combined in Figure 1.2, whose quadrants show typical examples of the resulting four signal classes.

A formally identical classification applies to the frequency representation of the signal. It too yields four classes in terms of discreteness of the frequency variable ω and of the periodicity of $F(\omega)$. Anticipating conclusions from Chapter 6, a duality between the discrete forms of the Fourier transform places the four classes of each classification in a one-to-one relationship to each other, as indicated by a second set of labels, shown in brackets in Figure 1.2. We exploit this remarkable duality to establish a framework that unifies the treatment of all the transforms, signals and systems covered in this book.

A further classification would apply in the context of **digital signal processing**, where also **signal levels are allowed only discrete values**, to match finite hardware precision. This falls outside the scope of this book.

1.1.3 System descriptions and solutions

A system reveals its existence by modifying signals. It can therefore be described, or characterized, in terms of its effects on signals. We briefly

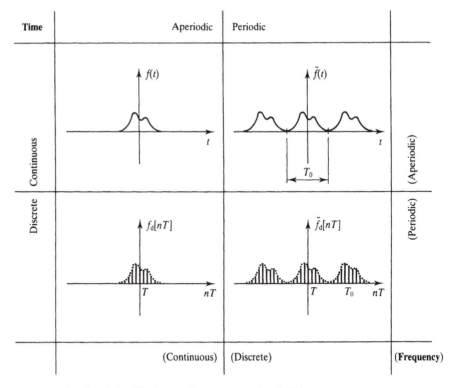

Figure 1.2 Signal classification by discreteness and periodicity.

outline what forms such descriptions take for continuous-time and discrete-time systems, how they relate to the applicable solution methods and what roles transforms play in both description and solution.

Continuous-time systems

The most immediate analytical description of a continuous-time system is a differential equation, typically of the form

$$a_2 y''(t) + a_1 y'(t) + a_0 y(t) = b_2 x''(t) + b_1 x'(t) + b_0 x(t)$$

taken from Chapter 4, which relates an arbitrary input signal $x(t)$ to the resulting output signal $y(t)$. The constant coefficients a_i and b_i are parameters associated with the system elements, which fully identify a given system. Such a differential equation is written explicitly in terms of the input signal. Integration **solves the system**, that is, yields the output $y(t)$ for the particular input $x(t)$.

If the system is to be solved repeatedly, for different input signals, it is desirable to characterize it in a form that is input independent. The system's impulse response $h(t)$ is typical. It represents the system's output when the input is the impulse $\delta(t)$ of Section 1.1.5. We will show in Chapter 5 that, when interpreting the arbitrary input $x(t)$ as a superposition of such impulses, the output $y(t)$ can be formulated as a **convolution of** $x(t)$ **with** $h(t)$, expressed by the integral

$$y(t) = \int_{-\infty}^{\infty} x(t - \tau) h(\tau) \, d\tau$$

to be introduced in Section 1.1.4. This solution method still requires an integration to be performed.

In contrast, replacing both the input signal $x(t)$ and the impulse response $h(t)$ by their frequency domain counterparts $X(s)$ and $H(s)$ by taking their Laplace transforms reduces the solution process to a mere algebraic multiplication,

$$Y(s) = X(s) H(s)$$

An inverse Laplace transformation then yields the desired time domain output $y(t)$. This is one of the cornerstones of transform methods.

Discrete-time systems

Formally similar concepts and descriptions apply to discrete-time systems. The immediate time domain description of the process now becomes a **difference equation**, typically of the form

$$a_2 y[nT - 2T] + a_1 y[nT - T] + a_0 y[nT]$$
$$= b_2 x[nT - 2T] + b_1 x[nT - T] + b_0 x[nT]$$

as found in Chapter 8. The parameters a_i b_i are again constant coefficients that identify the particular system, and functions of the form $x[nT - mT]$ indicate a version of $x[nT]$ delayed by m sampling intervals T.

The system is also characterized by its discrete-time impulse response $h[nT]$, conceptually similar to its continuous-time counterpart, obtained when the system's input is the discrete-time impulse $\delta[nT]$, also called the unit pulse. The associated solution method is formulated as a **discrete-time convolution of** $x[nT]$ **with** $h[nT]$, expressed by the sum

$$y[nT] = \sum_{m=-\infty}^{\infty} x[nT - mT] h[mT]$$

The z-transform of Chapter 7 yields the frequency domain counterparts $X(z)$ and $H(z)$ of the input signal and impulse response. It too has the property of reducing the infinite convolution sum to a frequency domain multiplication,

$$Y(z) = X(z)H(z)$$

thereby simplifying the solution process. Discrete-time transforms have a formally similar role here to that which continuous transforms have in continuous-time systems.

Links between continuous and discrete time

Most physical phenomena are essentially of continuous time, and this is reflected in the classical models used to represent them. To implement parts of a process in discrete time, signals are sampled and the process is approximated by a discrete-time system. This leads to the above-mentioned similarities, which we pursue and exploit in this book.

The link between continuous-time and discrete-time signals is established in the first half of Part 2, in the context of discrete-time transforms. Based on time domain and frequency domain sampling, it reaches a climax at the end of Chapter 7 where the transforms applicable to all signal classes are brought together on the framework offered by Figure 1.2.

Similarities of system characterizations are most revealing in the frequency domain, in terms of Laplace transform and z-transform representations. These are exploited in the second half of Part 2, where the associated solution methods owe their similarity to corresponding properties of those transforms, regarding convolution in time and multiplication in frequency.

But the main link between system classes is found in Chapter 11 where discrete-time system synthesis is formulated in terms of simulating a related continuous-time system. The underlying unity of the time-to-frequency transform family, headed by the Laplace transform, reveals a similar unity of concepts and methods that applies to systems from both classes.

1.1.4 Convolution

Convolution is an operation involving two functions, defined for continuous-time signals by an integral. For discrete-time signals it has a counterpart in a formally similar sum. We now introduce it by its analytical definition and give an appropriate graphical interpretation. As already mentioned, it plays a crucial role in the solution of systems and will be re-examined in that context in Chapters 5 and 9.

Continuous-time

Given two functions, or signals, $f(t)$ and $h(t)$, we form the integral

$$g(t) = \int_{-\infty}^{\infty} f(\tau)h(t - \tau)\,d\tau \tag{1.1}$$

whose result $g(t)$ is called the **convolution of $f(t)$ and $h(t)$**, expressed more concisely with the symbolic notation

$$g(t) = f(t) * h(t)$$

For example, in system analysis, the system's output $y(t)$ is the result of convolving the input $x(t)$ with the impulse response $h(t)$,

$$y(t) = x(t) * h(t)$$

Graphical interpretation

It is customary to interpret the integral 1.1 in the form shown in Figure 1.3. One of the functions, for example $f(t)$, is given the dummy integration variable τ, as shown at the top of the figure. The other is given the same dummy variable, indicated by $h(\tau)$ in the figure, and folded about the vertical axis, yielding $h(-\tau)$, a mirror image of $h(\tau)$. The folded function is then shifted to the time value $\tau = t$, for which the convolution integral is being evaluated, giving the function $h(t - \tau)$. The product of the functions thus formed, $f(\tau)h(t - \tau)$, represents the integrand of Equation 1.1, whose integral (shaded area) along the entire time axis τ yields one value of $g(t)$, as shown in Figure 1.3. Repeated for all values of t, this process builds up the resulting function $g(t)$.

The three-dimensional graphics of Figure 1.4 fully visualize the process by separating the actual time variable t from the integration variable τ. It highlights the point that every value $g(t)$ is the result of an infinite integral in the dummy variable τ.

An alternative visualization, more attuned to the **systems interpretation** of convolution, is given in Section 5.2.3.

Basic properties

We list, without immediate proof, the commutative and associative properties of convolution,

$$g(t) = f(t) * h(t) = h(t) * f(t) \tag{1.2}$$

hence

$$g(t) = \int_{-\infty}^{\infty} h(\tau)f(t - \tau)\,d\tau \tag{1.3}$$

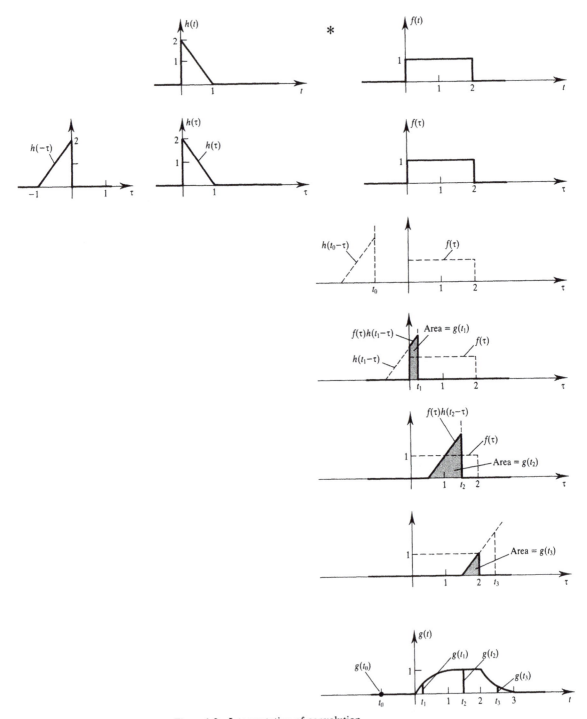

Figure 1.3 Interpretation of convolution.

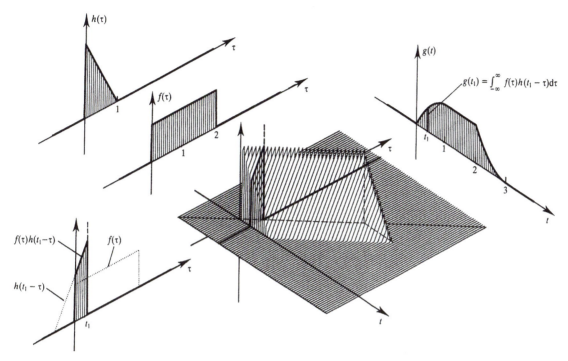

Figure 1.4   Full interpretation of convolution $g(t) = f(t) * h(t)$.

and

$$g(t) = [f(t) * h(t)] * e(t) = f(t) * [h(t) * e(t)] \qquad (1.4)$$

The justifications in terms of definition 1.1 are simple, but are even simpler using transform properties. These will be found in Section 2.4.5.

Convolution in frequency

The definition 1.1 is valid for functions in other variables. Thus, the frequency domain convolution of two functions $F(\omega)$ and $H(\omega)$ is expressed in terms of the actual frequency variable ω and the dummy integration variable v as

$$G(\omega) = \int_{-\infty}^{\infty} F(v) H(\omega - v) \, dv \qquad (1.5)$$

Discrete convolution

A formally similar definition, in terms of an infinite sum rather than an integral, applies to discrete-time or discrete-frequency functions,

$$g[nT] = f[nT] * h[nT] = \sum_{m=-\infty}^{\infty} f[mT]h[nT - mT] \tag{1.6}$$

which has a similar graphical interpretation and similar properties to the continuous form of Equation 1.1. We will explore their relationship in Chapter 11, in the context of the time-sampling method of simulation.

Causal signals

If one of the signals is causal, that is, it is zero-valued for negative time, the products $f(\tau)h(t - \tau)$ or $f[mT]h[nT - mT]$ are zero-valued outside some intervals and the integration or summation limits can be correspondingly reduced.

When both signals are causal, $f(\tau)$ is zero for negative time, $\tau < 0$, while $h(t - \tau)$ is zero for $\tau > t$, with the result

$$g(t) = \int_0^t f(\tau)h(t - \tau) \, d\tau \tag{1.7}$$

and

$$g[nT] = f[nT] * h[nT] = \sum_{m=0}^{n} f[mT]h[nT - mT] \tag{1.8}$$

Integration as convolution with unit step

When one of the functions of Equation 1.1 is the unit step $u(t)$, the integrand reduces to

$$f(\tau)u(t - \tau) = \begin{cases} f(\tau) & \tau < t \\ 0 & \tau > t \end{cases}$$

and leads to the interpretation of integration as a convolution with the unit step

$$g(t) = \int_{-\infty}^{t} f(\tau) \, d\tau = f(t) * u(t) \tag{1.9}$$

1.1.5 Impulse function

The impulse function $\delta(t)$ mentioned earlier, also called the delta function, is not a function in the usual analytical sense. For the purposes of system analysis, a **definition in terms of a convolution integral** yields all the required properties. For conceptual clarity, a simple intuitive model suffices. A rigorous mathematical formulation exists, but falls outside the requirements of this text.

Definition of impulse function

The impulse function $\delta(t)$ is defined by the integral properties of its product with an arbitrary function $f(t)$, as

$$\int_{-\infty}^{\infty} f(\tau)\delta(t - \tau)\, d\tau = f(t) \tag{1.10}$$

This is otherwise known as the **sifting property** of the impulse function, in that it sifts out the value of $f(\tau)$ at the instant $\tau = t$. An intuitive graphical development of this property is given in Section 5.2.2, where the function $f(t)$ is interpreted as **the convolution of itself with the impulse function**, $f(t) = f(t) * \delta(t)$. This, in turn, interprets the impulse function as the **identity element of convolution**, because, convolved with $f(t)$, it does not modify it.

Intuitive models

The impulse is interpreted intuitively as a pulse of infinitesimal duration, infinite height and unit area. This is expressed in the simpler form

$$\int_{-\infty}^{\infty} \delta(t)\, dt = 1$$
$$\delta(t) = 0 \quad \text{for } t \neq 0 \tag{1.11}$$

which does not specify a shape in the infinitesimal region surrounding the origin. Figure 1.5 gives two simple shapes of unit area. By a limiting process, $T \to 0$, each of these generates a valid impulse function. The actual shape of the generating function is unimportant because the impulse function is always applied in the context of an integration, so that only its integral properties are relevant.

Useful properties

The impulse is an even function, $\delta(t) = \delta(-t)$, so that Equation 1.10 can also be written

$$\int_{-\infty}^{\infty} f(\tau)\delta(\tau - t)\, d\tau = f(t)$$

while a simple change of integration variable yields the equivalent form

$$\int_{-\infty}^{\infty} f(t)\delta(\tau - t)\, dt = f(\tau)$$

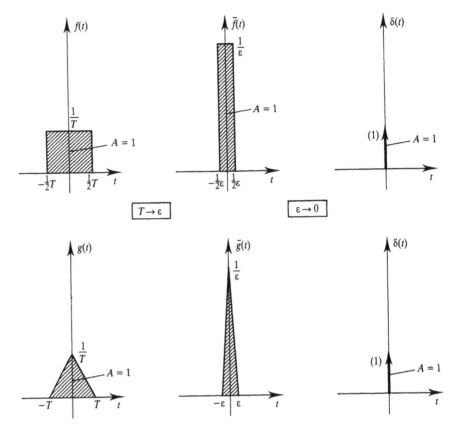

Figure 1.5 Impulse as limit of simple shape.

If $f(t)$ is continuous at the origin $t = 0$, it can be interpreted as being constant in the infinitesimal region surrounding the origin, giving rise to the equivalence of the product

$$f(t)\delta(t) = f(0)\delta(t) \tag{1.12}$$

Varying the upper integration limit of the simpler impulse definition 1.11, the integral is zero for $t < 0$ and unity for $t > 0$. This expresses the unit step function $u(t)$ as the integral of the impulse function

$$\int_{-\infty}^{t} \delta(\tau)\,d\tau = u(t) \tag{1.13}$$

Conversely, the impulse becomes the derivative of the unit step

$$\delta(t) = \frac{du(t)}{dt}$$

1.2   Fourier series as introduction to transforms

The familiar Fourier series is used here to introduce basic time domain and frequency domain concepts, the complex exponential and positive and negative frequencies, in anticipation of the more general Fourier transform in Chapter 2. A more systematic treatment of the Fourier series will be left to Chapter 6.

1.2.1   Trigonometric form

With the exception of some mathematical curiosities, any periodic signal of period T_0 can be expanded into a trigonometric series of the form

$$f(t) = a_0 + \sum_{k=1}^{\infty} (a_k \cos k\omega_0 t + b_k \sin k\omega_0 t) \tag{1.14}$$

where ω_0 is the fundamental frequency $\omega_0 = 2\pi/T_0$, and the coefficients a_0, a_k and b_k form a set of real numbers uniquely associated with the function $f(t)$. Each term

$$a_k \cos k\omega_0 t + b_k \sin k\omega_0 t$$

defines one harmonic of the function. These occur at integer multiples $k\omega_0$ of the fundamental frequency, and the term for $k = 1$ is called the fundamental. Thus, if all the frequency components a_k and b_k are known, Equation 1.14 is sufficient to synthesize the function $f(t)$.

Example 1.1

Given the fundamental frequency ω_0 and the Fourier coefficients $a_0 = 1.0$, $a_1 = 1.0$ and $b_3 = 0.5$, with all other coefficients zero, write the expression for the resulting time function $f(t)$.

As there are only three components, at $k = 0, 1, 3$, the direct application of Equation 1.14 yields

$$f(t) = 1 + \cos \omega_0 t + 0.5 \sin 3\omega_0 t$$

1.2.2   Graphical representation

The time domain and frequency domain components of the Fourier series in Equation 1.14 can be easily visualized. The individual terms $b_k \sin k\omega_0 t$ and $a_k \cos k\omega_0 t$ of the sum can be interpreted as projections

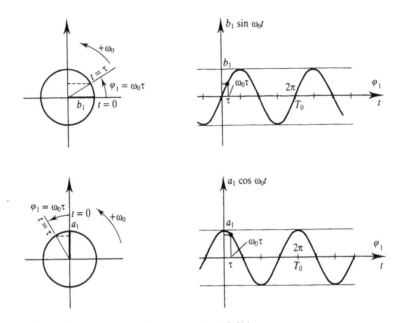

Figure 1.6   Sine and cosine components of $f(t)$.

on the vertical axis of segments of magnitude b_k and a_k, respectively, rotating with angular velocity $k\omega_0$. The case $k = 1$ is illustrated in Figure 1.6, where the initial positions of the segments (i.e. at $t = 0$) are indicated by bold segments.

To interpret the full expression 1.14, its cosine and sine components are placed as indicated in the frequency domain region of Figure 1.7, where the values of Example 1.1. are used for illustration. Two sets of frequency coordinates are provided for the coefficients a_k and b_k, oriented as shown to give the correct sine–cosine phase relationship. The constant component a_0 can be interpreted as a cosine of zero frequency, and included with the set a_k, as shown.

The frequency components are shown in their initial positions, at time zero. As time advances, each component rotates with the constant angular velocity (see arrow) determined by its position on the frequency axis. Their projections on the vertical plane, interpreted as functions of time, are shown in the foreground of the diagram. Thus, each frequency component, a_k or b_k, gives rise to a corresponding time domain component, and the sum of the latter represents the time function $f(t)$, as expressed by the Fourier series in Equation 1.14.

This establishes a graphical link, revealing the mechanism by which one representation is transformed into the other.

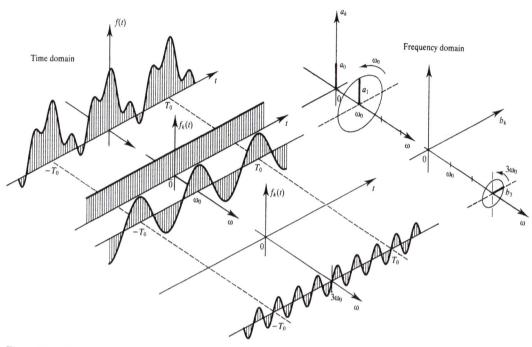

Figure 1.7   Representation of Fourier series, trigonometric form (Equation 1.14).

1.2.3   Inverse transformation

Expression 1.14 synthesizes the time domain function $f(t)$ from its frequency components, and is called the **synthesis equation of the Fourier series**. Conversely, the coefficients a_k and b_k, representing amplitudes of the associated cosine and sine waves, are obtained by multiplying $f(t)$ by similar waves, of unit magnitude, and integrating over one period T_0, i.e.

$$a_0 = \frac{1}{T_0}\int_0^{T_0} f(t)\,\mathrm{d}t$$

$$a_k = \frac{2}{T_0}\int_0^{T_0} f(t)\cos k\omega_0 t\,\mathrm{d}t \tag{1.15}$$

$$b_k = \frac{2}{T_0}\int_0^{T_0} f(t)\sin k\omega_0 t\,\mathrm{d}t$$

These are the **analysis equations of the Fourier series**, which exploit integral properties of products of sines and cosines, of the form

$$\cos k\omega_0 t \cos m\omega_0 t = \tfrac{1}{2}[\cos(k+m)\omega_0 t + \cos(k-m)\omega_0 t]$$

Since k and m are positive integers, the terms on the right are cosines having a number $n = |k \pm m|$ of complete cycles in one period T_0. The integral over such a period is zero, unless $n = 0$. Hence

$$\int_0^{T_0} \cos k\omega_0 t \cos m\omega_0 t = \begin{cases} \frac{1}{2} T_0 & (k = m) \\ 0 & (k \neq m) \end{cases}$$

Similar results apply to products of sines

$$\int_0^{T_0} \sin k\omega_0 t \sin m\omega_0 t = \begin{cases} \frac{1}{2} T_0 & (k = m) \\ 0 & (k \neq m) \end{cases}$$

whilst mixed products yield

$$\int_0^{T_0} \sin k\omega_0 t \cos m\omega_0 t = 0 \qquad \text{(all k and m)}$$

These integrals represent 'orthogonal properties' of sinusoidal functions. Collectively, the functions $\sin k\omega_0 t$ and $\cos k\omega_0 t$, with $k = 1, 2, 3, \ldots$ form an 'orthogonal set' in any interval of length T_0.

Example 1.2

We illustrate the basis of Equations 1.15 by means of the function of the preceding example,

$$f(t) = 1 + \cos 4t + 0.5 \sin 12t$$

The integration period T_0 is that of the fundamental. In this example the lowest frequency is also the fundamental, hence $\omega_0 = 4 \text{ rad/s}$ and

$$T_0 = 2\pi/\omega_0 = \tfrac{1}{2}\pi \text{ s}$$

The coefficients could be found by inspection, but our objective is to illustrate the use of the analysis equations. The first of Equations 1.15 yields

$$a_0 = \frac{1}{T_0} \int_0^{T_0} (1 + \cos \omega_0 t + \sin 3\omega_0 t) \, dt$$

whose last two terms represent integrals of one full cosine cycle and three full sine cycles, which vanish. The integral of the first term alone gives the value of the constant coefficient $a_0 = 1$.

Evaluated for $k = 1$, the second of Equations 1.15 gives

$$a_1 = \frac{2}{T_0} \int_0^{T_0} (\cos \omega_0 t + \cos^2 \omega_0 t + 0.5 \sin 3\omega_0 t \cos \omega_0 t) \, dt$$

With the exception of the second term, the integrals of all terms vanish, so that $a_1 = 1$.

The third of Equations 1.15, evaluated for $k = 1$, gives

$$b_1 = \frac{2}{T_0}\int_0^{T_0} (\sin \omega_0 t + \cos \omega_0 t \sin \omega_0 t + 0.5 \sin 3\omega_0 t \sin \omega_0 t)\, dt = 0$$

the integral of each term being zero. Proceeding similarly, it would be found that all other coefficients, except b_3 also vanish. The latter reduces to

$$b_3 = \frac{2}{T_0}\int_0^{T_0} 0.5 \sin^2 3\omega_0 t\, dt = 0.5$$

1.2.4   Exponential form

The trigonometric form of the Fourier series requires the specification of two independent quantities for each frequency component. A harmonic of index k is thus fully determined by the coefficients a_k and b_k:

$$f_k(t) = a_k \cos k\omega_0 t + b_k \sin k\omega_0 t \tag{1.16}$$

Alternative forms of the Fourier series also require the specification of two quantities for each frequency component. Complex number notation, and the use of complex exponentials, leads to a particularly compact form.

Expressing the sine and cosine of Equation 1.16 in terms of Euler's formulas (see Section 1.3.3), we have

$$f_k(t) = \frac{a_k}{2}(e^{jk\omega_0 t} + e^{-jk\omega_0 t}) + \frac{b_k}{2j}(e^{jk\omega_0 t} - e^{-jk\omega_0 t})$$

$$= \tfrac{1}{2}(a_k - jb_k)e^{jk\omega_0 t} + \tfrac{1}{2}(a_k + jb_k)e^{-jk\omega_0 t}$$

The coefficients associated with the exponentials of positive and negative frequency are identified as

$$c_k = \tfrac{1}{2}(a_k - jb_k) \qquad c_{-k} = \tfrac{1}{2}(a_k + jb_k) \quad (k > 0) \tag{1.17}$$

and assigning to the coefficient of index $k = 0$ the value $c_0 = a_0$, the Fourier series in Equation 1.14 can be rewritten as

$$f(t) = c_0 + \sum_{k=1}^{\infty} \{c_k e^{jk\omega_0 t} + c_{-k} e^{j(-k)\omega_0 t}\}$$

Noting that the sum of the last term does not change when the signs of the summation index $-k$ and its limits are changed, i.e.

$$\sum_{k=1}^{\infty} c_{-k}e^{j(-k)\omega_0 t} = \sum_{k=-1}^{-\infty} c_k e^{jk\omega_0 t}$$

and writing c_0 in a formally consistent way

$$c_0 = c_0 e^{j0}$$

leads to the compact expression

$$f(t) = \sum_{k=-\infty}^{\infty} c_k e^{jk\omega_0 t} \tag{1.18}$$

This is the **complex exponential form of the Fourier series synthesis equation**. By introducing the expressions 1.15 into the relationships 1.17, and using the Euler formulas

$$\cos x + j \sin x = e^{jx} \quad \text{and} \quad \cos x - j \sin x = e^{-jx}$$

it is easy to show that the three analysis equations (1.15) find their equivalent in the single **Fourier series analysis equation**

$$c_k = \frac{1}{T_0} \int_0^{T_0} f(t) e^{-jk\omega_0 t} \, dt \tag{1.19}$$

The expression 1.18 **synthesises** the time function $f(t)$ from known coefficients c_k. Conversely, Equation 1.19 **analyses** the known time function $f(t)$ for frequency coefficients c_k. Together they provide the bidirectional link between the two domains, expressed by the **Fourier series pair**

$$f(t) \overset{\text{FS}}{\longleftrightarrow} c_k$$

where FS indicates the Fourier series operator.

In deriving the above expressions we made use of complex number notation, complex exponentials and the concept of positive and negative frequencies. To proceed further we need to familiarize ourselves with these concepts.

1.3 Complex variables

Complex number notation is a convenient mathematical device for representing with one symbol certain quantities requiring two independ-

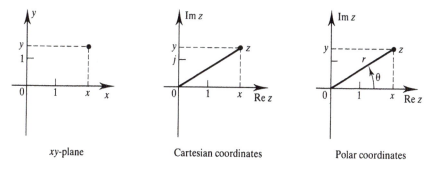

Figure 1.8   Geometrical interpretations of complex numbers.

ent values for their specification. In this section we will briefly review some of its properties, mainly to establish notation and to set the basis for three-dimensional geometrical representation of transforms.

A complex number z is written as a sum

$$z = x + jy = \text{Re}\, z + j\,\text{Im}\, z \tag{1.20}$$

where $x = \text{Re}\, z$ is the real part of z and $y = \text{Im}\, z$ is the imaginary part. Both x and y are real numbers and j is the imaginary unit.

The above definition leads to a natural one-to-one correspondence between a complex number z and a point (x, y) of the Cartesian xy-plane. The x-axis becomes the **real axis** and the y-axis the **imaginary axis** of the **complex plane** or z-plane, as indicated in Figure 1.8.

This Cartesian representation is taken further, by associating the complex number z with the directed line segment, or vector, from the origin O to the point representing z. The geometrical properties of vectors can thus be extended to represent complex numbers and their operations.

The complex number z can also be expressed in polar form and in exponential form. If r and θ are the polar coordinates of the point (x, y), see Figure 1.8, such that

$$x = r \cos \theta \qquad \text{and} \qquad y = r \sin \theta$$

the sum (1.20) takes the **polar form**

$$z = r(\cos \theta + j \sin \theta) \tag{1.21}$$

Using Euler's formula this can be written in the compact **exponential form**

$$z = re^{j\theta} \tag{1.22}$$

The real number r represents the **modulus** or **magnitude** of z and the angle θ represents its **argument**. These are related to the real and imaginary components of z by

$$r = |z| = \sqrt{x^2 + y^2} \quad \text{and} \quad \theta = \arg z = \arctan \frac{y}{x}$$

1.3.1 Functions of a complex variable

The definition of a complex function is formally similar to that of a real function $y = f(x)$. It is a rule that assigns to every point of the complex plane z a complex value $f(z)$. In general, such a function has a real part $\operatorname{Re} f(z)$ and an imaginary part $\operatorname{Im} f(z)$ associated with the independent variable z, expressed as

$$f(z) = \operatorname{Re} f(z) + j \operatorname{Im} f(z) \tag{1.23}$$

In signals and systems the variable z is found in one of three forms. It can be either a purely real variable, as in the case of time t, or a purely imaginary variable, when representing the angular frequency $j\omega$, or a complex variable, as in the case of the generalized frequency $s = \sigma + j\omega$.

One of the simplest complex functions, and arguably the most important in the interpretation of signals, is the **complex exponential** $e^{j\omega_0 t}$ encountered in the Fourier series expressions (1.18) and (1.19). If ω_0 is treated as a parameter, representing a constant angular frequency, the complex exponential becomes a function of time

$$f(t) = e^{j\omega_0 t}$$

This **complex function** is formally identical to the **complex number notation** (1.22). For each time value t, it defines a complex number of magnitude $r = 1$ and argument $\theta = \omega_0 t$, so that it can also be written in the form (1.21) as

$$f(t) = e^{j\omega_0 t} = \cos \omega_0 t + j \sin \omega_0 t \tag{1.24}$$

whose real and imaginary parts are

$$\operatorname{Re} f(t) = \cos \omega_0 t \quad \text{and} \quad \operatorname{Im} f(t) = \sin \omega_0 t$$

1.3.2 Graphical representation

The sine and cosine representations of Figure 1.6 are combined at the top left of Figure 1.9, with the cosine rotated by 90 degrees. As the vector OP rotates at constant angular velocity $+\omega_0$, whose positive direction is defined from the positive real axis towards the positive imaginary axis, it sweeps the angle $\theta = \omega_0 t$, as shown. The vector's projections on the real and imaginary axes are respectively cosine and

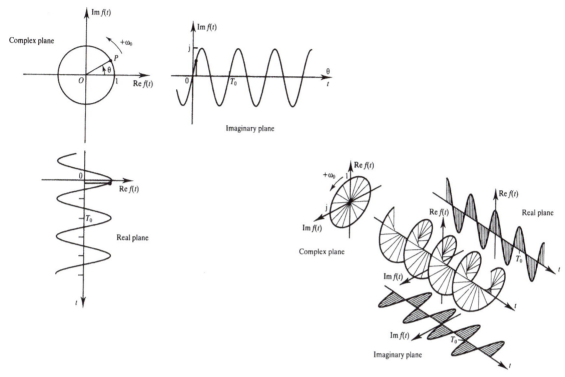

Figure 1.9   Representations of the complex exponential $f(t) = e^{j\omega_0 t}$.

sine functions of time, and represent the real and imaginary parts of the complex exponential function (1.24).

Placing the three planes thus defined orthogonally to each other, with the real plane vertical, as shown in the lower right of Figure 1.9, and adding corresponding values of the time functions according to Equation 1.23, leads to a full three-dimensional representation of the complex exponential. This is a right-handed helix of unit magnitude, a periodic function of time that repeats itself with each full turn of the vector, $\theta = \omega_0 t + 2\pi n$, as seen from

$$e^{j\omega_0 t} = e^{j(\omega_0 t + 2\pi n)}$$

In terms of the time variable, the corresponding period is $T_0 = 2\pi/\omega_0$, so that

$$f(t) = f(t + nT_0)$$

It can be verified that the projections of $f(t)$ on the three orthogonal planes represent the real and imaginary components of Equation 1.24 as related in the complex plane by Equation 1.23.

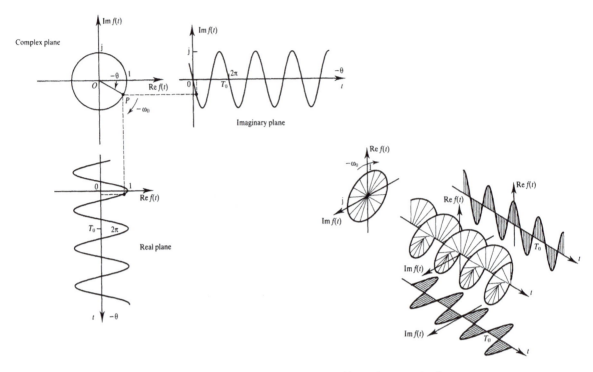

Figure 1.10 Complex exponential, $f(t) = e^{-j\omega_0 t}$, negative frequency.

Negative frequency

Had the vector OP of Figure 1.9 been rotating in the opposite direction, that is, with negative angular velocity $-\omega_0$, where ω_0 is a positive real number, it would simply sweep an angle θ of progressively larger negative value. Its projections on the real and imaginary axes would then be those shown in the upper left of Figure 1.10.

This is expressed analytically by substituting $-\omega_0$ for ω_0 in Equation 1.24, as

$$g(t) = e^{j(-\omega_0)t} = \cos(-\omega_0)t + j\sin(-\omega_0)t = \cos\omega_0 t - j\sin\omega_0 t \quad (1.25)$$

whose real part is identical to that of Equation 1.24, while the two imaginary parts are of opposite sign, that is, in anti-phase.

The full graphical representation of this function is the **left-handed** helix of the lower right of Figure 1.10.

1.3.3   Basic operations with complex variables

The Cartesian notation of Equation 1.20 is ideally suited for adding and subtracting complex numbers, as it involves only the separate addition or subtraction of the corresponding real and imaginary parts. Thus, addition of the functions

$$f(t) = \text{Re} f(t) + \text{j} \text{Im} f(t) \quad \text{and} \quad g(t) = \text{Re} g(t) + \text{j} \text{Im} g(t)$$

produces the function

$$h(t) = f(t) + g(t) = [\text{Re} f(t) + \text{Re} g(t)] + \text{j}[\text{Im} f(t) + \text{Im} g(t)]$$

The exponentials 1.24 and 1.25 of this section provide two useful examples. Their sum gives a purely real function, while their difference results in a pure imaginary function, namely the **Euler expressions**

$$e^{\text{j}\omega_0 t} + e^{-\text{j}\omega_0 t} = 2\cos \omega_0 t$$

$$e^{\text{j}\omega_0 t} - e^{-\text{j}\omega_0 t} = 2\text{j}\sin \omega_0 t \tag{1.26}$$

In the complex plane these operations correspond to additions of counter-rotating vectors, as shown in Figure 1.11. In the case of the sum, the upper diagram shows that the two imaginary parts are opposed and cancel, thus leaving only real parts, which double. The case of subtraction, shown in the lower half of Figure 1.11 as an addition of the diametrally opposed vector, cancels the real components and doubles the imaginary parts.

This interpretation is easily extended to the three-dimensional representations of Figures 1.9 and 1.10. While a direct addition of the helices would call for a vivid imagination, there are no such difficulties in separately adding or subtracting the real planes and the imaginary planes. The results simply confirm expressions 1.26 and the earlier interpretations of Figure 1.11. But they show clearly that the result of subtraction lies in the imaginary plane.

Complex conjugation

Two complex numbers z and z^* are said to be **complex conjugates** when they have the same real parts and opposite imaginary parts. Thus the numbers

$$z = x + \text{j}y \quad \text{and} \quad z^* = x - \text{j}y$$

form a conjugate pair. Inherent in this definition is the property that the sum of such a pair is real and the difference is imaginary,

$$z + z^* = 2x \quad \text{and} \quad z - z^* = 2\text{j}y$$

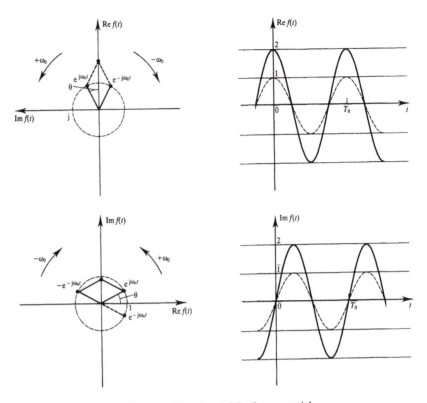

Figure 1.11   Sum and difference (Equation 1.26) of exponentials.

Similarly, two complex functions are said to form a conjugate pair, when for every value of the independent variable the corresponding complex values are conjugate. The above complex exponentials with positive and negative frequency provide an example of such a pair.

Conversely, a function that is a sum of two complex functions can only be real valued if the latter form a complex conjugate pair. This is an essential requirement of real functions of time.

If at the origin of time, $t = 0$, the argument θ of the rotating vector has an initial value θ_0, we write

$$z = re^{j(\omega_0 t + \theta_0)}$$

If we wish the resultant of two such counter-rotating vectors to be real, then, not only need their angular velocities be equal and opposite, but their initial phases must also be equal and opposite, i.e.

$$z^* = re^{-j(\omega_0 t + \theta_0)}$$

Multiplication

Division

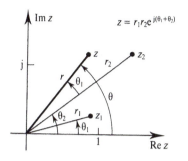

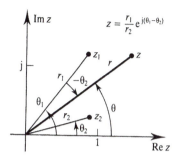

$z = r_1 r_2 e^{j(\theta_1 + \theta_2)}$

$z = \dfrac{r_1}{r_2} e^{j(\theta_1 - \theta_2)}$

Figure 1.12 Multiplication and division of complex numbers.

Multiplication and division

While Cartesian notation is ideal for addition and subtraction, it becomes cumbersome for other operations, where the exponential form 1.22 provides compact notation and simple geometric interpretations.

The rules of multiplication are formally similar to those of real exponentials. Thus, the product of two complex numbers

$$z_1 = r_1 e^{j\theta_1} \quad \text{and} \quad z_2 = r_2 e^{j\theta_2}$$

is another complex number z

$$z = z_1 z_2 = r_1 r_2 e^{j(\theta_1 + \theta_2)} = r e^{j\theta} \tag{1.27}$$

of modulus $r = r_1 r_2$ and argument $\theta = \theta_1 + \theta_2$. Similarly, division of two complex numbers yields a complex number

$$z = \frac{z_1}{z_2} = \frac{r_1}{r_2} e^{j(\theta_1 - \theta_2)} \tag{1.28}$$

of modulus $r = r_1/r_2$ and argument $\theta = \theta_1 - \theta_2$. These results have the graphical interpretations of Figure 1.12.

Multiplication by the imaginary unit j forms an important special case. Interpreted in exponential form, the imaginary unit is a complex number of modulus $r = 1$ and argument $\theta = \frac{1}{2}\pi$, i.e.

$$j = 1 e^{j\pi/2}$$

Applying the rules of Equations 1.27 and 1.28, it follows that multiplication of a complex number by j is equivalent to a pure rotation $+\frac{1}{2}\pi$, while division by j represents a pure rotation $-\frac{1}{2}\pi$, as shown in Figure 1.13. Clearly, division by j is equivalent to multiplication by $-j$.

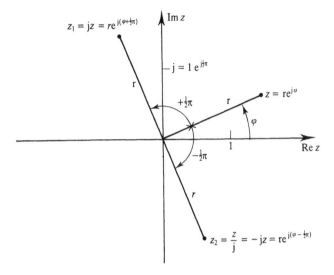

Figure 1.13   Multiplication and division by j.

To multiply a complex function by a complex constant involves multiplication of each point of the function by the constant. For example, multiplying the exponential function $e^{j\omega_0 t}$ of Figure 1.9 by $-j$ simply rotates every point of the helix in a clockwise direction. The result is shown in the left half of Figure 1.14, where the whole helix is rotated back, with the effect of shifting its real and imaginary parts **forward in time**, by one quarter of the period T_0.

Another example is shown in the right half of Figure 1.14, where the function $e^{-j\omega_0 t}$ of Figure 1.10 is multiplied by j, that is, rotated anti-clockwise. The sum of the two functions is a sinusoid located in the real plane, another manifestation of Euler's sine equation

$$\sin \omega_0 t = \frac{1}{2}\left(\frac{e^{j\omega_0 t}}{j} - \frac{e^{-j\omega_0 t}}{j}\right) = \frac{1}{2}\left(-je^{j\omega_0 t} + je^{-j\omega_0 t}\right)$$

This relationship was almost reached earlier in Equation 1.26, where it only remained to rotate the result $2j \sin \omega_0 t$ back into the real plane and to scale it, by dividing by the constant 2j.

1.4   Fourier series, exponential form

The preceding section introduced three important concepts. Firstly, there is no fundamental difference between positive and negative frequencies. It is a matter of convention as to which is called positive.

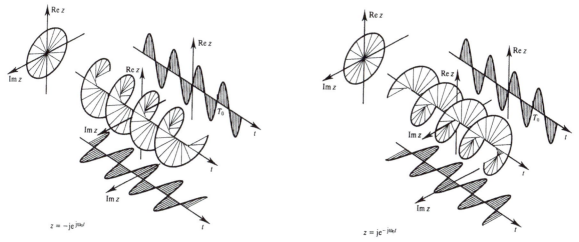

Figure 1.14 Exponentials rotated by $-\frac{1}{2}\pi$ and $+\frac{1}{2}\pi$.

Secondly, labelling signal components as **real** and **imaginary** enables the use of the notation and properties of complex numbers to express and manipulate variables described by two independent quantities.

Thirdly, to produce real-valued time signals, frequency components must appear in complex conjugate pairs. We will elaborate in Chapter 2.

The expressions 1.18 and 1.19 of the exponential form of the Fourier series, repeated here

$$f(t) = \sum_{k=-\infty}^{\infty} c_k e^{jk\omega_0 t} \tag{1.29}$$

$$c_k = \frac{1}{T_0} \int_0^{T_0} f(t) e^{-jk\omega_0 t} \, dt \tag{1.30}$$

can be re-examined in the light of these findings. The synthesis equation 1.29 is now seen as the sum of complex functions of time, namely of the scaled complex exponentials $c_k e^{jk\omega_0 t}$. The magnitude and initial angle, or phase, of each exponential are those of the magnitude $|c_k|$ and the argument φ_k of the associated frequency coefficient

$$c_k = |c_k| e^{j\varphi_k}$$

In addition to compactness, the exponential form of the Fourier series is particularly useful when multiplication and division are involved, and for logarithmic representation. Furthermore, this notation is consistent with those of the Fourier, Laplace and z transforms, making it ideal for deriving them from each other, so that the unity of the whole subject is emphasized.

Example 1.3

The function of Figure 1.15, consisting of the periodic repetition, at intervals T_0, of a rectangular pulse of width τ, is a versatile function for illustrating transforms. We will find its Fourier coefficients.

Since the function is symmetrical about the origin, we use the integration interval $(-\frac{1}{2}T_0, \frac{1}{2}T_0)$, and noting that the function is zero outside the interval $(-\frac{1}{2}\tau, \frac{1}{2}\tau)$, we have

$$c_k = \frac{1}{T_0}\int_{-T_0/2}^{T_0/2} f(t)e^{-jk\omega_0 t}\,dt = -\frac{1}{jk\omega_0 T_0}\,e^{-jk\omega_0 t}\,\Big|_{-\tau/2}^{\tau/2}$$

$$c_k = \frac{1}{jk\omega_0 T_0}\,(e^{jk\omega_0 \tau/2} - e^{-jk\omega_0 \tau/2})$$

Multiplying and dividing by $\frac{1}{2}\tau$, and using Euler's sine equation, this reduces to the real-valued function

$$c_k = \frac{\tau}{T_0}\,\frac{\sin\left(\frac{1}{2}k\omega_0\tau\right)}{\frac{1}{2}k\omega_0\tau} \tag{1.31}$$

The coefficients c_k, all of them real, can thus be interpreted as discrete evaluations, at points $\omega = k\omega_0$ of the continuous-frequency envelope

$$\frac{\tau}{T_0}\,\frac{\sin\left(\frac{1}{2}\omega\tau\right)}{\frac{1}{2}\omega\tau} \tag{1.32}$$

indicated by a dotted line in Figure 1.15.

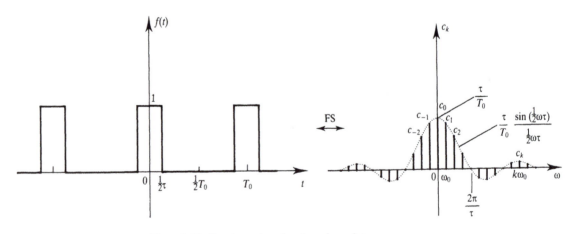

Figure 1.15 Fourier series of rectangular pulses.

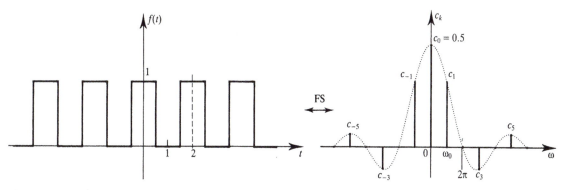

Figure 1.16 Fourier coefficients of square wave.

The last expression shows that the **shape of the envelope**, of the form $\sin x/x$, is fully determined by the pulse width τ, as evidenced by the crossings of the frequency axis, which occur at multiples of $2\pi/\tau$. The **envelope amplitude** τ/T_0 depends on both pulse width and period, while the spacing between coefficients depends only on the period T_0.

Example 1.4

We calculate the numerical values of the Fourier coefficients of the square wave of Figure 1.16, a special case of the preceding example, with period $T_0 = 2$ s and $\tau = 1$ s.

With a fundamental frequency $\omega_0 = 2\pi/T_0 = \pi$ rad/s, the Fourier coefficients are given by Equation 1.31 as

$$c_k = 0.5 \, \frac{\sin\left(\frac{1}{2}k\pi\right)}{\frac{1}{2}k\pi}$$

For $k = 0$ the $\sin x/x$ function is unity, hence $c_0 = 0.5$. Even indices k involve sines of integer multiples of π, giving zero-valued coefficients. Odd k give alternating sine values $+1$ and -1. Hence all coefficients are real, those nearest the origin taking the values

$$c_0 = 0.5$$
$$c_1 = c_{-1} = 1/\pi \simeq 0.3183$$
$$c_3 = c_{-3} = -1/(3\pi) \simeq -0.1061$$
$$c_5 = c_{-5} = 1/(5\pi) \simeq 0.063\,66$$

These are interpreted, according to Equation 1.32 as samples of a $\sin x/x$ envelope of amplitude $\tau/T_0 = 0.5$ whose zero crossings, at integer multiples of 2π rad/s, coincide with the samples of even k, as seen in Figure 1.16.

The square wave is thus represented by the infinite series

$$f(t) = \ldots + \frac{1}{5\pi}e^{-j5\pi t} - \frac{1}{3\pi}e^{-j3\pi t} + \frac{1}{\pi}e^{-j\pi t} + 0.5$$

$$+ \frac{1}{\pi}e^{j\pi t} - \frac{1}{3\pi}e^{j3\pi t} + \frac{1}{5\pi}e^{j5\pi t} - \ldots$$

1.4.1 Relationship to trigonometric form

The relationship between the complex coefficients c_k and the coefficients a_k and b_k of the trigonometric form was given in Equation 1.17 as

$$c_k = \tfrac{1}{2}(a_k - jb_k) \qquad c_{-k} = \tfrac{1}{2}(a_k + jb_k) \quad (k > 0)$$

so that,

$$a_k = c_k + c_{-k} \qquad b_k = j(c_k - c_{-k})$$

Most commonly $f(t)$ is a real-valued function, with real coefficients a_k and b_k, in which case the coefficients c_k and c_{-k} form complex conjugate pairs,

$$c_k = |c_k|e^{j\varphi_k} \qquad \text{and} \qquad c_{-k} = |c_{-k}|e^{j\varphi_{-k}}$$

of identical magnitudes, $|c_{-k}| = |c_k|$, and opposite phases, $\varphi_{-k} = -\varphi_k$. These relationships are illustrated in Figure 1.17.

It will be seen later that the time domain description of a tangible real-world signal usually takes the form of a real-valued function, though not always. In contrast, the frequency domain description of the same signal usually contains real and imaginary parts.

1.4.2 Simplifications for real functions of time

The exponentials of the sum 1.29 as well as their coefficients are usually complex. The term of index k can thus be expanded into products of real and imaginary parts as

$$f_k(t) = c_k e^{jk\omega_0 t}$$

$$= (\text{Re } c_k + j\,\text{Im } c_k)(\cos k\omega_0 t + j\sin k\omega_0 t)$$

$$= (\text{Re } c_k \cos k\omega_0 t - \text{Im } c_k \sin k\omega_0 t)$$

$$+ j(\text{Re } c_k \sin k\omega_0 t + \text{Im } c_k \cos k\omega_0 t)$$

where the two terms in parentheses represent the real and imaginary parts of $f_k(t)$. If $f(t)$ is real, the sum 1.29 only contains such real terms,

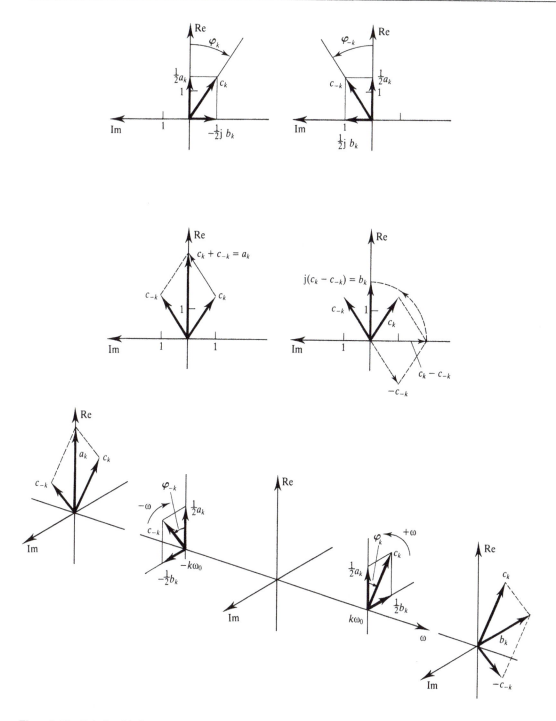

Figure 1.17 Relationship between exponential and trigonometric Fourier coefficients (e.g. $a_k = 3$, $b_k = 2$).

and can be interpreted by two sums

$$\operatorname{Re} f(t) = \sum_{k=-\infty}^{\infty} \operatorname{Re} c_k \cos k\omega_0 t - \sum_{k=-\infty}^{\infty} \operatorname{Im} c_k \sin k\omega_0 t \qquad (1.33)$$

which neatly separates the real and imaginary parts of the coefficients and bundles together their effects.

Thus, if $f(t)$ is known to be real, and in addition all the coefficients c_k are also real, the first sum of Equation 1.33 is sufficient to evaluate $f(t)$. Similarly, the second sum is sufficient for purely imaginary coefficients.

It was shown earlier that real $f(t)$ implies complex conjugate frequency coefficients $c_{-k} = c_k^*$. This in turn implies some simplifying symmetries, namely the real parts of the coefficients must be even-symmetrical, while imaginary parts must be odd-symmetrical. These symmetries and a generalization to time functions $f(t)$ containing both real and imaginary parts will be examined further in Section 2.3, in the context of the Fourier transform.

1.4.3 Graphical interpretation

The three-dimensional interpretation of the trigonometric form of the Fourier series, shown earlier in Figure 1.7, will now be extended to the exponential form. This is done not simply to visualize mathematical expressions, but to develop simple graphical aids (e.g. time shifting) that will help to introduce new concepts.

To simplify presentation, in this section only real-valued time functions will be considered, starting with real coefficients, which come in even-symmetrical pairs, i.e. $c_{-k} = c_k$.

One such pair is shown in the frequency domain region of Figure 1.18, located in planes identified by their frequency values $+\omega_k$ and $-\omega_k$. As these components rotate at the appropriate angular velocities, their projections on the vertical plane describe the real time functions

$$f_k(t) = c_k \cos \omega_k t \qquad \text{and} \qquad f_{-k}(t) = c_{-k} \cos(-\omega_k) t$$

shown in the foreground of Figure 1.18. Their sum gives the contribution of a real pair c_k, c_{-k} to the function $f(t)$. This diagram provides the core of our interpretations of the Fourier, Laplace and z transforms.

For a full interpretation of a function, the effects of all the coefficient pairs must be added according to Equation 1.33. This is shown in Figure 1.19, using, for illustration, the coefficients obtained in Example 1.4.

The coefficients are represented in the frequency domain region of Figure 1.19, together with their envelope shown by dotted lines. As

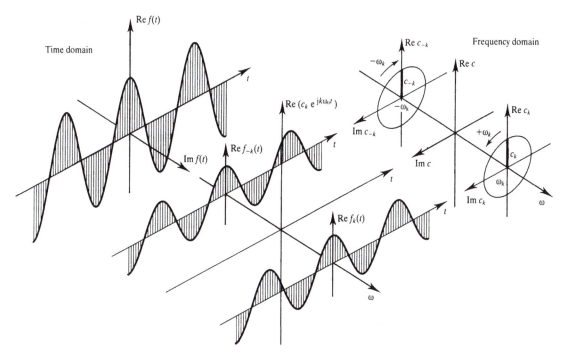

Figure 1.18 Real parts of time domain contributions from a coefficient pair c_k, c_{-k}.

time progresses, each component, rotating at the appropriate positive or negative frequency, describes a time function. If all the infinite components were added, as prescribed by the first sum of Equation 1.33, they would reconstitute the original square wave of Figure 1.16. Even the sum of the few harmonics shown in Figure 1.19 suggests a convergence towards the square wave.

1.4.4   Time shifting

The square wave of the time domain of Figure 1.19 and the set of coefficients c_k of the frequency domain offer two complete and equivalent representations of the same function. The construction in the foreground contains time domain representations of given frequency components, establishing a link to the time domain.

The rotating vector imagery used in this construction can be usefully extended to obtain the coefficients of a time shifted function $f(t - \tau)$. If the coefficients c_k of Figure 1.19 are interpreted as a

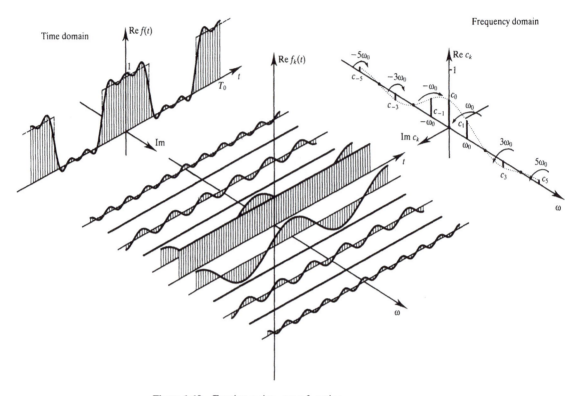

Figure 1.19 Fourier series, even function.

snapshot of the rotating vectors in their initial positions, at time $t = 0$, consecutive time-lapsed exposures at times $t = t_1,\ t_2,\ t_3,\ t_4,\ \ldots$ would show them in the configurations indicated in the right-hand background of Figure 1.20.

To relate the two domains quantitatively, we note that the fundamental c_1 requires a time interval of one period T_0 to describe one full turn of 2π radians. Thus, at time $t = T_0$ all harmonics will have rotated a proportional number of positive or negative turns, only to end up in their original angular relationships.

In Figure 1.20 the spacing between displayed instants was set to $t_1 = \frac{1}{12} T_0$, so that successive positions of the fundamental differ by $\frac{1}{6}\pi$, those of the third harmonic by $\frac{1}{2}\pi$, and so on. The vertical projections of these components can be found on the respective cosines of the foreground, and their sums give the corresponding values of $f(t_1)$, $f(t_2)$, etc.

The instant t_3 presents an interesting case, where all the vectors, except the average value c_0, find themselves in the imaginary plane, so that $f(t_3) = c_0$.

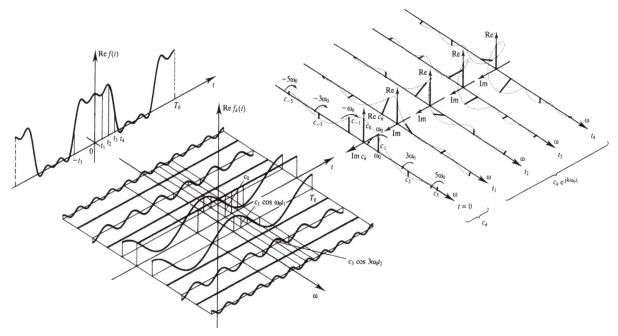

Figure 1.20   Time-lapse exposures of rotating vectors.

We emphasize that of all the diagrams in the frequency domain sequence of Figure 1.20 only the first, that for $t = 0$, represents the Fourier series coefficients of the square wave shown in the time domain! The others simply show time-lapsed configurations of the rotating vectors.

Change of time origin

The time origin of a function is an arbitrary reference point, chosen for a particular purpose, e.g. to take advantage of even symmetry. A change of origin does not change the shape of the signal, only its description. We will here establish how this is reflected in the frequency description.

For instance, if we were to make the instant t_3 of Figure 1.20 our new time origin, what so far was only a snapshot, at time t_3, of the original configuration would now represent the new Fourier coefficients. The signal would still be the same, but, allowing for the d.c. component c_0, its time domain would now be described by an odd function, while its frequency description would contain only imaginary components.

To express this process analytically, if a component of the original signal

$$f_k(t) = c_k e^{jk\omega_0 t}$$

is delayed by an amount τ, the delayed signal is expressed as

$$f_k(t - \tau) = c_k e^{jk\omega_0(t-\tau)} = [c_k e^{-jk\omega_0 \tau}] e^{jk\omega_0 t}$$

The product in brackets on the right represents a backward rotation, proportional to the delay τ, of the coefficient c_k. This yields the new coefficient

$$c'_k = c_k e^{-jk\omega_0 \tau} = c_k e^{-jk2\pi\tau/T_0} \tag{1.34}$$

of the delayed component

$$f_k(t - \tau) = c'_k e^{jk\omega_0 t}$$

If the overall function $f(t)$ is real, a pure delay can only produce another real function, and the simplified expression 1.33 applies,

$$f(t - \tau) = \sum_{k=-\infty}^{\infty} \mathrm{Re}\, c'_k \cos k\omega_0 t - \sum_{k=-\infty}^{\infty} \mathrm{Im}\, c'_k \sin k\omega_0 t$$

Clearly, the only frequency domain effect is a rotation of the coefficients c_k. This rotation is a linear function of the frequency $k\omega_0$, the rate being determined by the delay τ. In the context of systems we will interpret this as an addition of **linear phase** to the original components.

The time-lapsed diagrams of Figure 1.20 thus interpret the process of shifting the time origin of a function. This greatly simplifies the process of finding the transform of some functions, as illustrated in the following example.

Example 1.5

Consider the basically odd function $g(t)$ of Figure 1.21, which is a time-shifted version of the square wave $f(t)$ of the preceding example. We will derive the Fourier coefficients of $g(t)$ from those of $f(t)$, by shifting the time origin.

Recall from Figure 1.20 that, to find the value of $f(t)$ at the positive time value $t = t_3$, we let c_1 rotate with positive frequency $+\omega_0$ for the stipulated time, at which instant all the harmonics were in the imaginary plane, in the configuration labeled t_3.

Similarly, to find the function value at time $t = -t_3$ of Figure 1.20, we wind c_1 and all the harmonics in the opposite directions.

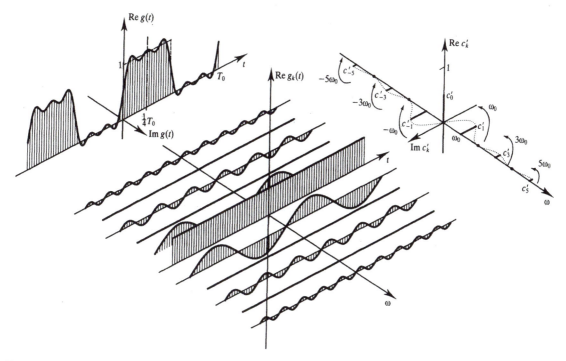

Figure 1.21 Fourier series, odd function with d.c. component.

Again, all the harmonics end up in the imaginary plane, but pointing in the opposite directions, as indicated in the frequency domain of Figure 1.21. If we now define $t = -t_3$ of Figure 1.20 as our new time origin, the resulting time function is identical to $g(t)$, i.e. $f(t - \frac{1}{4}T_0) = g(t)$, and the transform pair of Figure 1.21 is the desired result.

The numerical values of the new coefficients can be obtained by rotating those of Example 1.4 according to Equation 1.34. For a delay $\tau = \frac{1}{4}T_0$ we have

$$c_k' = c_k e^{-jk\pi/2} = 0.5 \frac{\sin\left(\frac{1}{2}k\pi\right)}{\frac{1}{2}k\pi} e^{-jk\pi/2}$$

which confirms that the d.c. value $c_0' = 0.5$ remains real, while all other components become imaginary, as each is rotated by an odd multiple of $\frac{1}{2}\pi$.

Having derived the Fourier coefficients c_k', these reconstruct the time function $g(t)$ as shown in Figure 1.21, which interprets Equation 1.33 as

$$g(t) = c_0' - \sum_{k=-\infty}^{\infty} \operatorname{Im} c_k' \sin k\omega_0 t$$

1.4.5 Period doubling

We now apply the imagery developed for time-shifted functions to examine, in purely graphical terms, the frequency implications of doubling the period of a function. In the next chapter this forms the basis for our derivation of the Fourier transform.

Consider the function $f_1(t)$, of the form of Example 1.3, shown in both domains at the top of Figure 1.22. The coefficients c_k' are values of Equation 1.31 for a pulse of width $\tau = 1$ s and period $T_0 = 4$ s, and represent samples, spaced at intervals $\omega_0 = 2\pi/T_0 = \frac{1}{2}\pi$ rad/s, of an envelope of amplitude $c_0' = 0.25$.

Shifting $f_1(t)$ by half its period, we generate the second function

$$f_2(t) = f_1(t - \tfrac{1}{2}T_0)$$

shown in the middle row in Figure 1.22. Its coefficients c_k'' are derived from those of $f_1(t)$ by using the imagery developed in Figures 1.20 and 1.21 for time-shifted functions. This involves a half turn of the fundamental c_1', corresponding turns of the harmonics and an appropriate change of the time origin.

We now add the two signals, as shown in the lower row of Figure 1.22. In the time domain this yields a denser function

$$f(t) = f_1(t) + f_2(t)$$

with twice the number of pulses. In the frequency domain, even-numbered components are the same for both signals, whilst odd-numbered ones are opposite. Their sum is

$$c_k = c_k' + c_k'' = \begin{cases} 0 & (k \text{ odd}) \\ 2c_k' & (k \text{ even}) \end{cases}$$

so that the fundamental and all its odd harmonics are lost, and a new fundamental component c_1 emerges at twice the original frequency. In short, the number of frequency components is halved, but those that remain double their value. The d.c. value c_0 provides a familiar landmark for consolidating ideas.

Conversely, given a periodic function of time, a doubling of its period, obtained by inserting zero-valued segments in the appropriate places, has the effect of halving the amplitude of the coefficient envelope and doubling the number of components.

There is an inverse duality of the domains of the Fourier transform, paraphrased as 'expansion in one domain means contraction in the other'. In the present context this relates the period in time to the fundamental frequency, and 'thinning out one domain means crowding

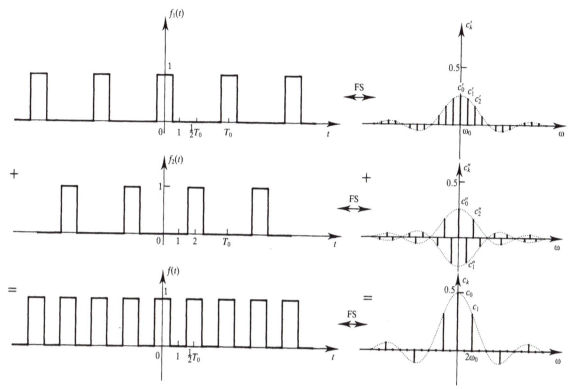

Figure 1.22   Doubling or halving the period.

the other'. We will mention other aspects of this rule in various parts of the book.

Exercises

1.1 Using the construction of Figure 1.3 plot the convolution $g(t) = f(t) * h(t)$ of the functions shown in Figure 1.23.

1.2 Justify analytically the commutative property of convolution (Equation 1.2), the associative property (Equation 1.4) and the distributive property

$$f(t) * [g(t) + h(t)] = f(t) * g(t) + f(t) * h(t)$$

1.3 Using the sifting property (Equation 1.10) of the impulse function in the form

$$\int_{-\infty}^{\infty} f(t)\, \delta(t - \tau)\, dt = f(\tau)$$

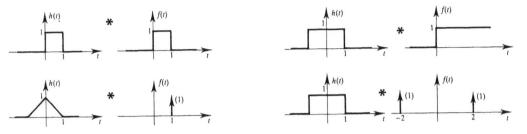

Figure 1.23 Functions for Exercise 1.1.

evaluate the following integrals, without performing the integration

$$\int_{-\infty}^{\infty} \sin t \, \delta(t - \pi/2) \, dt \qquad\qquad \int_{-\infty}^{\infty} e^{-j\omega t} \, \delta(t) \, dt$$

$$\int_{-\infty}^{\infty} e^{2t} \, \delta(t - 1) \, dt \qquad\qquad \int_{0}^{\infty} \cos t \, \delta(t + 1) \, dt$$

1.4 Sketch the real exponential functions e^{at} and e^{-at}, where a is a real positive number. Also sketch their semi-sum and semi-difference,

$$\cosh at = \tfrac{1}{2}(e^{at} + e^{-at})$$

$$\sinh at = \tfrac{1}{2}(e^{at} - e^{-at})$$

and contrast the results with those seen in Section 1.3.3 for complex exponentials.

1.5 Show that the time derivative of the complex exponential function $f(t) = e^{j\omega_0 t}$ (Equation 1.24) is

$$g(t) = \frac{d}{dt} e^{j\omega_0 t} = j\omega_0 e^{j\omega_0 t}$$

Sketch its real and imaginary parts and compare them with those of Figure 1.9.

1.6 Based on the construction of Figure 1.7, build up the functions $f(t)$ and $g(t)$ described respectively by the Fourier series coefficients

$$f(t): a_1 = 1, \; a_3 = -\tfrac{1}{3}, \; a_5 = \tfrac{1}{5}$$

$$g(t): b_1 = 1, \; b_2 = \tfrac{1}{2}, \; b_3 = \tfrac{1}{3}$$

1.7 Convert the coefficients a_k and b_k of Exercise 1.6 to equivalent complex coefficients c_k and build up the functions $f(t)$ and $g(t)$ by the constructions of Figures 1.19 or 1.21, as applicable.

1.8 Find the Fourier series coefficients of the periodic functions shown in Figure 1.24.

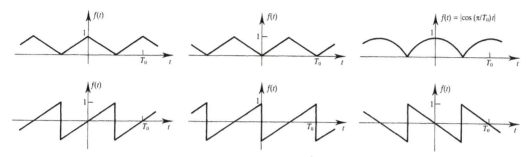

Figure 1.24 Functions for Exercise 1.8.

1.9 Show that the Fourier series (Equation 1.18) represents a linear operator (see Section 2.4.1 for linearity conditions).

1.10 Making use of linearity, generate a new Fourier pair by graphically adding the two functions of Figures 1.19 and 1.21 and their transforms. In the time domain, consider only the limiting rectangular shapes. Add the frequency components graphically, as vectors.

1.11 The resulting frequency components of Exercise 1.10 are complex, with real and imaginary parts. Using the time-shifting imagery of Section 1.4.4, graphically rotate the fundamental by $+\pi/4$ and show that all frequency components end up in the real plane, that this represents a time shift by $-T_0/8$ and that the resulting time-shifted function is real and even in both domains.

1.12 Discuss the full time-domain and frequency-domain implications of rotating the fundamental frequency component c_1 of a periodic function by one full revolution in the positive direction, and similarly for one full revolution in the negative direction and for any integer number of turns.

PART 1

Analysis of Continuous-time Signals and Systems

We now analyse the class of continuous and aperiodic signals and the appropriate systems. In the first two chapters we derive the analytical tools applicable to this signal class, namely the Fourier and Laplace transforms. The Fourier transform is derived as an extension of the Fourier series, while the Laplace transform is treated as a generalization of the Fourier transform.

Systems capable of handling signals of this class are then analysed with the aid of these tools. General purpose characterizations are derived and related in terms of such transforms and so are the associated solution methods.

Together with Chapter 10 this part gives a self-contained treatment of continuous-time signals and systems. But it is structured with the requirements of discrete-time signals and systems of Part 2 and system synthesis of Part 3 in mind. It thus serves the purpose of introducing concepts and methods valid for all classes.

CHAPTER 2

Fourier Transform

The Fourier transform is a generalization of the Fourier series. In a strict sense it applies to the class of continuous and aperiodic functions, but the use of impulse functions permits extending it to discrete signal forms. It also represents a subset of the Laplace transform, of which it provides a simple interpretation. These relationships make the Fourier transform a key link between all the time-to-frequency transforms.

In this key capacity we will examine the Fourier transform in depth, as most of the associated concepts and properties can be extended to all the other transforms.

2.1 Fourier series to Fourier transform

The Fourier series of the preceding chapter provides the basis for deriving the Fourier transform as its extension to aperiodic functions, as suggested in Figure 2.1 in terms of the signal classification chart of Figure 1.2.

We use the time and frequency implications of doubling the period of a function, as seen in Section 1.4.5, to extend the Fourier series Equations 1.29 and 1.30 to functions that are not periodic in time.

2.1.1 Graphical link

Consider the periodic function $\tilde{f}(t)$, of period T_0 and frequency coefficients c_k, shown in the top row of Figure 2.2. Successive doubling

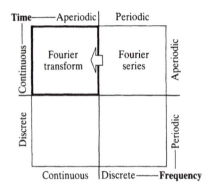

Figure 2.1   Fourier series to Fourier transform, on signal classification chart.

of the period to $T_1, T_2, \ldots$, generates a proportionally larger number of coefficients of diminishing amplitude, as indicated in the middle column of Figure 2.2. To counteract the diminution and eventual vanishing of coefficients, we scale their amplitudes by the current period, as shown in the last column of the figure, where the envelope of the product $T_0 c_k$ remains unaffected by period doubling.

When T_i approaches infinity, in the limit, it leads to an infinite number of infinitesimally close frequency components, as shown in the bottom row of Figure 2.2. Values of the form $T_0 c_k$, which until then were defined at points of the discrete frequency variable $k\omega_0$, merge into a continuum, to become a function $F(\omega)$ of the continuous frequency variable ω.

The function $F(\omega)$ now provides a description of the envelope of products of the form $T_0 c_k$. These can be interpreted as samples, at points $\omega = k\omega_0$, of that envelope, that is,

$$T_0 c_k = F(\omega)\big|_{\omega = k\omega_0} \tag{2.1}$$

The transition to infinite period is an important concept that links two functions of fundamentally different frequency variables. We elaborate in Chapter 6 in the context of sampling.

2.1.2 Analytical link

Using symmetrical integration limits for convenience, the Fourier analysis Equation 1.30 relates the products $T_0 c_k$ to $\tilde{f}(t)$ as

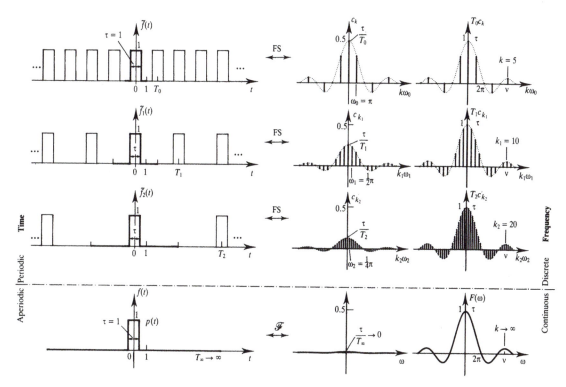

Figure 2.2   Fourier series to Fourier transform.

$$T_0 c_k = \int_{-T_0/2}^{T_0/2} \tilde{f}(t) e^{-jk\omega_0 t}\, \mathrm{d}t$$

Comparing the functions of the top and bottom rows of Figure 2.2, within the integration interval their values are identical, $f(t) = \tilde{f}(t)$, so that they can be interchanged. Also, the resulting aperiodic function $f(t)$ is zero outside that interval, therefore it is valid to extend the integration limits to infinity,

$$T_0 c_k = \int_{-\infty}^{\infty} f(t) e^{-jk\omega_0 t}\, \mathrm{d}t$$

For a given index k the value $k\omega_0$ of the exponential can be interpreted as a value of the continuous variable ω, and with the interpretation of Equation 2.1 we have

$$F(\omega) = \int_{-\infty}^{\infty} f(t) e^{-j\omega t}\, \mathrm{d}t \tag{2.2}$$

This expression gives a continuous-frequency representation $F(\omega)$ of the continuous and aperiodic function $f(t)$.

2.1.3   Inverse transformation

With the insights of Figure 2.2 we derive the inversion formula for Equation 2.2. Starting from the Fourier series synthesis equation (1.30), repeated here

$$\tilde{f}(t) = \sum_{k=-\infty}^{\infty} c_k e^{jk\omega_0 t}$$

its coefficients c_k can be interpreted by means of Equation 2.1 as scaled samples of the envelope $F(\omega)$, as

$$c_k = \frac{1}{T_0} F(k\omega_0) = \frac{\omega_0}{2\pi} F(k\omega_0)$$

so that the above Fourier series becomes

$$\tilde{f}(t) = \frac{1}{2\pi} \sum_{k=-\infty}^{\infty} F(k\omega_0) e^{jk\omega_0 t} \omega_0 \qquad (2.3)$$

The fundamental frequency ω_0 of this expression can also be interpreted as the frequency increment $\Delta\omega$ between adjacent samples, which decreases in inverse proportion to the period. Since the density of frequency components increases in direct proportion to the period, the typical product values $k\omega_0 = k_1\omega_1 = k_2\omega_2 = v$ indicate the same frequency value $\omega = v$.

In the limit, as the period tends to infinity, the time function $\tilde{f}(t)$ tends to the related aperiodic function $f(t)$, the frequency increment $\Delta\omega$ becomes the differential $d\omega$ and the discrete values $k\omega_0$ merge into the continuous frequency ω. As a result of these simultaneous changes the sum in Equation 2.3 becomes the integral

$$f(t) = \frac{1}{2\pi} \int_{-\infty}^{\infty} F(\omega) e^{j\omega t} \, d\omega \qquad (2.4)$$

2.1.4   Forward and inverse Fourier transforms

The pair of Equations 2.2 and 2.4 relate the time domain and frequency domain representations of a function. To simplify initial presentation we treated the frequency ω as a real variable. In the next chapter, in the context of the Laplace transform, we will use the complex frequency variable $s = \sigma + j\omega$, of which ω represents the imaginary part. There we will interpret the Fourier transform as the subset $F(j\omega)$ of the Laplace transform $F(s)$. For overall consistency, the remainder of this book will denote the Fourier transform with $j\omega$ as the frequency variable. With this notation Equations 2.2 and 2.4 become

$$F(j\omega) = \int_{-\infty}^{\infty} f(t)e^{-j\omega t}\,dt \tag{2.5}$$

$$f(t) = \frac{1}{2\pi} \int_{-\infty}^{\infty} F(j\omega)e^{j\omega t}\,d\omega \tag{2.6}$$

The first represents the **Fourier transform** of $f(t)$, otherwise known as its analysis equation, or the Fourier integral. The second defines the **inverse Fourier transform** of $F(j\omega)$, also called its synthesis equation.

Discontinuities

If a function $f(t)$ is discontinuous at some point $t = \tau$, as in Figure 2.3, the integral of Equation 2.6 converges to the average value at the discontinuity

$$f(\tau) = \tfrac{1}{2}[f(\tau^+) + f(\tau^-)]$$

where $f(\tau^+)$ and $f(\tau^-)$ represent the values taken by the function as the discontinuity is approached from the positive and from the negative side, respectively. Taking this as the definition of $f(t)$ at discontinuities, the expression 2.6 is valid for all t.

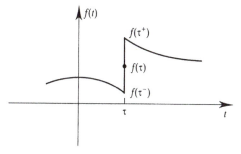

Figure 2.3 Function value at discontinuity.

Example 2.1

We find the Fourier transform of the rectangular pulse of Figure 2.4. Analytically, the function is defined as

$$f(t) = p_\tau(t) = \begin{cases} 1 & |t| < \tfrac{1}{2}\tau \\ \\ 0 & |t| > \tfrac{1}{2}\tau \end{cases} \qquad \tau = 1\ \text{s}$$

Applying the definition 2.5 we have

$$F(j\omega) = \int_{-\infty}^{\infty} p_\tau(t) e^{-j\omega t}\, dt$$

Since $p_\tau(t)$ is zero outside the interval $|t| = \tfrac{1}{2}\tau$, the integration limits can be reduced accordingly,

$$F(j\omega) = \int_{-\tau/2}^{\tau/2} e^{-j\omega t}\, dt = \left. \frac{-e^{-j\omega t}}{j\omega} \right|_{-\tau/2}^{\tau/2} = \frac{e^{j\omega \tau/2} - e^{-j\omega \tau/2}}{j\omega}$$

Multiplying numerator and denominator by $\tfrac{1}{2}\tau$, and using Euler's equation, the Fourier transform becomes

$$F(j\omega) = \tau\, \frac{\sin \tfrac{1}{2}\omega\tau}{\tfrac{1}{2}\omega\tau}$$

This is the familiar $\sin x/x$ function, of amplitude τ at the origin and zero crossings at multiples of $2\pi/\tau$, shown on the right side of Figure 2.4.

It is a real and even function. Considered in the context of Figure 2.2 and duly scaled, it represents the envelope to the Fourier coefficients of the related periodic functions, as expressed in Equation 1.32 of Example 1.3.

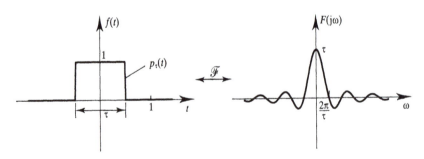

Figure 2.4   Fourier transform of a rectangular pulse.

2.2 Graphical interpretation and developments

We now extend the imagery of the Fourier series to the Fourier transform. If the frequency coefficients c_k of the earlier Figure 1.19 are scaled by the period T_0, the same figure serves to interpret the synthesis of the function $\tilde{f}(t)$ of the top row of Figure 2.2, from its coefficients $T_0 c_k$.

A similar interpretation of the function $\tilde{f}_1(t)$ of Figure 2.2, with period $T_1 = 2T_0$, retains the same cosine components $\tilde{f}_k(t)$, and has intermediate cosines inserted midway between these, in correspondence with the frequencies and magnitudes of the additional coefficients of $T_1 c_{k1}$.

At each stage of the period doubling process the number of intermediate cosines doubles, without affecting previous ones. In the limit, as the coefficients merge into the continuous function $F(\omega)$ of Figure 2.2, the corresponding cosines merge into the continuous surface of the ωt-plane of Figure 2.5.

2.2.1 Inverse transform

Taken as a limiting case, Figure 2.5 leads to the interpretation of the inverse Fourier transform of the frequency spectrum $F(j\omega)$, as derived in Example 2.1. Being real, this function is located in the real plane of

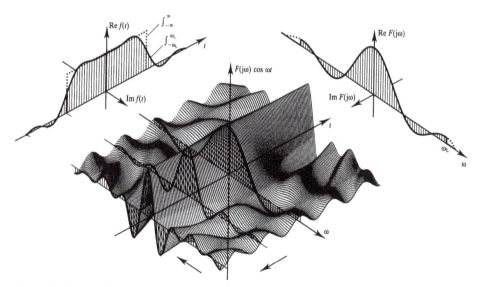

Figure 2.5 Inverse Fourier transform of a real even function.

the frequency domain, and the surface of the foreground represents the real part of the integrand of Equation 2.6, i.e.

$$F(j\omega)\cos\omega t = \text{Re}\,[F(j\omega)e^{j\omega t}]$$

To find a value of $f(t)$ at some arbitrary time $t = \tau$ we take a slice of constant τ through this surface and integrate the intersected function according to Equation 2.6, i.e.

$$f(\tau) = \frac{1}{2\pi}\int_{-\infty}^{\infty} F(j\omega)\cos\omega\tau d\omega$$

Three values are illustrated in Figure 2.5. The case $\tau = 0$ is the simplest, as it makes $\cos\omega\tau = 1$ for all ω, so that the integral

$$f(0) = \frac{1}{2\pi}\int_{-\infty}^{\infty} F(j\omega)\,d\omega$$

represents the area of the slice $t = 0$, which is a replica of $F(j\omega)$. The other values are represented by areas of more elaborate curves. This integration process is the counterpart of the summing process in the Fourier series of Figure 1.19.

Two alternative results are superimposed in the time domain of Figure 2.5. The full line belongs to $f(t)$ when $F(j\omega)$ is truncated at the edge ω_L of the illustration, when $F(j\omega) = 0$ for $|\omega| > \omega_L$. The dotted line indicates values attained by $f(t)$ when $F(j\omega)$ is allowed to extend to infinity.

This diagram provides a complete interpretation when $F(j\omega)$ is real and even. This can be justified on grounds of symmetry, as in Section 1.4.3. We will elaborate later in this section.

2.2.2   Forward transform

The forward transform of Equation 2.5 of a real and even function of time can be similarly interpreted. Consider the rectangular time pulse $f(t)$ of Example 2.1, shown in the time domain of Figure 2.6. For every point (ω, t) of the ωt-plane we can calculate and plot a value

$$f(t)\cos\omega t = \text{Re}\,[f(t)e^{-j\omega t}]$$

By analogy to the preceding case, taking a slice of constant $\omega = v$ through this surface and integrating in time gives the value $F(jv)$ of the transform

$$F(jv) = \int_{-\infty}^{\infty} f(t)\cos vt\,dt$$

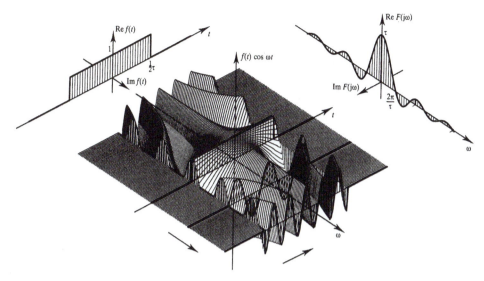

Figure 2.6 Forward transform of a real and even function.

Repeating this for every value of ω builds up the expected $\sin x/x$ function of the frequency domain. Three values are illustrated in the figure.

2.2.3 Duality of forward and inverse transforms

We next consider a function whose frequency domain representation is the rectangular pulse of width $\pm\omega_L$ illustrated in Figure 2.7. In the context of systems, this shape would describe an ideal filter of bandwidth ω_L. The whole of Figure 2.7, which gives the graphical interpretation of the inverse Fourier transform of this function, is a mirror image of Figure 2.6, and its time domain is a $\sin x/x$ function.

Example 2.2

We examine analytically the result obtained in Figure 2.7. Proceeding as in Example 2.1, we express the frequency domain pulse as

$$F(j\omega) = \begin{cases} 1 & |\omega| < \omega_L \\ 0 & \text{elsewhere} \end{cases}$$

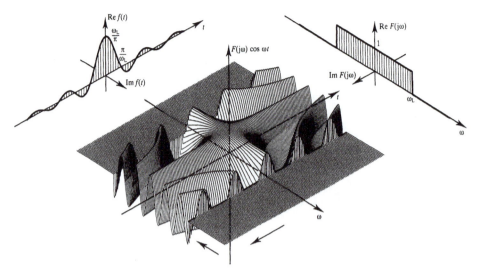

Figure 2.7   Inverse transform of a real even pulse.

for which the inverse Fourier transform (2.6) reduces to

$$f(t) = \frac{1}{2\pi} \int_{-\omega_L}^{\omega_L} e^{j\omega t}\, d\omega = \frac{1}{2\pi} \frac{e^{j\omega_L t} - e^{-j\omega_L t}}{jt}$$

Multiplying and dividing by ω_L and using Euler's equation, we have

$$f(t) = \frac{\omega_L}{\pi} \frac{\sin \omega_L t}{\omega_L t}$$

which confirms the $\sin x/x$ shape and provides its scaling factor.

The similarity observed here is a manifestation of the **duality property** of the Fourier transform, to be formalized in Section 2.4.3. Its basis lies in the symmetry of the exponentials in the forward and inverse Fourier transform expressions and is partly revealed in the surfaces of the two ωt-planes of Figures 2.6 and 2.7. We will now reveal the surface further.

2.2.4   Transform of an impulse

Increasing the bandwidth of the function $F(j\omega)$ of Figure 2.7, first to ω_L of Figure 2.8 and then to infinity, suggests a sequence that leads to the transform of the impulse $\delta(t)$.

Being wider, the pulse of Figure 2.8 produces a proportionally taller and narrower $\sin x/x$ pulse in the time domain, which has the

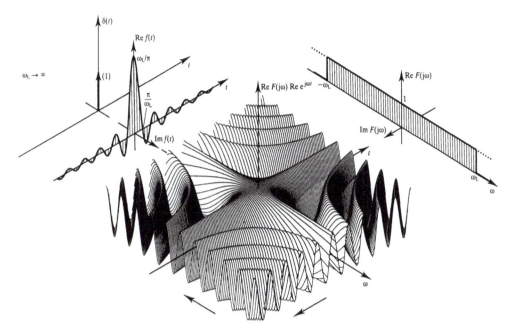

Figure 2.8   Inverse transform of a unit height pulse.

same area as that of Figure 2.7. As the bandwidth extends to infinity, so that $F(j\omega) = 1$ for all ω, the height of the pulse $f(t)$ grows to infinity, while its width becomes infinitesimal, retaining a constant area quantified by the unchanging d.c. value $F(0) = 1$.

Intuition tells us that, in the limit, the pulse becomes an impulse $\delta(t)$, as indicated in the background of the figure's time domain. Figure 2.8 thus represents the inverse transform of the real constant $F(j\omega) = 1$, i.e.

$$\mathcal{F}^{-1}\{1\} = \delta(t)$$

But intuition is not sufficient when dealing with infinity, the above result is substantiated analytically in Example 2.3.

Example 2.3

We derive the transform of the impulse $\delta(t)$. Inserting $f(t) = \delta(t)$ into the Fourier transform expression (2.5) we have

$$\mathcal{F}\{\delta(t)\} = \int_{-\infty}^{\infty} \delta(t)e^{-j\omega t}\, dt \tag{2.7}$$

But the integral on the right can be interpreted as the formal definition

(Equation 1.10) of an impulse, in the form

$$\int_{-\infty}^{\infty} \delta(t - \tau) f(t) \, dt = f(\tau)$$

With $f(t) = e^{-j\omega t}$ and $\tau = 0$, the integral of (2.7) takes the value

$$\int_{-\infty}^{\infty} \delta(t) e^{-j\omega t} \, dt = e^{-j\omega \tau} = 1$$

thus yielding the transform pair

$$\delta(t) \xleftrightarrow{\;\mathcal{F}\;} 1 \tag{2.8}$$

which confirms the earlier graphical deductions.

The symmetry of the surface in Figure 2.8 suggests exchanging the functions of the two domains. Thus, if we were to make $f(t) = 1$ and represent its Fourier transform, we would obtain a mirror image of Figure 2.8, such that the surfaces of the two ωt-planes would be identical. The result would be a frequency domain impulse $\delta(\omega)$, scaled by the factor 2π, that spoils the otherwise complete symmetry between the forward and inverse Fourier expressions, i.e.

$$\mathcal{F}\{1\} = 2\pi \delta(\omega)$$

This is confirmed in Example 2.4.

Example 2.4

We apply the same arguments as those of the last example to its dual case, the frequency domain impluse $\delta(\omega)$, whose inverse Fourier transform is

$$\mathcal{F}^{-1}\{\delta(\omega)\} = \frac{1}{2\pi} \int_{-\infty}^{\infty} \delta(\omega) e^{j\omega t} \, d\omega$$

The impulse definition (Equation 1.10) for frequency variables is

$$\int_{-\infty}^{\infty} \delta(\omega - v)\phi(\omega) \, d\omega = \phi(v)$$

Making $\phi(\omega) = e^{j\omega t}$ and $v = 0$, we have

$$\int_{-\infty}^{\infty} \delta(\omega) e^{j\omega t} \, dt = e^{jvt} = 1$$

which leads to the Fourier pair

$$1 \xleftrightarrow{\;\mathcal{F}\;} 2\pi \delta(\omega) \tag{2.9}$$

2.2.5   Time-lapse sequence

The earlier Figure 1.20 showed a time-lapse sequence of rotating vectors associated with the Fourier series coefficients of a function. This was subsequently used to examine the Fourier series of time-shifted functions. We now extend these concepts to the Fourier transform.

Consider the continuous frequency representation $F(j\omega)$ of Figure 2.8 to be made up of infinitesimal frequency components. This was suggested earlier in Figure 2.2 and will be formalized in Section 5.3.3. Interpreting these components as rotating vectors, the function $F(j\omega)$ represents their original configuration, a snapshot at the time origin $t = 0$.

Imagine now that the function $F(j\omega)$ of Figure 2.8 is released, so that the infinitesimal frequency components are allowed to rotate, each at its own constant angular velocity ω. As time progresses to $t = t_1, t_2, t_3, \ldots$ of Figure 2.9, the original function $F(j\omega)$ twists into an ever tightening helix, suggested by the time-lapse sequence in the frequency domain region.

The helices are samples of the integrand $F(j\omega)e^{j\omega t_i}$ of the inverse Fourier transform (2.6), and the cosines shown on the ωt-plane represent their real parts. Because $F(j\omega)$ of this example is even, the areas of the cosines suffice to determine values of $f(t)$, as indicated in the time domain of Figure 2.9.

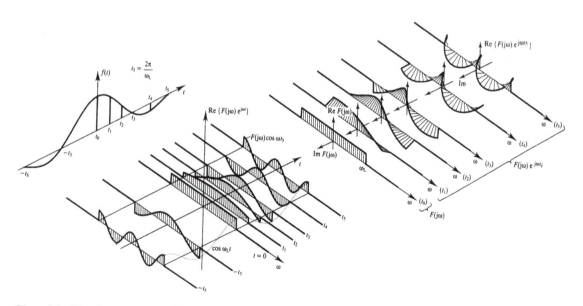

Figure 2.9   Time-lapse sequence of frequency pulse.

Those time values for which the truncated frequency domain helix has an integer number of turns yield $f(t) = 0$. For instance, at $t = t_5$ the frequency component at band-edge $\omega = \omega_L$ describes one full revolution $\omega_L t_5 = 2\pi$ (which makes $t_5 = 2\pi/\omega_L$) so that the integral of the cosine is zero.

One fundamental difference to the Fourier series case of Figure 1.20 is that the function $F(j\omega)$ is continuous, so that the whole concept of fundamental frequency and of associated harmonics is missing.

2.2.6 Transform of time-shifted function

The above time-lapse sequence gives an insight into the changes imposed on a transform when the original function is shifted in time. It also provides a tool for deriving such transforms. This concept is an extension of that presented in Section 1.4.4 for periodic functions.

If we were to evaluate $f(t)$ of Figure 2.9 for a negative time value, for instance $t = -t_5$, we would twist the helix by the same amount as in the snapshot at $t = t_5$, but in the opposite direction. Having reached this configuration, instead of evaluating $f(t)$, we define $t = -t_5$ as our new time origin, so that the resulting helix becomes the transform of the shifted function $f(t - t_5)$ thus defined.

This is illustrated in Figure 2.10 for a function $g(t) = f(t - \tau)$, where the function $f(t)$ is that of the preceding two figures and $\tau = t_5$. The result can be verified graphically by evaluating $g(t)$ at $t = \tau$. We allow the infinitesimal frequency components to rotate forward until $t = \tau$. At this instant they all reach the vertical position, and the integrand becomes

$$G(j\omega)e^{j\omega\tau} = F(j\omega) = 1$$

If the helix $G(j\omega)$ is not truncated at $|\omega| = \omega_L$, but extends to infinity, the construction of Figure 2.10 builds up to the shifted impulse $g(t) = \delta(t - \tau)$ shown in the background of the time domain.

Being the transform of the complex exponential, the shifted impulse plays a crucial role in signals and systems, as it permits writing the Fourier transform of sinusoidal functions and, by extension, of periodic functions. In this capacity it provides the analytical basis for time and frequency sampling. It is also fundamental to the impulse response method of Chapter 5, for solving systems and, by duality properties of the Fourier transform, also to the frequency response method.

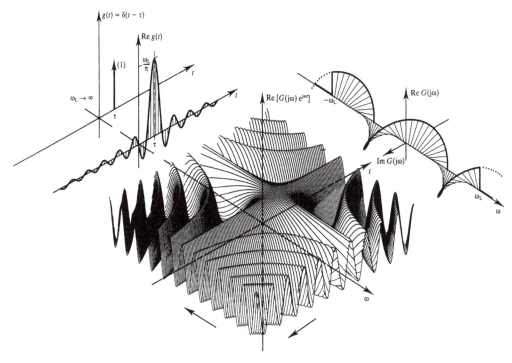

Figure 2.10   Time-shifted function.

Example 2.5

We confirm the above result by applying the Fourier transform definition to the shifted impulse $\delta(t - \tau)$

$$\mathcal{F}\{\delta(t - \tau)\} = \int_{-\infty}^{\infty} \delta(t - \tau)e^{-j\omega\tau}\,dt$$

This is similar to Example 2.3, except that this integral fits one variant of the impulse definition (Equation 1.10) exactly. We conclude that

$$\delta(t - \tau) \quad \overset{\mathcal{F}}{\longleftrightarrow} \quad e^{-j\omega\tau} \tag{2.10}$$

This result is expressed graphically in Figure 2.11, where the frequency domain is represented in terms of magnitude and phase. Note that the magnitude $A(\omega)$ is unity, the same as for the impulse at the origin $\delta(t)$, while the phase $\phi(\omega)$ varies linearly with frequency, at a rate determined by the displacement τ. Verify that this representation agrees with that provided by the helix of Figure 2.10.

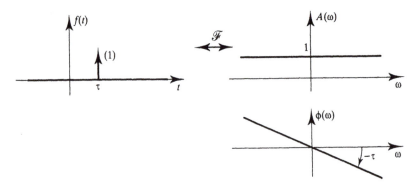

Figure 2.11 Transform of shifted impulse $\delta(t - \tau)$.

Example 2.6

The dual case of the frequency shifted impulse $\delta(\omega - v)$ gives similar results. Proceeding as in Example 2.4 we have

$$\mathscr{F}^{-1}\{\delta(\omega - v)\} = \frac{1}{2\pi} \int_{-\infty}^{\infty} \delta(\omega - v)e^{j\omega t} \, d\omega$$

which leads to the pair

$$e^{jvt} \quad \overset{\mathscr{F}}{\longleftrightarrow} \quad 2\pi\delta(\omega - v) \tag{2.11}$$

The symmetry, or duality, with Equation 2.10 is evident.

2.2.7 Odd-symmetric function

The functions encountered so far in this chapter have been real and even in both domains. The exception was the shifted impulse, where the real impulse ceased to be even and the transform also acquired an imaginary odd part. We now interpret graphically the transform of an odd-symmetric function.

Consider the function of Figure 2.12, defined in the frequency domain by an imaginary odd function $F(j\omega)$, band limited to $|\omega| = \omega_L$. To visualize its time domain, we release the frequency components from their imaginary plane, letting each rotate at its own angular velocity ω.

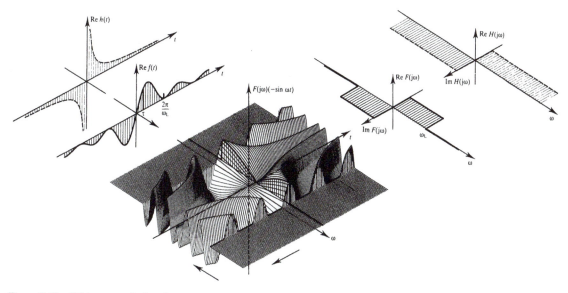

Figure 2.12 Odd-symmetric function.

The projections on the real plane Re $F(j\omega)$ describe sine functions in the time domain, as shown on the ωt-plane of Figure 2.12. Slicing the resulting surface with a plane of constant $t = \tau$, and integrating over frequency, gives one value $f(\tau)$. Repeating this for all values of t produces the complete time function $f(t)$ shown in the figure. We will see in Section 2.3.2 that this is a real odd function.

The time and frequency domain backgrounds show the same function not limited in frequency. In Section 10.3 this function is associated with the Hilbert transformer.

This example suggests that the transform of an arbitrary function involving even and odd components with real and imaginary parts calls for more elaborate interpretations.

2.3 Transform of general complex function

A general complex function has a real part and an imaginary part. Each of these can be interpreted as a sum of an even and an odd component. This gives four elemental function types with distinctive behaviour under the Fourier transform, which help to visualize the general shape of a transformed function and to predict its properties.

2.3.1   Forward transform

We first analyse the Fourier transform of a general function $f(t)$ in terms of symmetry. Writing $f(t)$ and $e^{-j\omega t}$ as sums of their even and odd components,

$$f(t) = f_e(t) + f_o(t)$$

$$e^{-j\omega t} = \cos \omega t - j \sin \omega t$$

the Fourier expression 2.5 expands into the sum

$$F(j\omega) = \int_{-\infty}^{\infty} f_e(t) \cos \omega t \, dt - j \int_{-\infty}^{\infty} f_o(t) \sin \omega t \, dt$$

$$+ \int_{-\infty}^{\infty} f_o(t) \cos \omega t \, dt - j \int_{-\infty}^{\infty} f_e(t) \sin \omega t \, dt$$

But the last two terms are zero on grounds of symmetry. In each of these the integrand, being the product of an even and an odd function, is an odd function of time, so that its infinite integral vanishes. This leaves the generally valid expression

$$F(j\omega) = \int_{-\infty}^{\infty} f_e(t) \cos \omega t \, dt - j \int_{-\infty}^{\infty} f_o(t) \sin \omega t \, dt \qquad (2.12)$$

Separating $f_e(t)$ and $f_o(t)$ into their real and imaginary parts, for instance,

$$f_e(t) = \mathrm{Re}\, f_e(t) + j \,\mathrm{Im}\, f_e(t)$$

we have

$$F(j\omega) = \int_{-\infty}^{\infty} \mathrm{Re}\, f_e(t) \cos \omega t \, dt + j \int_{-\infty}^{\infty} \mathrm{Im}\, f_e(t) \cos \omega t \, dt$$

$$- j \int_{-\infty}^{\infty} \mathrm{Re}\, f_o(t) \sin \omega t \, dt + \int_{-\infty}^{\infty} \mathrm{Im}\, f_o(t) \sin \omega t \, dt \qquad (2.13)$$

where all four integrands are real and even functions of time, so that the integrals are also real.

The function $F(j\omega)$ can be broken up, in its own right, into elemental components

$$F(j\omega) = \mathrm{Re}\, F_e(j\omega) + j \,\mathrm{Im}\, F_e(j\omega) + \mathrm{Re}\, F_o(j\omega) + j \,\mathrm{Im}\, F_o(j\omega)$$

where even and odd symmetry now relate to the frequency variable ω. Equating terms with Equation 2.13, where only the factors $\cos \omega t$ and $\sin \omega t$ are functions of ω, thus determining the symmetry of the integrals, we find

$$\mathrm{Re}\,F_e(j\omega) = \int_{-\infty}^{\infty} \mathrm{Re}\,f_e(t)\cos\omega t\,dt$$

$$\mathrm{Im}\,F_e(j\omega) = \int_{-\infty}^{\infty} \mathrm{Im}\,f_e(t)\cos\omega t\,dt \tag{2.14}$$

$$\mathrm{Re}\,F_o(j\omega) = \int_{-\infty}^{\infty} \mathrm{Im}\,f_o(t)\sin\omega t\,dt$$

$$\mathrm{Im}\,F_o(j\omega) = -\int_{-\infty}^{\infty} \mathrm{Re}\,f_o(t)\sin\omega t\,dt$$

which gives a one-to-one correspondence between elements of the two domains.

2.3.2   Inverse transform

The symmetry of the Fourier transform suggests that the same procedure, applied to the inverse transform 2.6, leads to similar elemental relationships. Indeed, with the slight asymmetry caused by the sign of the exponential, the counterpart of Equation 2.12 takes the form

$$f(t) = \frac{1}{2\pi} \int_{-\infty}^{\infty} F_e(j\omega)\cos\omega t\,d\omega + \frac{j}{2\pi} \int_{-\infty}^{\infty} F_o(j\omega)\sin\omega t\,d\omega$$

and that of (2.13) the form

$$f(t) = \frac{1}{2\pi} \int_{-\infty}^{\infty} \mathrm{Re}\,F_e(j\omega)\cos\omega t\,d\omega + \frac{j}{2\pi} \int_{-\infty}^{\infty} \mathrm{Im}\,F_e(j\omega)\cos\omega t\,d\omega$$

$$+ \frac{j}{2\pi} \int_{-\infty}^{\infty} \mathrm{Re}\,F_o(j\omega)\sin\omega t\,d\omega$$

$$- \frac{1}{2\pi} \int_{-\infty}^{\infty} \mathrm{Im}\,F_o(j\omega)\sin\omega t\,d\omega$$

which yields the inverse relationships between the functions in Equation 2.14,

$$\mathrm{Re}\,f_e(t) = \frac{1}{2\pi} \int_{-\infty}^{\infty} \mathrm{Re}\,F_e(j\omega)\cos\omega t\,d\omega$$

$$\mathrm{Im}\,f_e(t) = \frac{1}{2\pi} \int_{-\infty}^{\infty} \mathrm{Im}\,F_e(j\omega)\cos\omega t\,d\omega \tag{2.15}$$

$$\mathrm{Re}\,f_o(t) = -\frac{1}{2\pi} \int_{-\infty}^{\infty} \mathrm{Im}\,F_o(j\omega)\sin\omega t\,d\omega$$

$$\mathrm{Im}\,f_o(t) = \frac{1}{2\pi} \int_{-\infty}^{\infty} \mathrm{Re}\,F_o(j\omega)\sin\omega t\,d\omega$$

Figure 2.8 provides a typical example of the first expression, while Figure 2.12 gives a typical example of the third.

2.3.3   Function symmetries and real functions of time

We conclude that symmetry plays a decisive role in the destination of the elemental components: each element of one domain has a specific counterpart in the transformed domain. Symmetry is preserved as even functions remain even and odd functions remain odd. Regarding real and imaginary types, if the function is even, it remains of the same type, but if the function is odd, its transform is of the alternative type.

The subdivision of the time domain and frequency domain descriptions of a function into elemental components, and the relationships (2.14) and (2.15) between such components, can be summarized as follows

$$f(t)$$

$$
\begin{array}{ccc}
\overbrace{f_e(t)} & + & \overbrace{f_o(t)}
\end{array}
$$

$$
\begin{array}{ccccc}
\overbrace{\mathrm{Re}\, f_e(t) \quad + \quad j\,\mathrm{Im}\, f_e(t)} & + & \overbrace{\mathrm{Re}\, f_o(t) \quad + \quad j\,\mathrm{Im}\, f_o(t)}
\end{array}
$$

$$
\begin{array}{cc}
\updownarrow (\cos \omega t) \quad \updownarrow (\cos \omega t) & (\sin \omega t) \diagdown \diagup (-\sin \omega t)
\end{array}
$$

$$
\begin{array}{ccccc}
\underbrace{\mathrm{Re}\, F_e(j\omega) \quad + \quad j\,\mathrm{Im}\, F_e(j\omega)} & + & \underbrace{\mathrm{Re}\, F_o(j\omega) \quad + \quad j\,\mathrm{Im}\, F_o(j\omega)}
\end{array}
$$

$$
\begin{array}{ccc}
\underbrace{F_e(j\omega)} & + & \underbrace{F_o(j\omega)}
\end{array}
$$

$$F(j\omega)$$

$$(2.16)$$

These relationships contain some important special cases. If $f(t)$ is a real function of time, the above schematic reduces to

$$\mathrm{Re}\, f(t)$$

$$
\begin{array}{ccc}
\overbrace{\mathrm{Re}\, f_e(t)} & + & \overbrace{\mathrm{Re}\, f_o(t)} \\
\updownarrow (\cos \omega t) & & \updownarrow (-\sin \omega t) \\
\underbrace{\mathrm{Re}\, F_e(j\omega)} & + & j\,\mathrm{Im}\, F_o(j\omega)
\end{array}
$$

$$(2.17)$$

$$F(j\omega)$$

The frequency domain representation has a real even part, derived from the even part of $f(t)$, and an imaginary odd part, derived from the odd part of $f(t)$. Functions with this type of symmetry are called Hermitian.

Similarly, the transform of a pure imaginary function of time is given by

$$
\overbrace{
\begin{array}{ccc}
\mathrm{j\,Im}\,f_e(t) & + & \mathrm{j\,Im}\,f_o(t) \\
\Big\downarrow (\cos \omega t) & & \Big\downarrow (\sin \omega t) \\
\mathrm{j\,Im}\,F_e(\mathrm{j}\omega) & + & \mathrm{Re}\,F_o(\mathrm{j}\omega)
\end{array}
}^{\mathrm{j\,Im}\,f(t)}
\tag{2.18}
$$
$$
\underbrace{\phantom{\mathrm{j\,Im}\,F_e(\mathrm{j}\omega) \qquad + \qquad \mathrm{Re}\,F_o(\mathrm{j}\omega)}}_{F(\mathrm{j}\omega)}
$$

whose real and imaginary parts are respectively odd and even, and this is called an anti-Hermitian function.

Example 2.7

We derive the transform of the real causal exponential function

$$f(t) = e^{at} u(t) \quad a < 0$$

The transform definition 2.5 yields

$$F(\mathrm{j}\omega) = \int_{-\infty}^{\infty} e^{at} u(t) e^{-\mathrm{j}\omega t}\, dt = \int_{0}^{\infty} e^{(a-\mathrm{j}\omega)t}\, dt = \frac{1}{\mathrm{j}\omega - a}$$

and the Fourier pair

$$e^{at} u(t) \quad \overset{\mathcal{F}}{\longleftrightarrow} \quad \frac{1}{\mathrm{j}\omega - a} \quad a < 0$$

The function $F(\mathrm{j}\omega)$ can be expresed in terms of its real and imaginary components

$$F(\mathrm{j}\omega) = \mathrm{Re}\,F(\mathrm{j}\omega) + \mathrm{j\,Im}\,F(\mathrm{j}\omega) = \frac{-a}{\omega^2 + a^2} + \mathrm{j}\frac{-\omega}{\omega^2 + a^2}$$

which correspond to the even and odd components of the scheme 2.17 as

$$f_e(t) = \tfrac{1}{2}e^{at} u(t) + \tfrac{1}{2}e^{-at} u(-t) \quad \overset{\mathcal{F}}{\longleftrightarrow} \quad \mathrm{Re}\,F_e(\mathrm{j}\omega) = \frac{-a}{\omega^2 + a^2}$$

and

$$f_o(t) = \tfrac{1}{2}e^{at} u(t) - \tfrac{1}{2}e^{-at} u(-t) \quad \overset{\mathcal{F}}{\longleftrightarrow} \quad \mathrm{j\,Im}\,F_o(\mathrm{j}\omega) = \mathrm{j}\frac{-\omega}{\omega^2 + a^2}$$

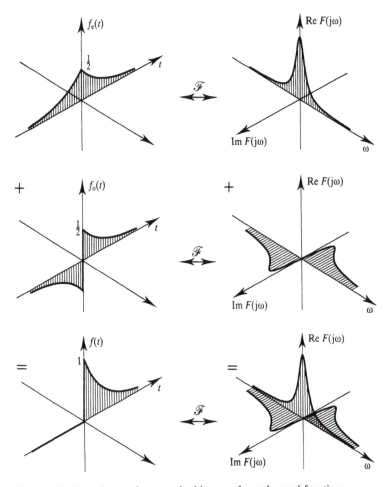

Figure 2.13   Transforms of even and odd parts of a real causal function.

These components and their sums are shown related in Figure 2.13

For completeness we write $F(j\omega)$ in terms of magnitude and phase as

$$F(j\omega) = A(\omega)e^{j\phi(\omega)}$$

where

$$A(\omega) = \frac{1}{\sqrt{\omega^2 + a^2}} \quad \text{and} \quad \phi(\omega) = \arctan\frac{\omega}{a}$$

2.3.4 Full graphical interpretation

We can now re-examine earlier graphical interpretations in the light of the above results, and then extend them to general complex functions.

Even function

Figure 2.8 holds the key to the interpretation. It represents the inverse Fourier transform of a rectangular pulse of width ω_L. But, with ω_L increased to infinity, it also shows the inverse transform of $F(j\omega) = 1$ that leads to the impulse $\delta(t)$.

In this latter capacity, the symmetrical three-dimensional surface on the ωt-plane must be thought of as extending to infinity, in both directions. We call this the **cosine surface**. It exhibits a remarkable regularity and a symmetry in the variables ω and t, in each of which it is also even symmetrical.

Since $F(j\omega) = 1$ is real and even, the first of expressions 2.15 applies, so that in Figure 2.8 we are visualizing the integral

$$f(t) = \frac{1}{2\pi} \int_{-\infty}^{\infty} \cos \omega t \, d\omega$$

This is the case of simplest $F(j\omega)$, where the integrand is the $\cos \omega t$ surface itself. For other real and even functions, depending on whether it is a forward or inverse transform, the integrand on the ωt-plane is one of the products

$$f(t) \cos \omega t \quad \text{or} \quad F(j\omega) \cos \omega t$$

To simplify explanations, we borrow the terminology of **amplitude modulated signals**, where a sinusoidal carrier is multiplied by a modulating signal. We loosely extend this to two dimensions by associating the term **carrier** with the cosine surface and the term **modulating function** with an irregular cylinder whose cross-section is the given function $f(t)$ or $F(j\omega)$. At each point of the ωt-plane the integrand is the product of the carrier surface and the modulating function.

With this interpretation, the simple rectangular pulse of width ω_L of Figure 2.8 defines a cylindrical envelope that truncates the cosine carrier surface at $\omega = \pm\omega_L$. This surface is then sliced and integrated.

Similarly, in Figure 2.7 the cosine carrier surface is modulated by a narrower cylinder, leading to a wider pulse in the time domain. In contrast, in the forward transform of Figure 2.6 the same cosine carrier surface is cylindrically modulated, by a similar rectangular time window $f(t)$, but now in the direction of the frequency axis.

In the case of Figure 2.5, the variability of $F(j\omega)$, which includes sign changes, complicates the shape of the integrand surface. But there is still a faint suggestion of cylindrical modulation in the direction of the time axis.

Odd function

Just as even functions were modulating the even cosine carrier surface of Figure 2.8 to give even integrands, so odd functions modulate the odd sine carrier surface.

This surface is revealed in the process of representing the real part of the transform of $F(j\omega) = -j$. This is shown in Figure 2.14, which also confirms that the real part of the transform of an imaginary even function is zero. There is again a symmetry in the variables ω and t, in each of which it is also odd symmetric.

The two three-dimensional carrier surfaces, $\cos \omega t$ and $\sin \omega t$, are the real even and imaginary odd parts of a four-dimensional surface representing the complex exponential $e^{j\omega t} = \cos \omega t + j \sin \omega t$. Their symmetries account for many of the dual properties of Fourier transforms, to be seen in Section 2.4.

The integrand on the ωt-plane of Figure 2.12 can now be reinterpreted as the result of modulating the sine carrier surface of Figure 2.14 with the imaginary odd part of $F(j\omega)$. That case represents the third of Equations 2.15, where the product $F(j\omega)(-\sin \omega t)$ is a real even function of frequency, with real integral.

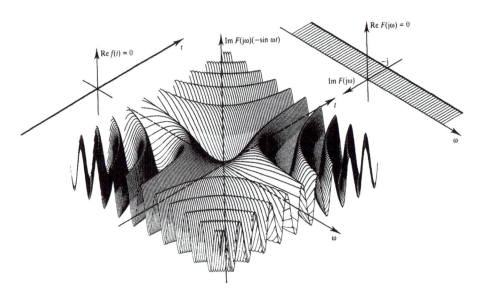

Figure 2.14　Sine carrier surface.

Real function with even and odd components

Physical signals are in general real functions of time, and so are their idealized models. Their even and odd components are thus transformed according to the scheme 2.17.

We illustrate this important case in the sequence of Figures 2.15 to 2.18, where we re-examine an even and an odd function, and then show the transform of the sum. Each figure combines two diagrams involving the cosine and sine carriers. Using the same layout introduces some redundancy, but this helps to clarify some earlier concepts.

The real even function $f_e(t)$ of Figure 2.15 modulates the cosine carrier to give an integrand surface that is even in time, thus yielding a real even transform. The same function, modulating the odd sine carrier, produces an integrand that is odd in time, with zero integral.

Similarly, the real odd function $f_o(t)$ of Figure 2.16 produces an odd integrand with the cosine carrier and an even integrand with the sine carrier, so that the overall transform is imaginary odd. The shape of this transform is of the form $(1 - \cos x)/x$, to be confirmed later, in Example 2.18.

The sum of these two functions is the rectangular pulse

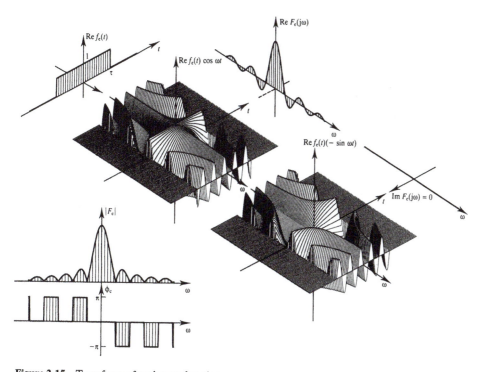

Figure 2.15 Transform of real even function.

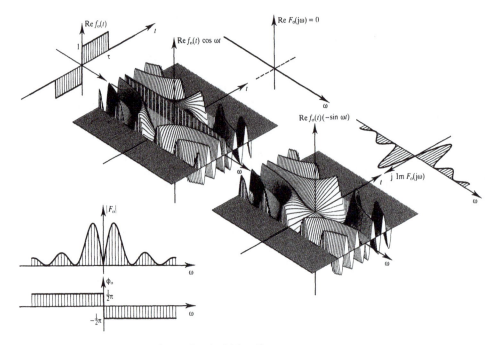

Figure 2.16   Transform of real odd function.

$f(t) = f_e(t) + f_o(t)$ of Figure 2.17, which modulates the two carriers as shown. Note that in this example, in which $f(t)$ is causal, the integrand surfaces replicate the positive-time halves of the corresponding even surfaces of the preceding two figures, but at twice the magnitude. The integrals are thus the same, and their complex sum represents the transform of $f(t)$, as expressed in the scheme 2.17.

This last example graphically confirms the general expression 2.12, showing that the cosine carrier extracts the even part of the Fourier transform of an arbitrary real function, while the sine carrier extracts the odd part. The two odd surfaces of Figures 2.15 and 2.16 simply illustrate the symmetry considerations that led to the result 2.12.

The real and imaginary parts of the frequency domain of Figure 2.17 are combined in Figure 2.18 to show their contributions to magnitude and phase.

The graphical representations introduced here will be extended in Chapter 3 to represent the Laplace transform, and later the z-transform.

General complex function

We conclude these interpretations by relating the four elemental components of a general complex function $f(t)$ to their frequency domain counterparts, as expressed in the relationships 2.16. Note that abund-

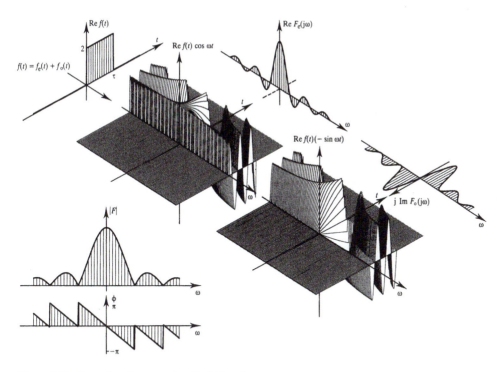

Figure 2.17   Sum of real even and real odd functions.

ance of detail does not signify complexity.

It is easier to visualize the subdivision of $f(t)$ into component elements by first representing its real and imaginary parts, as shown at the centre of Figure 2.19, and subdividing these into even and odd components as shown.

To reconstitute $F(j\omega)$ from its elemental components, we follow a similar inverse process, as indicated in the central cluster of Figure 2.20. Each of the frequency components is derived from the appropriate time component, as expressed in scheme 2.16. These are represented by the four satellite clusters of Figure 2.20, each of which visualizes one of the expressions 2.14. The capricious orientations of the time domain components are dictated by the eventual orientation of the frequency components.

2.3.5   Fourier sine and cosine transforms

The generally valid Fourier transform expression 2.12, repeated here,

$$F(j\omega) = \int_{-\infty}^{\infty} f_e(t) \cos \omega t \, dt - j \int_{-\infty}^{\infty} f_o(t) \sin \omega t \, dt \qquad (2.19)$$

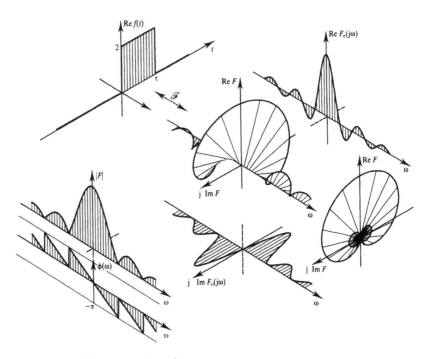

Figure 2.18 Full representation of sum.

acknowledges that any arbitrary function $f(t)$ can be interpreted as the sum of an even and an odd component, expressed as

$$f(t) = f_e(t) + f_o(t)$$

where

$$f_e(t) = \tfrac{1}{2}[f(t) + f(-t)] \quad \text{and} \quad f_o(t) = \tfrac{1}{2}[f(t) - f(-t)]$$

Each of these components may have real and imaginary parts, as seen earlier. Two special forms of the Fourier transform are associated with these symmetries, namely the **Fourier cosine transform** of $f(t)$, denoted by

$$f(t) \xleftrightarrow{\mathscr{F}_c} F_c(\omega)$$

and defined by the forward and inverse expressions

$$F_c(\omega) = 2\int_0^\infty f(t) \cos \omega t \, dt \tag{2.20}$$

$$f(t) = \frac{1}{\pi} \int_0^\infty F_c(\omega) \cos \omega t \, d\omega$$

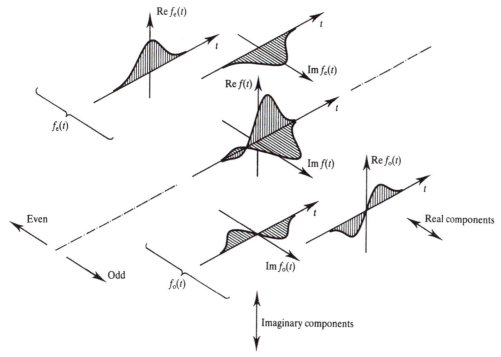

Figure 2.19   Elemental subdivision of general time function.

and the **Fourier sine transform** of $f(t)$, denoted by

$$f(t) \xleftrightarrow{\mathscr{F}_s} F_s(\omega)$$

and defined as

$$F_s(\omega) = 2\int_0^\infty f(t)\sin\omega t \, dt$$

$$f(t) = \frac{1}{\pi} \int_0^\infty F_s(\omega)\sin\omega t \, d\omega$$

(2.21)

These definitions effectively ignore any function values of negative time, tacitly implying that these values are, respectively, even-symmetric or odd-symmetric extensions of the corresponding positive time values.

Given the even symmetry of the integrands of Equation 2.19, with these definitions we can write

$$F(j\omega) = F_c(\omega) - jF_s(\omega)$$

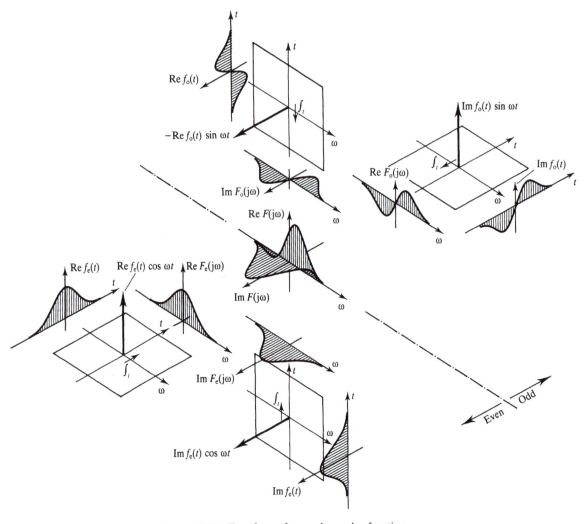

Figure 2.20   Transform of general complex function.

so that

$$\mathcal{F}\{f(t)\} = \mathcal{F}_c\{f_e(t)\} - j\mathcal{F}_s\{f_o(t)\}$$

$$\mathcal{F}^{-1}\{F(j\omega)\} = \mathcal{F}_c^{-1}\{F_c(\omega)\} + j\mathcal{F}_s^{-1}\{F_s(\omega)\}$$

(2.22)

The great merit of this notation lies in the vast number of Fourier sine and cosine pairs that can be found tabulated in specialized handbooks, e.g. Oberhettinger (1957). The Fourier transform of a function $f(t)$ can thus be found by separating it into its even and odd parts, looking up their cosine and sine transforms, and adding these as in Equations 2.22.

The earlier set of Figures 2.15 to 2.18 illustrates this process. If $f(t)$ of Figure 2.17 is given, then $f_e(t)$ and $f_o(t)$ of Figures 2.15 and 2.16 are its even and odd components. Their transforms relate to the cosine and sine transforms as $F_c(\omega) = \operatorname{Re} F_e(j\omega)$ and $F_s(\omega) = -\operatorname{Im} F_o(j\omega)$ and the modulated cosine and sine surfaces of Figure 2.17 represent the integrands of Equations 2.20 and 2.21.

Nowadays, with the advent of the fast Fourier transform and the extensive use of numerical evaluations, such tables tend to be over-looked, but they remain invaluable in the study of many theoretical problems involving analytical manipulations.

2.4 Transform properties

We have already met some transform properties in the context of the earlier graphical developments. Here we formalize their presentation and indicate ways of deriving new transform pairs from known ones.

Most of these properties are valid, in slightly modified forms, for the Laplace transform, the discrete classes of the Fourier transform and the z-transform. For this reason we present' them in some detail, at the same time developing examples for use in later chapters.

The Fourier integral 2.5 does not converge for all functions. The Dirichlet conditions,

- the function $f(t)$ is absolutely integrable, that is $\int_{-\infty}^{\infty}|f(t)|\mathrm{d}t < \infty$,
- $f(t)$ has a finite number of maxima and minima and a finite number of discontinuities in any finite interval,

provide a set of sufficient conditions for the existence of the Fourier transform $F(j\omega)$. We examine this problem in more detail in Section 3.3, in the context of the region of convergence of the Laplace transform.

Functions that do not meet the Dirichlet conditions may still have a Fourier transform. These include periodic functions, whose transforms consist of impulses, and functions whose Fourier integral only converges as a limit. We illustrate the latter concept with an example.

Example 2.8

We wish to find the Fourier transform of the signum function defined as

$$f(t) = \operatorname{sgn}(t) = \begin{cases} -1 & t < 0 \\ +1 & t > 0 \end{cases}$$

which is not absolutely integrable and its integral (2.5) does not converge. We form the auxiliary function

$$g(t) = \begin{cases} -e^{\varepsilon t} & t < 0 \\ e^{-\varepsilon t} & t > 0 \end{cases}$$

illustrated in Figure 2.21, which yields $f(t)$ as the limit

$$f(t) = \lim_{\varepsilon \to 0} g(t)$$

The Fourier transform of $g(t)$ is the sum

$$\mathcal{F}\{g(t)\} = \int_{-\infty}^{0} -e^{\varepsilon t} e^{-j\omega t}\, dt + \int_{0}^{\infty} e^{-\varepsilon t} e^{-j\omega t}\, dt$$

$$= -\int_{-\infty}^{0} e^{(\varepsilon - j\omega)t}\, dt + \int_{0}^{\infty} e^{-(\varepsilon + j\omega)t}\, dt$$

$$= -\frac{1}{\varepsilon - j\omega} + \frac{1}{\varepsilon + j\omega}$$

and the transform of $f(t)$ is obtained as the limit

$$F(j\omega) = \mathcal{F}\{\text{sgn}(t)\} = \lim_{\varepsilon \to 0} \mathcal{F}\{g(t)\} = \frac{2}{j\omega} = -j\frac{2}{\omega}$$

This yields the transform pair

$$\text{sgn}(t) \xleftrightarrow{\;\mathcal{F}\;} -j\frac{2}{\omega} \tag{2.23}$$

which is real odd in time, hence imaginary odd in frequency, as shown in Figure 2.21.

2.4.1 Linearity

An operator L is said to be linear, if the two conditions

$$L\{x + y\} = L\{x\} + L\{y\} \quad \text{and} \quad L\{cx\} = cL\{x\}$$

hold for all x and y for which $L\{x\}$ and $L\{y\}$ are defined, and for every constant c. This leads to the more general condition

$$L\{ax + by\} = aL\{x\} + bL\{y\}$$

where a and b are arbitrary constants.

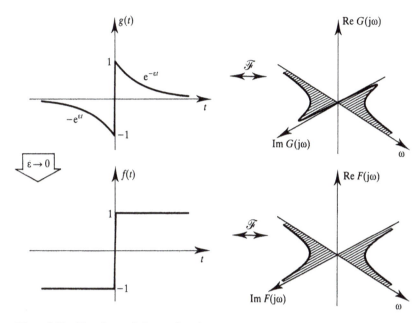

Figure 2.21   Transform of signum function.

The Fourier transform meets these requirements. Denoting two time functions and their transforms by the pairs

$$f_1(t) \overset{\mathcal{F}}{\longleftrightarrow} F_1(j\omega) \quad \text{and} \quad f_2(t) \overset{\mathcal{F}}{\longleftrightarrow} F_2(j\omega)$$

we have

$$\boxed{a f_1(t) + b f_2(t) \overset{\mathcal{F}}{\longleftrightarrow} a F_1(j\omega) + b F_2(j\omega)} \tag{2.24}$$

This can be verified by applying the Fourier transform definition 2.5 to the sum, and by the linearity of the integral operator.

Note that both the functions in expression 2.24 as well as the constants a and b can take complex values, so that addition involves vector quantities.

Example 2.9

We find the transform of the unit step $u(t)$ from the transforms of its even and odd components,

$$u(t) = \tfrac{1}{2} + \tfrac{1}{2} \, \text{sgn}(t)$$

shown in Figure 2.22. The first term, a real constant representing the d.c. component of the step, transforms to a real impulse of strength π. Combined with the result 2.23 it yields the transform pair

$$u(t) \quad \overset{\mathcal{F}}{\longleftrightarrow} \quad U(j\omega) = \frac{1}{j\omega} + \pi\delta(\omega) \qquad (2.25)$$

Other examples are found in Sections 2.3 and 2.4, which interpret arbitrary functions as sums of elemental components. For instance, the

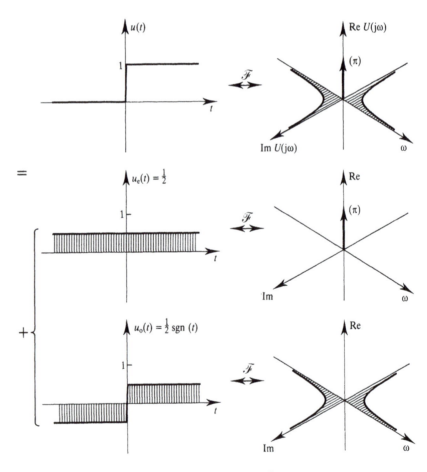

Figure 2.22 Transform of $u(t)$ as sum of even and odd parts.

sequence of Figures 2.15 to 2.18 illustrates linearity when sums of even and odd functions and real and imaginary functions are involved.

2.4.2 Time and frequency scaling

This property represents an inverse relationship between time and frequency domain representations. It relates the transforms of functions that only differ in the scaling of the independent variable. It should not be confused with amplitude scaling, which simply involves linearity.

Time-scaling

Given a function $f(t)$ with transform $F(j\omega)$, the transform of the time-scaled function $f(at)$ is

$$\mathcal{F}\{f(at)\} = \int_{-\infty}^{\infty} f(at)e^{-j\omega t}\,dt$$

The change of variable $x = at$, which implies $t = x/a$ and $dt = dx/a$, yields

$$\mathcal{F}\{f(at)\} = \frac{1}{a} \int_{-\infty}^{\infty} f(x)e^{-j\frac{\omega}{a}x}\,dx$$

This is valid when $a > 0$. For $a < 0$ the integration limits are inverted,

$$\mathcal{F}\{f(at)\} = \frac{1}{a} \int_{\infty}^{-\infty} f(x)e^{-j\frac{\omega}{a}x}\,dx = -\frac{1}{a} \int_{-\infty}^{\infty} f(x)e^{-j\frac{\omega}{a}x}\,dx$$

These are combined into the single pair

$$\boxed{f(at) \quad \overset{\mathcal{F}}{\longleftrightarrow} \quad \frac{1}{|a|} F\left(j\frac{\omega}{a}\right)} \tag{2.26}$$

Thus, expansion of the time scale (or time duration) leads to compression of the frequency scale (or bandwidth) and vice versa. And this is accompanied by an inverse scaling of the amplitude.

Example 2.10

Given the transform pair of Example 2.1,

$$f(t) = p_\tau(t) \quad \overset{\mathcal{F}}{\longleftrightarrow} \quad F(j\omega) = \tau \frac{\sin\frac{1}{2}\omega\tau}{\frac{1}{2}\omega\tau}$$

illustrated with $\tau = 1$ in the middle row of Figure 2.23, we derive two related pairs by scaling the time variable by $a_1 = 2$ and $a_2 = \frac{1}{2}$. Applying expression 2.26, we have

$$f_1(t) = p_\tau(2t) = p_{\tau/2}(t) \quad \overset{\mathscr{F}}{\longleftrightarrow} \quad F_1(j\omega) = \frac{1}{2}\tau \, \frac{\sin\frac{1}{4}\omega\tau}{\frac{1}{4}\omega\tau}$$

and

$$f_2(t) = p_\tau(\tfrac{1}{2}t) = p_{2\tau}(t) \quad \overset{\mathscr{F}}{\longleftrightarrow} \quad F_2(j\omega) = 2\tau \, \frac{\sin\omega\tau}{\omega\tau}$$

shown in the top and bottom rows of Figure 2.23 respectively.

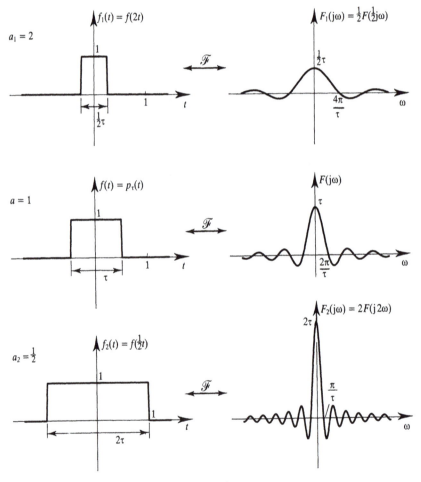

Figure 2.23 Time scaling, even function, $a > 0$.

Had we used a negative scaling factor in the above example, for instance $a = -1$, the results would have been identical, since the function was even symmetrical. This is not so when the function contains an odd part.

Example 2.11

Consider the exponential $f_1(t) = e^{-t}u(t)$ which, according to Example 2.7, has the transform

$$f_1(t) = e^{-t}u(t) \quad \overset{\mathcal{F}}{\longleftrightarrow} \quad F_1(j\omega) = \frac{1}{j\omega + 1}$$

We obtain the transform of the related function $f_2(t) = e^{2t}u(-t)$ by interpreting it as $f_2(t) = f_1(-2t)$ where $u(-2t) = u(-t)$. The property 2.15 yields

$$f_2(t) = e^{2t}u(-t) \quad \overset{\mathcal{F}}{\longleftrightarrow} \quad F_2(j\omega) = \frac{1}{-j\omega + 2}$$

The relationship between the two functions is seen in Figure 2.24.

Frequency scaling

Scaling the frequency variable by a factor b has a similar effect to time scaling. Making the substitution $b = 1/a$ in expression 2.26 yields the expression

$$\boxed{\frac{1}{|b|} f\left(\frac{t}{b}\right) \quad \overset{\mathcal{F}}{\longleftrightarrow} \quad F(jb\omega)} \tag{2.27}$$

which is completely symmetrical with expression 2.26.

Example 2.12

We construct the dual counterpart of Example 2.10, using for reference the result of Example 2.2

$$f(t) = \frac{v}{\pi} \frac{\sin vt}{vt} \quad \overset{\mathcal{F}}{\longleftrightarrow} \quad F(j\omega) = p_{2v}(j\omega)$$

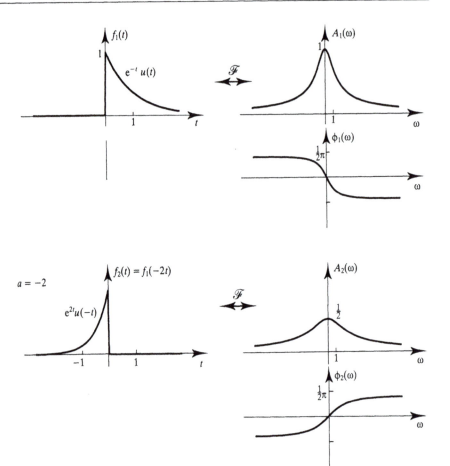

Figure 2.24 Time scaling, arbitrary function, $a < 0$.

shown in the middle row of Figure 2.25. The scaling factors $b_1 = 2$ and $b_2 = \frac{1}{2}$ lead to the results shown in the top and bottom rows of that figure, respectively. These results are strongly symmetrical with those of Figure 2.23.

The property 2.27 yields an important equivalence of impulses. Frequency scaling the pair

$$1 \overset{\mathcal{F}}{\longleftrightarrow} 2\pi\delta(\omega)$$

by the factor $b = j$ gives

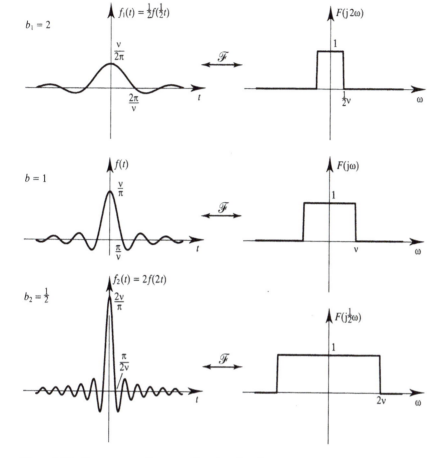

Figure 2.25 Frequency scaling, even function, $b > 0$.

$$\frac{1}{|j|} = 1 \quad \overset{\mathcal{F}}{\longleftrightarrow} \quad 2\pi\delta(j\omega)$$

hence $\delta(\omega) = \delta(j\omega)$.

2.4.3 Transform duality

The examples in Figures 2.23 and 2.25 reiterate the duality property of the Fourier transform, which was introduced graphically in Section 2.2.3. We observed there that the symmetries of the cosine and sine surfaces of Figures 2.8 and 2.14 caused similar shapes from alternative domains to be transformed to similar shapes. We now formalize those results.

Give a transform pair

$$f(t) \overset{\mathcal{F}}{\longleftrightarrow} g(j\omega)$$

we write the **inverse** Fourier transform definition 2.6 as

$$2\pi f(t) = \int_{-\infty}^{\infty} g(j\omega)e^{j\omega t}\, d\omega$$

Making a double change of variables, $t = -\tau$ and $\omega = v$, yields

$$2\pi f(-\tau) = \int_{-\infty}^{\infty} g(jv)e^{-jv\tau}\, dv$$

A second exchange of variables, now across domains $\tau = \omega$ and $v = t$, gives

$$2\pi f(-\omega) = \int_{-\infty}^{\infty} g(jt)e^{-j\omega t}\, dt$$

The right side represents the **forward** Fourier transform of $g(jt)$, which leads to the conclusion

$$g(jt) \overset{\mathcal{F}}{\longleftrightarrow} 2\pi f(-\omega) \tag{2.28}$$

This represents one possibility for expressing the **dual property** of the Fourier transform. A more useful form results from scaling the frequency domain according to Equation 2.27, with the scaling factor $b = j$, with the result

$$\boxed{g(t) \overset{\mathcal{F}}{\longleftrightarrow} 2\pi f(-j\omega)} \tag{2.29}$$

The dual property greatly simplifies the derivation of a transform pair, whenever the dual pair is known. We illustrate with two earlier examples.

Example 2.13

Derive the time domain representation of the frequency impulse $\delta(\omega)$, as the dual counterpart of the pair 2.8,

$$\delta(t) \overset{\mathcal{F}}{\longleftrightarrow} 1$$

By the property 2.29 and the even symmetry of the signal $\delta(-\omega) = \delta(\omega)$, the result is simply

$$1 \overset{\mathcal{F}}{\longleftrightarrow} 2\pi\delta(\omega)$$

which confirms expression 2.9.

The two pairs of the preceding example were derived separately in Section 2.2.4, by applying the forward and inverse Fourier transform definitions and the sifting property of the impulse function. Using the dual property simplifies one of these derivations. The illustration of Figure 2.26 stresses this symmetry.

The duality property implies that taking a second **forward transform** of the result $F(j\omega)$ of a first **forward transform** of $f(t)$ leads to a time-reversed version of the original function, expressed by

$$f(t) \underset{\mathscr{F}^{-1}}{\overset{\mathscr{F}}{\rightleftharpoons}} \quad F(j\omega) \underset{\mathscr{F}^{-1}}{\overset{\mathscr{F}}{\rightleftharpoons}} \quad 2\pi f(-t)$$

Example 2.14

We find the inverse transform of a rectangular frequency domain pulse

$$F(j\omega) = p_\nu(\omega) = \begin{cases} 1 & |\omega| < \tfrac{1}{2}\nu \\ 0 & \text{elsewhere} \end{cases}$$

from the knowledge of the symmetrical case of Example 2.1, i.e. the pair

$$p_\tau(t) \quad \overset{\mathscr{F}}{\longleftrightarrow} \quad \tau \frac{\sin \tfrac{1}{2}\omega\tau}{\tfrac{1}{2}\omega\tau}$$

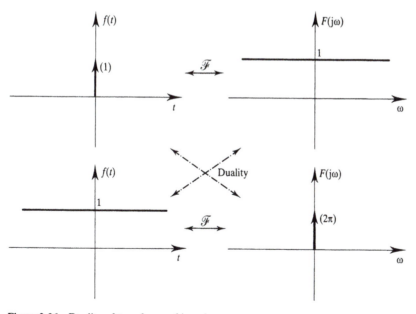

Figure 2.26 Duality of transforms of impulses.

which we illustrate at the top of Figure 2.27. By the dual property 2.29, and the evenness of the pulse $p_v(-\omega) = p_v(\omega)$, the desired result is

$$v\frac{\sin\frac{1}{2}vt}{\frac{1}{2}vt} \xleftrightarrow{\mathscr{F}} 2\pi p_v(\omega)$$

which we illustrate at the bottom of Figure 2.27.

The result of the preceding example is formally identical to that of Example 2.2, obtained by directly applying the transform definition. Scaling differences reflect the pulse widths of the two examples.

Gaussian function

The Gaussian function

$$f(t) = e^{-at^2}$$

although not unique in this, represents the ultimate in symmetry and duality, as it is symmetrical and has another Gaussian function as its Fourier transform. The Fourier definition gives

$$F(j\omega) = \int_{-\infty}^{\infty} e^{-at^2}e^{-j\omega t}\, dt$$

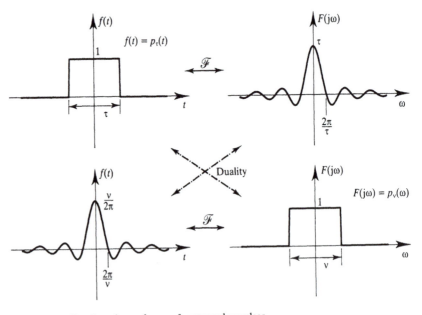

Figure 2.27   Duality of transforms of rectangular pulses.

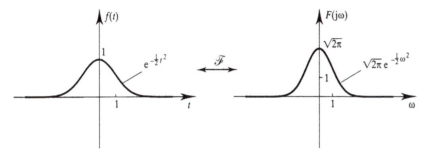

Figure 2.28 Gaussian function and its transform.

We complete the square of the exponent by multiplying and dividing by $e^{\omega^2/4a}$

$$F(j\omega) = e^{-\omega^2/4a} \int_{-\infty}^{\infty} e^{-(\sqrt{a}\,t + j\omega/2\sqrt{a})^2}\,dt$$

Changing the integration variable to $x = \sqrt{a}\,t + j\omega/2\sqrt{a}$, hence $dx = \sqrt{a}\,dt$, gives

$$F(j\omega) = \frac{e^{-\omega^2/4a}}{\sqrt{a}} \int_{-\infty}^{\infty} e^{-x^2}\,dx = \sqrt{\frac{\pi}{a}}\, e^{-\omega^2/4a}$$

because $\int_{-\infty}^{\infty} e^{-x^2}\,dx = \sqrt{\pi}$. This yields the transform pair

$$e^{-at^2} \quad \overset{\mathcal{F}}{\longleftrightarrow} \quad \sqrt{\frac{\pi}{a}}\, e^{\omega^2/4a}$$

and the special symmetrical case for $a = \tfrac{1}{2}$,

$$e^{-t^2/2} \quad \overset{\mathcal{F}}{\longleftrightarrow} \quad \sqrt{2\pi}\, e^{-\omega^2/2} \qquad\qquad (2.30)$$

illustrated in Figure 2.28.

2.4.4 Time and frequency shifting

We briefly touched on this property in Section 2.2.6, when deriving the transform of the shifted impulse $\delta(t - t_0)$, where we used the symbol τ for t_0. We wound the frequency components back in time until the value $t = -t_0$ was reached, and then defined this value as the new origin of time, effectively shifting the impulse $\delta(t)$ to $t = t_0$. This process is valid for arbitrary functions $f(t)$, as we now show analytically.

Time shifting

Given the transform pair $f(t) \leftrightarrow F(j\omega)$, the transform of the shifted function $g(t) = f(t - t_0)$ is, from the Fourier definition,

$$G(j\omega) = \int_{-\infty}^{\infty} f(t - t_0)e^{-j\omega t}\, dt$$

A change of time variable $\tau = t - t_0$ implies $d\tau = dt$ and yields

$$G(j\omega) = \int_{-\infty}^{\infty} f(\tau)e^{-j\omega(\tau + t_0)}\, d\tau = e^{-j\omega t_0} \int_{-\infty}^{\infty} f(\tau)e^{-j\omega\tau}\, d\tau$$

The last integral represents the transform $F(j\omega)$ of the original function $f(t)$, and leads to the property

$$\boxed{f(t - t_0) \quad \overset{\mathcal{F}}{\longleftrightarrow} \quad e^{-j\omega t_0} F(j\omega)} \tag{2.31}$$

The exponential has unit magnitude, so that its sole contribution to the product is its angle $\theta = -\omega t_0$, linear with ω and proportional to the time shift t_0, which is added to the phase of the original function $F(j\omega)$,

$$\angle G(j\omega) = \angle F(j\omega) - \omega t_0$$

Graphical interpretation

At this point the reader should re-examine the sequence of Figures 2.8 to 2.10, to confirm that time shifting a function is equivalent to adding linear phase to its frequency representation, which in turn is equivalent to multiplying $F(j\omega)$ by the factor $e^{-j\omega t_0}$. The two concurrent examples (finite and infinite bandwidth) processed in those figures are extracted and repeated in Figure 2.29, where the frequency representations are in terms of magnitude and phase, to stress phase linearity.

The infinite bandwidth case, that of the shifted impulse

$$\delta(t - t_0) \quad \overset{\mathcal{F}}{\longleftrightarrow} \quad e^{-j\omega t_0} \tag{2.32}$$

forms the basis for interpreting time shifting of an arbitrary function as a convolution with a shifted impulse, to be developed in Section 2.4.8.

Frequency shifting

The same derivation applied to the dual case, that is, shifting a frequency representation $F(j\omega)$ by an amount ω_0, leads to a formally similar property

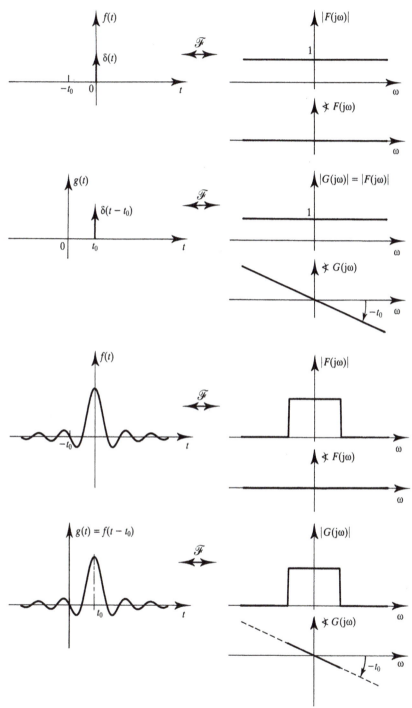

Figure 2.29 Two time-shifted functions and their transforms.

$$\boxed{e^{j\omega_0 t} f(t) \quad \overset{\mathcal{F}}{\longleftrightarrow} \quad F(j\omega - j\omega_0)} \tag{2.33}$$

This has the effect of adding the angle $\theta = -\omega_0 t$ to the time representation. Thus, if the function $f(t)$ was originally real, a shift in frequency makes it acquire an imaginary part.

Example 2.15

We find the transform of the shifted frequency impulse $\delta(\omega - \omega_0)$. Applying the expression 2.33 to the pair

$$\frac{1}{2\pi} \quad \overset{\mathcal{F}}{\longleftrightarrow} \quad \delta(\omega)$$

gives the desired result

$$\frac{1}{2\pi} e^{j\omega_0 t} \quad \overset{\mathcal{F}}{\longleftrightarrow} \quad \delta(\omega - \omega_0) \tag{2.34}$$

shown in Figure 2.30 for positive and negative ω_0. The first of these is the dual equivalent of the displaced time impulse of the previous figure.

Example 2.16

A further example is given is Figure 2.31, which highlights the effects on phase when a function, in this case the rectangular pulse $p_\tau(t)$, is shifted by different amounts, and in opposite directions. The magnitudes of the transforms remain constant, only phase varies.

The curious shapes adopted by the phase representations are the result of arbitrarily assigning phase values $+\pi$ or $-\pi$ to the negative lobes of the original transform (second row from the top), and to the convenience of using modulo 2π representation. The latter is always associated with phase algorithms, which return phase values in the range $-\pi < \phi < +\pi$.

Despite the jagged shapes, the time shifts can be recognized in the slopes of constant rate. The direction of the slope is opposite to that of the time shift, and the rate indicates the amount of shift. In the figure the shift values t_0 were chosen arbitrarily as multiples of $\frac{1}{2}\tau$ to give relatively simple phase diagrams.

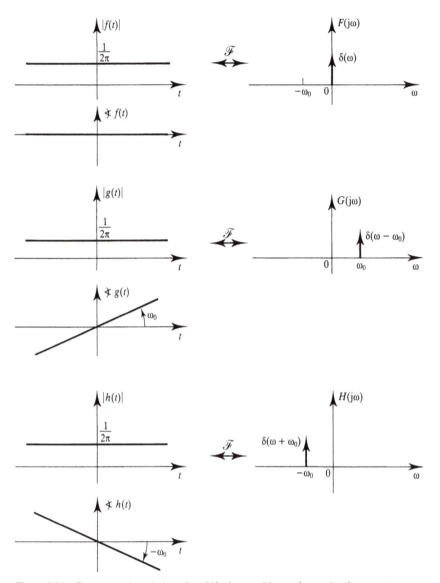

Figure 2.30   Frequency domain impulse shifted to positive and negative frequencies.

2.4.5   Convolution properties

The study of systems often involves either multiplication or convolution of two functions. We now show that multiplication in one domain involves convolution of the transforms in the other domain.

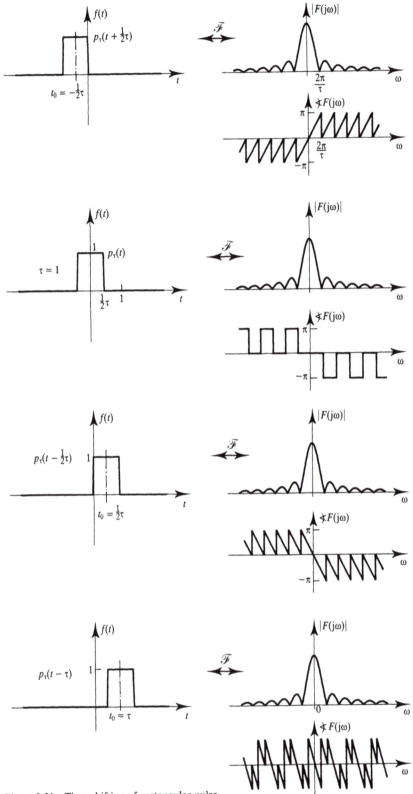

Figure 2.31 Time shifting of rectangular pulse.

Time domain convolution

Given two functions $x(t)$ and $h(t)$ with transforms

$$x(t) \overset{\mathcal{F}}{\longleftrightarrow} X(j\omega)$$

$$h(t) \overset{\mathcal{F}}{\longleftrightarrow} H(j\omega)$$

their convolution produces another function $y(t)$ expressed as the integral

$$y(t) = x(t) * h(t) = \int_{-\infty}^{\infty} x(\tau)h(t - \tau)\,d\tau$$

Taking the Fourier transform of both sides gives

$$Y(j\omega) = \int_{-\infty}^{\infty} \left[\int_{-\infty}^{\infty} x(\tau)h(t - \tau)\,d\tau \right] e^{-j\omega t}\,dt$$

Exchanging the integration order, and noting that $x(\tau)$ is not a function of t, we have

$$Y(j\omega) = \int_{-\infty}^{\infty} x(\tau) \left[\int_{-\infty}^{\infty} h(t - \tau)e^{-j\omega t}\,dt \right] d\tau$$

A change of variable $p = t - \tau$, which implies $dp = dt$ and $t = p + \tau$, yields

$$Y(j\omega) = \int_{-\infty}^{\infty} x(\tau) \left[\int_{-\infty}^{\infty} h(p)e^{-j\omega(p+\tau)}\,dp \right] d\tau$$

Since $e^{-j\omega\tau}$ is not a function of p the double integration can be decoupled as the product of two integrals

$$Y(j\omega) = \int_{-\infty}^{\infty} x(\tau)e^{-j\omega\tau}\,d\tau \cdot \int_{-\infty}^{\infty} h(p)e^{-j\omega p}\,dp$$

which shows that convolution in time means multiplication in frequency,

$$\boxed{\; y(t) = x(t) * h(t) \overset{\mathcal{F}}{\longleftrightarrow} Y(j\omega) = X(j\omega) \cdot H(j\omega) \;}$$ (2.35)

Example 2.17

We derive the transform of a triangular pulse using the convolution property as indicated in Figure 2.32. The time domain of the triangular pulse of width 2τ can be interpreted as the convolution of two rectangular pulses $p_\tau(t)$ of width τ. With the pair

$$p_\tau(t) \overset{\mathcal{F}}{\longleftrightarrow} P_\tau(j\omega) = \tau \, \frac{\sin \frac{1}{2}\omega\tau}{\frac{1}{2}\omega\tau}$$

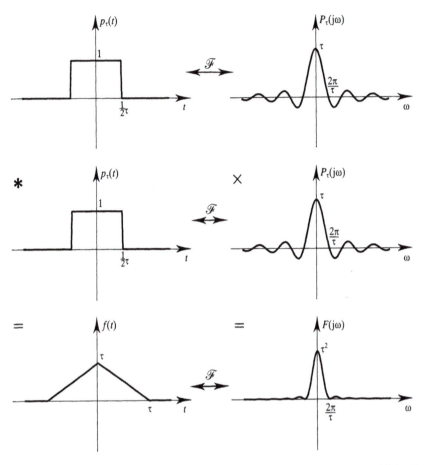

Figure 2.32   Transform of triangular pulse as convolution of rectangular pulse with itself.

and the property 2.35, we have

$$f(t) = p_\tau(t) * p_\tau(t) \quad \overset{\mathcal{F}}{\longleftrightarrow} \quad F(j\omega) = P_\tau^2(j\omega) = \tau^2 \left(\frac{\sin \frac{1}{2}\omega\tau}{\frac{1}{2}\omega\tau} \right)^2$$

We thus avoided a cumbersome integration leading to the result of the form $(\sin x / x)^2$.

The frequency domain product of property 2.35 is commutative, hence

$$y(t) = h(t) * x(t) \quad \overset{\mathcal{F}}{\longleftrightarrow} \quad Y(j\omega) = H(j\omega) X(j\omega)$$

which provides the proof of the convolution property 1.2 and can be simply extended to justify 1.4.

Convolution in frequency

A similar procedure, carried out in the frequency domain on the convolution $G(j\omega) = 1/2\pi X(j\omega) * H(j\omega)$, would lead to the dual result

$$g(t) = x(t)h(t) \quad \overset{\mathcal{F}}{\longleftrightarrow} \quad G(j\omega) = \frac{1}{2\pi} X(j\omega) * H(j\omega) \qquad (2.36)$$

But we will start with the time domain product $g(t) = x(t)h(t)$ and use the frequency-shifting property to show an alternative path, also valid for the preceding derivation. Using the dummy frequency variable v, we write $h(t)$ as the inverse transform of $H(j\omega)$,

$$g(t) = x(t) \frac{1}{2\pi} \int_{-\infty}^{\infty} H(jv)e^{jvt} \, dv$$

Since $x(t)$ is not a function of v it can go inside the integral. Both sides of this expression are functions of time, and can be Fourier transformed, now using the frequency variable ω, to yield

$$G(j\omega) = \int_{-\infty}^{\infty} \left[\frac{1}{2\pi} \int_{-\infty}^{\infty} H(jv)e^{jvt}x(t) \, dv \right] e^{-j\omega t} \, dt$$

Interchanging the integration order, and noting that $e^{-j\omega t}$ and $H(jv)$ are not functions of v and t, respectively, we have

$$G(j\omega) = \frac{1}{2\pi} \int_{-\infty}^{\infty} H(jv) \left[\int_{-\infty}^{\infty} \{e^{jvt}x(t)\}e^{-j\omega t} \, dt \right] dv$$

But the bracketed integral is the forward Fourier transform of the function $\{e^{jvt}x(t)\}$. According to expression 2.33 it represents the frequency domain of the frequency-shifted function

$$e^{jvt}x(t) \quad \overset{\mathcal{F}}{\longleftrightarrow} \quad X(j\omega - jv)$$

Consequently,

$$G(j\omega) = \frac{1}{2\pi} \int_{-\infty}^{\infty} H(jv)X(j\omega - jv) \, dv$$

This integral is the formal definition of the frequency domain convolution of $H(j\omega)$ and $X(j\omega)$, and leads to the desired transform pair 2.36.

2.4.6 Differentiation and integration

Given the transform pair $f(t) \leftrightarrow F(j\omega)$ we will now show that, in essence, the transform of the derivative (or integral) of $f(t)$ equals

$F(j\omega)$ multiplied (or divided) by the linear factor $j\omega$. A similar property applies to the derivative (or integral) of $F(j\omega)$.

Differentiation

For the time domain derivative, we differentiate both sides of the inverse Fourier transform expression 2.6,

$$f(t) = \frac{1}{2\pi} \int_{-\infty}^{\infty} F(j\omega)e^{j\omega t} \, d\omega$$

with respect to time. Noting that, under the integral, only $e^{j\omega t}$ is a function of time, we have

$$\frac{df(t)}{dt} = \frac{1}{2\pi} \int_{-\infty}^{\infty} [j\omega F(j\omega)]e^{j\omega t} \, d\omega$$

where the right side has the form of the inverse transform of the function $[j\omega F(j\omega)]$. Consequently,

$$\boxed{\frac{df(t)}{dt} \quad \overset{\mathcal{F}}{\longleftrightarrow} \quad j\omega F(j\omega)} \tag{2.37}$$

A time domain differentiation therefore causes a frequency domain rotation $+\frac{1}{2}\pi$ of $F(j\omega)$, and a linear scaling by ω.

Repeated differentiation of expression 2.6 leads to the general result

$$\boxed{\frac{d^n f(t)}{dt^n} \quad \overset{\mathcal{F}}{\longleftrightarrow} \quad (j\omega)^n F(j\omega)} \tag{2.38}$$

Similarly, differentiation in frequency of the forward transform (2.5) leads to the dual counterpart of expression 2.38,

$$\boxed{(-jt)^n f(t) \quad \overset{\mathcal{F}}{\longleftrightarrow} \quad \frac{d^n F(j\omega)}{d\omega^n}} \tag{2.39}$$

Example 2.18

The odd-symmetric rectangular function $g(t)$ of Figure 2.33 is the derivative of the triangular pulse $f(t)$ of Example 2.17. With

$$f(t) \quad \overset{\mathcal{F}}{\longleftrightarrow} \quad F(j\omega) = \tau^2 \left(\frac{\sin \frac{1}{2}\omega\tau}{\frac{1}{2}\omega\tau} \right)^2$$

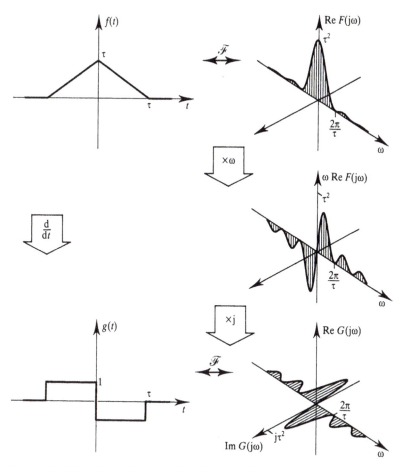

Figure 2.33   Transform of the derivative of a triangular function.

the transform of $g(t)$ is obtained according to expression 2.37 by multiplying $F(j\omega)$ by $j\omega$, to yield

$$g(t) \quad \overset{\mathscr{F}}{\longleftrightarrow} \quad G(j\omega) = j4\,\frac{\sin^2 \tfrac{1}{2}\omega\tau}{\omega}$$

This multiplication process is illustrated in two stages in Figure 2.33, first a scaling by ω, followed by a rotation into the imaginary plane.

Using the equivalence $\sin^2 x = \tfrac{1}{2}(1 - \cos 2x)$ the above result becomes

$$g(t) \quad \overset{\mathscr{F}}{\longleftrightarrow} \quad G(j\omega) = j2\tau\,\frac{1 - \cos \omega\tau}{\omega\tau} \tag{2.40}$$

which is of the more familiar form $(1 - \cos x)/x$.

Example 2.19

We find the transform of the derivative of the Gaussian function. Applying expression 2.37 to both sides of expression 2.30 yields functions of the same shape,

$$f(t) = -te^{-t^2/2} \quad \overset{\mathcal{F}}{\longleftrightarrow} \quad F(j\omega) = j\sqrt{2\pi}\,\omega e^{-\omega^2/2}$$

Being the derivative of a real even function, the time domain is real odd, so that the frequency domain is imaginary odd, as shown in Figure 2.34.

Integration

Given a function

$$f(t) \quad \overset{\mathcal{F}}{\longleftrightarrow} \quad F(j\omega)$$

we interpret its integral as a convolution with the unit step $u(t)$, as in Equation 1.9,

$$g(t) = \int_{-\infty}^{t} f(\tau)\,d\tau = u(t) * f(t)$$

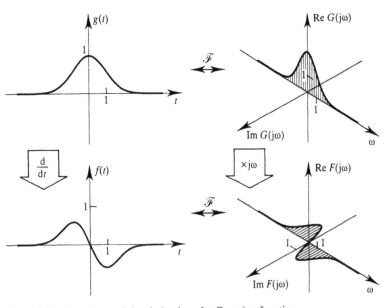

Figure 2.34 Transform of the derivative of a Gaussian function.

Invoking the convolution property 2.35 and using the result 2.25

$$u(t) \quad \overset{\mathcal{F}}{\longleftrightarrow} \quad U(j\omega) = \frac{1}{j\omega} + \pi\delta(\omega)$$

we write

$$G(j\omega) = U(j\omega)F(j\omega) = \frac{1}{j\omega} F(j\omega) + \pi\delta(\omega)F(j\omega)$$

But the impulse property 1.12, interpreted for an impulse of the frequency domain as $\delta(\omega)F(j\omega) = \delta(\omega)F(0)$, yields

$$\boxed{\int_{-\infty}^{t} f(\tau)\,d\tau \quad \overset{\mathcal{F}}{\longleftrightarrow} \quad \frac{1}{j\omega} F(j\omega) + \pi\delta(\omega)F(0)} \qquad (2.41)$$

The first term on the right, $F(j\omega)/j\omega$, represents the inverse of the differentiation property 2.37. If $F(0) = 0$, as was the case in Figures 2.33 and 2.34, then the two properties are fully reversible, in the sense that the function $G(j\omega)$ can be recovered from $F(j\omega)$ through division by $j\omega$.

In contrast, if the function $f(t)$ contains a non-zero d.c. component, represented by a non-zero value of $F(0)$, the transform of its integral contains an additional impulse of strength $\pi F(0)$ at the origin.

Example 2.20

We transform the integral of the function (see Example 2.12)

$$f(t) = \frac{v}{\pi} \frac{\sin vt}{vt} \quad \overset{\mathcal{F}}{\longleftrightarrow} \quad F(j\omega) = p_{2v}(j\omega)$$

shown in the upper half of Figure 2.35. The d.c. value of this function is $F(0) = 1$, and the property 2.41 yields the transform of the integral as

$$g(t) = \int_{-\infty}^{t} f(\tau)\,d\tau \quad \overset{\mathcal{F}}{\longleftrightarrow} \quad G(j\omega) = -\frac{j}{\omega} p_{2v}(j\omega) + \pi\delta(\omega)$$

The frequency representation consists of a hyperbola in the imaginary plane, truncated by the pulse $p_{2v}(j\omega)$, and a real impulse at the origin of magnitude π. This impulse transforms back to the time domain as the constant $\frac{1}{2}$, which represents the d.c. value of $g(t)$. Even symmetry of $f(t)$ makes this value coincide with the value at the origin, $g(0) = \frac{1}{2}$, and leads to the asymptotic value $g(\infty) = 1$.

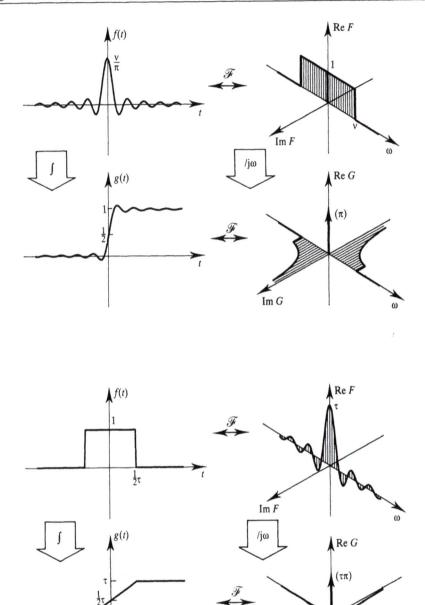

Figure 2.35 Transform of integral.

Example 2.21

The transform of the integral of the time domain pulse

$$f(t) = p_\tau(t) \overset{\mathcal{F}}{\longleftrightarrow} F(j\omega) = \tau \frac{\sin \frac{1}{2}\omega\tau}{\frac{1}{2}\omega\tau}$$

shown in the lower half of Figure 2.35 is similarly obtained. With $F(0) = \tau$, the transform is

$$g(t) = \int_{-\infty}^{t} f(\tau)\,d\tau \overset{\mathcal{F}}{\longleftrightarrow} G(j\omega) = -j2 \frac{\sin \frac{1}{2}\omega\tau}{\omega^2} + \pi\tau\delta(\omega)$$

The integral of $f(t)$ is a truncated ramp, whose value at the origin $g(0) = \frac{1}{2}\tau$ corresponds to the function's d.c. value and relates to the frequency domain impulse of strength $\pi\tau$.

2.4.7 Energy and power

Given two interacting functions $f(t)$ and $g(t)$, the integral of their product represents an energy,

$$E = \int_{-\infty}^{\infty} f(t)g(t)\,dt \tag{2.42}$$

For example, if the two functions represent the voltage $v(t)$ and current $i(t)$ associated with a source, then the product $p(t) = v(t)i(t)$ represents the instantaneous power delivered by the source and its integral E gives the energy delivered over the life span of the source.

We now express the energy E in terms of the frequency representations of the signals, given by

$$f(t) \overset{\mathcal{F}}{\longleftrightarrow} F(j\omega) \quad \text{and} \quad g(t) \overset{\mathcal{F}}{\longleftrightarrow} G(j\omega)$$

Representing $f(t)$ of Equation 2.42 by the inverse transform of $F(j\omega)$ yields

$$E = \int_{-\infty}^{\infty} \left[\frac{1}{2\pi} \int_{-\infty}^{\infty} F(j\omega)e^{j\omega t}\,d\omega \right] g(t)\,dt$$

and interchanging the order of integration gives

$$E = \frac{1}{2\pi} \int_{-\infty}^{\infty} F(j\omega) \left[\int_{-\infty}^{\infty} g(t)e^{j\omega t}\,dt \right] d\omega$$

But $g(t)$ of the bracketed integral can be interpreted as the inverse Fourier transform of the function $G(-j\omega)$, so that

$$E = \int_{-\infty}^{\infty} f(t)g(t)\,dt = \frac{1}{2\pi} \int_{-\infty}^{\infty} F(j\omega)G(-j\omega)\,d\omega \tag{2.43}$$

which expresses the energy associated with those signals, in both domains.

This result can be expressed more symmetrically by interpreting $g(t)$ as the complex conjugate of an auxiliary function $h(t)$, i.e. $g(t) = h^*(t)$. But the transform of the conjugate of a function is

$$h^*(t) \xleftrightarrow{\mathcal{F}} H^*(-j\omega)$$

Thus, if $g(t)$ of the left side of Equation 2.43 is replaced by $h^*(t)$, then the factor $H^*(j\omega)$ takes the place of $G(-j\omega)$ of the right side, yielding

$$\int_{-\infty}^{\infty} f(t)h^*(t)\,dt = \frac{1}{2\pi} \int_{-\infty}^{\infty} F(j\omega)H^*(j\omega)\,d\omega \tag{2.44}$$

Since $g(t)$ was an arbitrary function, the function $h(t)$ too is arbitrary, and the self-contained relationship 2.44 is valid if $g^*(t)$ takes the place of $h^*(t)$ and $G^*(j\omega)$ that of $H^*(j\omega)$.

Special cases

If $f(t)$ and $g(t)$ represent the same signal, the energy equation (2.43) expresses the energy contents of the signal

$$\int_{-\infty}^{\infty} |f(t)|^2\,dt = \frac{1}{2\pi} \int_{-\infty}^{\infty} F(j\omega)F(-j\omega)\,d\omega \tag{2.45}$$

If in addition $f(t)$ is real, then $F(-j\omega) = F^*(j\omega)$ and both of these have the same magnitude as $F(j\omega)$, but opposite phase, so that the energy equation simplifies to the more commonly used form

$$E = \int_{-\infty}^{\infty} f^2(t)\,dt = \frac{1}{2\pi} \int_{-\infty}^{\infty} |F(j\omega)|^2\,d\omega \tag{2.46}$$

This is known as Parseval's formula and Equation 2.44 represents its general form. By expressing the same energy E on both sides, expression 2.46 provides a useful integral relationship between the magnitude values of a real signal $f(t)$ and the magnitude values of its equivalent frequency representation $F(j\omega)$.

2.4.8 Transforms involving shifted impulses

The transform pairs 2.32 and 2.34, which relate a shifted time domain impulse and a shifted frequency domain impulse to their respective transforms, are of the utmost importance in Fourier analysis. In Chapter

5 we will derive system solution methods by interpreting arbitrary input signals as either time domain or frequency domain superpositions of infinitesimal impulses. In Part 2 shifted impulses provide the analytical link between functions of continuous variable and discrete variable.

We conclude this section with examples of periodic functions, which involve such impulses in the alternative domain. We show how these functions are related by earlier transform properties and also interpret the shifting property for arbitrary functions as a convolution with a shifted impulse.

Transforms of sinusoidal functions

We will show in Chapter 6 that the continuous Fourier transform of a periodic function is a sum of shifted impulses. Sinusoids provide simple examples.

Example 2.22

We derive the Fourier transforms of the time domain functions $f_1(t) = \cos \omega_1 t$ and $g_1(t) = \sin \omega_1 t$. Interpreting the transform pair 2.34 for the displacements ω_1 and $-\omega_1$ as

$$\frac{1}{2}e^{j\omega_1 t} \quad \overset{\mathscr{F}}{\longleftrightarrow} \quad \pi\delta(\omega - \omega_1)$$
$$\frac{1}{2}e^{-j\omega_1 t} \quad \overset{\mathscr{F}}{\longleftrightarrow} \quad \pi\delta(\omega + \omega_1)$$
(2.47)

and adding both sides yields the transform pair

$$f_1(t) = \cos \omega_1 t \quad \overset{\mathscr{F}}{\longleftrightarrow} \quad F_1(j\omega) = \pi\delta(\omega - \omega_1) + \pi\delta(\omega + \omega_1)$$
(2.48)

The frequency representation of the cosine function thus consists of two impulses of strength π, located in the real plane, as shown in the upper half of Figure 2.36.

Similarly, dividing both sides of the expressions 2.47 by j and subtracting gives the pair

$$g_1(t) = \sin \omega_1 t \quad \overset{\mathscr{F}}{\longleftrightarrow} \quad G_1(j\omega) = -j\pi\delta(\omega - \omega_1) + j\pi\delta(\omega + \omega_1)$$
(2.49)

The frequency domain impulses are those of the cosine function, rotated through multiplication by −j and j onto the imaginary plane, as shown in the upper half of Figure 2.36. As expected, the real odd sine function has imaginary odd frequency components.

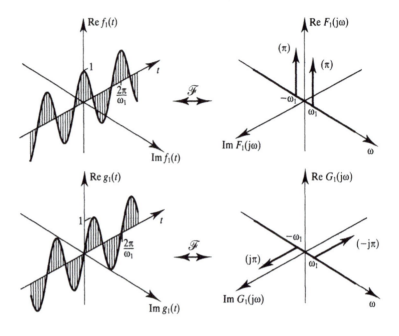

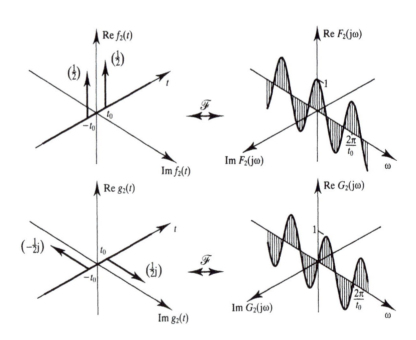

Figure 2.36 Transforms of time domain and frequency domain sinusoids.

Had we interpreted the function $f_1(t) = \cos \omega_1 t$ as an ordinary function, the forward Fourier transform definition 2.5 would not have been very helpful. Periodic functions do not decay at infinity and are therefore not absolutely integrable, thus violating the Dirichlet sufficiency conditions mentioned at the beginning of Section 2.4. It is still possible to find the transform of the cosine by means of the definition 2.5 through the device illustrated in Example 2.8, that is, by slightly modifying the function to be transformed and using the concept of 'convergence in the limit'.

But this is where the impulse definition integral 1.10 and its frequency domain counterpart make their great contribution. They embody the mathematical niceties required for simple, yet rigorous derivations of the transforms 2.32 and 2.34 of shifted impulses. These results are then incorporated into the transforms of functions containing impulses in one or the other domain, thus circumventing the need for 'integration in the limit'.

Example 2.23

We derive the transforms of the frequency domain sinusoids $F_2(j\omega) = \cos \omega t_0$ and $G_2(j\omega) = \sin \omega t_0$. Writing the pair 2.32 for two impulses displaced to $t = t_0$ and $t = -t_0$ as

$$\tfrac{1}{2}\delta(t - t_0) \xleftrightarrow{\ \mathcal{F}\ } \tfrac{1}{2}e^{-j\omega t_0}$$
$$\tfrac{1}{2}\delta(t + t_0) \xleftrightarrow{\ \mathcal{F}\ } \tfrac{1}{2}e^{j\omega t_0} \qquad (2.50)$$

and adding or subtracting as in the preceding example leads to the dual frequency domain equivalents

$$f_2(t) = \tfrac{1}{2}\delta(t - t_0) + \tfrac{1}{2}\delta(t + t_0) \xleftrightarrow{\ \mathcal{F}\ } F_2(j\omega) = \cos \omega t_0$$
$$g_2(t) = \tfrac{1}{2}j\delta(t - t_0) - \tfrac{1}{2}j\delta(t + t_0) \xleftrightarrow{\ \mathcal{F}\ } G_2(j\omega) = \sin \omega t_0 \qquad (2.51)$$

represented in the lower half of Figure 2.36.

Shifting as convolution with shifted impulse

The transform of the time domain convolution of a function $f(t)$ with a shifted impulse $\delta(t - t_0)$ can be interpreted according to the property 2.35 as

$$f(t) \xleftrightarrow{\ \mathcal{F}\ } F(j\omega)$$
$$\delta(t - t_0) \xleftrightarrow{\ \mathcal{F}\ } e^{-j\omega t_0}$$
$$\delta(t - t_0) * f(t) \xleftrightarrow{\ \mathcal{F}\ } e^{-j\omega t_0} F(j\omega)$$

Comparing the frequency representation with that of the shifting property 2.31 we conclude that time shifting a function is equivalent to convolving it with a similarly shifted impulse, that is,

$$g(t) = f(t - t_0) = \delta(t - t_0) * f(t) \tag{2.52}$$

This relationship is shown explicitly in Figure 2.37 by relating the two examples of Figure 2.29. It confirms that time shifting does not modify

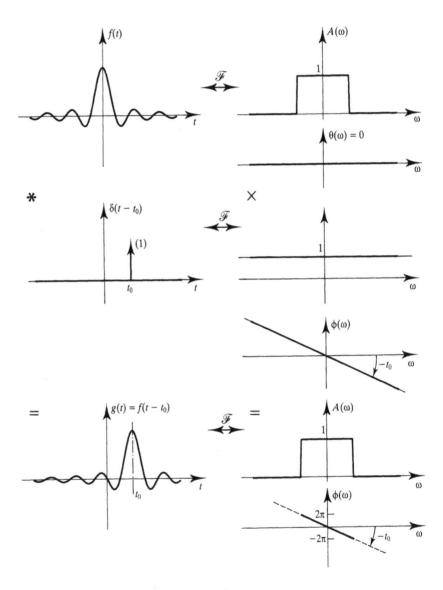

Figure 2.37   Shifting as convolution with shifted impulse.

the frequency domain amplitude $A(\omega)$ of $F(j\omega)$, it simply adds the linear phase $\phi(\omega) = -\omega t_0$ associated with the shifted impulse to the phase $\theta(\omega)$ of the original function $F(j\omega)$,

$$G(j\omega) = A(\omega)e^{j[\theta(\omega)+\phi(\omega)]}$$

In the example of Figure 2.37 we have $\theta(\omega) = 0$.

A similar interpretation is given to the dual case, the frequency shifting property 2.33, by using the convolution property 2.36 on the function $F(j\omega)$ and the impulse $\delta(\omega - \omega_1)$

$$f(t) \xleftrightarrow{\;\mathcal{F}\;} F(j\omega)$$

$$\frac{1}{2\pi} e^{j\omega_1 t} \xleftrightarrow{\;\mathcal{F}\;} \delta(\omega - \omega_1)$$

$$e^{j\omega_1 t} f(t) \xleftrightarrow{\;\mathcal{F}\;} \delta(\omega - \omega_1) * F(j\omega)$$

where the angle of the time domain exponential, see Figure 2.30, is added to that of $f(t)$.

Relationships involving transform properties

We now interrelate the functions of Figure 2.36 in terms of various transform properties. If the sine function $g_1(t)$ is interpreted as the cosine $f_1(t)$, right-shifted by a quarter of their wavelength $T_0 = 2\pi/\omega_1$

$$g_1(t) = f_1(t - \tfrac{1}{4}T_0) = \cos(\omega_1 t - \tfrac{1}{2}\pi) = \sin \omega_1 t$$

then its frequency representation $G_1(j\omega)$ results from rotating the two impulses of $F_1(j\omega)$ by $-\tfrac{1}{2}\pi$ and $+\tfrac{1}{2}\pi$, as expressed by the shifting property 2.31.

This is consistent with the imagery developed in Section 1.4.4 for the Fourier series, in particular with the concepts of Figure 1.20. Rotating the impulses of $F_1(j\omega)$ by $\tfrac{1}{4}$ revolution, as shown in Figure 2.36, represents a $\tfrac{1}{4}$ wavelength displacement of the time pointer towards negative time. Taking this as the new time origin yields the pair $g_1(t) \leftrightarrow G_1(j\omega)$.

Alternatively, if we interpret $g_1(t)$ as the convolution

$$g_1(t) = \delta(t - \tfrac{1}{4}T_0) * f_1(t)$$

then the interpretation 2.52 leads to the same conclusions.

We can also use the differential relationship

$$\frac{dg_1(t)}{dt} = \omega_1 \cos \omega_1 t = \omega_1 f_1(t) \xleftrightarrow{\;\mathcal{F}\;} \omega_1 F_1(j\omega)$$

and by the derivative property 2.37

$$\frac{dg_1(t)}{dt} \xleftrightarrow{\;\mathcal{F}\;} j\omega G_1(j\omega)$$

so that we can interpret $F_1(j\omega)$ in terms of $G_1(j\omega)$ as

$$F_1(j\omega) = j \frac{\omega}{\omega_1} G_1(j\omega)$$

The impulses of $G_1(j\omega)$ located at $\omega = \omega_1$ and $\omega = -\omega_1$ are therefore multiplied by j and $-j$ respectively, to give the corresponding impulses of $F_1(j\omega)$.

Similar interpretations can be given to the relationships between the frequency domain sinusoids of the lower half of Figure 2.36. These represent the dual counterparts of the functions of the upper half. Transposing the sinusoidal shapes from time to frequency yields suitably scaled and time-reversed impulses in the time domain, as expressed by the duality property 2.29.

Exercises

2.1 Referring to the graphics of Figure 2.6, draw slices of the integrand at integer multiples of $\omega = \pi/\tau$. Show that slices at even multiples have zero area, while those located at odd multiples have alternating positive and negative areas, which decrease in inverse proportion to frequency.

2.2 Show that the five truncated helices shown in the frequency domain of Figure 2.9 represent the Fourier transforms of the same shape $f(t)$ of the time domain, but with its time origin displaced respectively to the values $t = t_1$, t_2, t_3, t_4 and t_5. Plot the magnitude and angle representations of the five helices.

2.3 Using the Fourier transform definition 2.5 derive the frequency representation of the triangular pulse

$$f(t) = \begin{cases} 1 - |t|/\tau & |t| < \tau \\ 0 & \text{elsewhere} \end{cases}$$

2.4 Derive the rules governing the multiplication of even and odd functions and illustrate them graphically.

2.5 Show that the Fourier transform is a linear operator (property 2.24).

2.6 Verify the results of Figure 2.18 analytically, by finding the even and odd parts of the time domain pulse $f(t)$, transforming each part by the Fourier integral 2.5, and combining the complex results.

2.7 Given the transform of the time domain signum function (Equation

2.23), find the Fourier transform of the Hilbert transformer $h(t) = 1/(\pi t)$ by using transform dualities.

2.8 Derive the convolution properties of the Fourier transform (Equations 2.35 and 2.36) by proceeding in the alternative domain to that of the text, i.e. by respectively transforming the functions

$$f(t) = \mathcal{F}^{-1}\{G(j\omega) H(j\omega)\} \quad \text{and} \quad f(t) = \mathcal{F}^{-1}\{G(j\omega) * H(j\omega)\}$$

2.9 Using the convolution properties of the Fourier transform show that the convolution of two odd symmetric functions is even symmetric.

2.10 Find the Fourier transforms of the shifted unit step $f(t) = u(t - t_0)$ and of the shifted signum function $g(t) = \text{sgn}(t - t_0)$.

2.11 Derive the transform of the even rectangular pulse $f(t) = p_\tau(t)$ by interpreting it as the sum of two shifted unit steps and, again, as the sum of two shifted signum functions.

2.12 Derive the Fourier property of differentiation in frequency

$$-j\,t\,f(t) \quad \overset{\mathcal{F}}{\longleftrightarrow} \quad \frac{dF(j\omega)}{d\omega}$$

2.13 Given the transform of the function $f(t) = \cos\omega_1 t$ (Equation 2.48), use the layout of Figure 2.36 to sketch both domains of

(a) the rotated sequence: $f(t), jf(t), j^2 f(t), j^3 f(t), j^4 f(t)$

(b) the shifted sequence: $\cos(\omega_1 t - \theta)$, with $\theta = \pi/4, \pi/2, \pi$ and 2π.

2.14 Find the result of the time domain convolution indicated in Figure 2.38 and sketch both domains of the process on the layout of Figure 2.32. (The frequency domain will need a three-dimensional representation.)

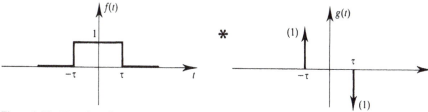

Figure 2.38 Functions for Exercise 2.14.

CHAPTER 3

Laplace Transform

While the Fourier transform is valuable in the study of continuous-time signals and systems, in many situations the general Laplace transform is more versatile. Firstly, the Laplace transform covers a wider range of functions, including some that do not decay with increasing time, such as the increasing exponential and the unit ramp. Secondly, it describes a function in greater depth, for instance revealing the location of poles and zeros. Thirdly, the simpler notation s for the Laplace frequency variable simplifies algebriac manipulations, a welcome bonus. Finally, the unilateral form of the Laplace transform simplifies the study of systems with initial conditions.

In a strict sense the Laplace transform applies to the class of continuous and aperiodic functions, the same class as the Fourier transform, as indicated in Figure 3.1 on the signal classification chart of Figure 1.2. In Chapter 7 we extend its validity to discrete-time signals, by interpreting discrete-time samples in terms of equivalent impulses.

3.1 Laplace transform as extension of Fourier transform

We first give an analytical definition of the Laplace transform that provides a basis for interpretation as an extension of the Fourier transform. The main concepts are then introduced graphically.

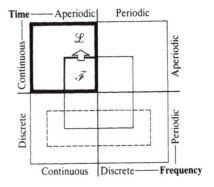

Figure 3.1   Relationship of Laplace transform, Fourier transform and signal classes.

3.1.1   Analytical link

It is expedient to define the Laplace transform $F(s)$ of a function $f(t)$ by generalizing the Fourier expression 2.5, as

$$F(s) = \int_{-\infty}^{\infty} f(t)e^{-st}\,dt \tag{3.1}$$

also expressed as $F(s) = \mathcal{L}\{f(t)\}$, where $\mathcal{L}$ is the Laplace operator, or by the transform pair

$$f(t) \xleftrightarrow{\mathcal{L}} F(s)$$

This definition associates the Fourier frequency ω with the imaginary part of the more general complex frequency variable $s = \sigma + j\omega$. This frees the frequency domain from being restricted to the imaginary axis $j\omega$ of the associated s-plane.

Writing the complex variable of Equation 3.1 in full yields

$$F(\sigma + j\omega) = \int_{-\infty}^{\infty} f(t)e^{-(\sigma+j\omega)t}\,dt = \int_{-\infty}^{\infty} [f(t)e^{-\sigma t}]e^{-j\omega t}\,dt \tag{3.2}$$

where the last integral represents the Fourier transform of the bracketed function, so that

$$F(\sigma + j\omega) = \mathcal{F}\{f(t)e^{-\sigma t}\} \qquad (3.3)$$

For any desired value of σ, the expression 3.3 interprets the Laplace transform of a function $f(t)$ as the Fourier transform of the **modified function** $f(t)e^{-\sigma t}$. This is valid for all values of the parameter σ for which the Fourier integral converges. In particular, the value $\sigma = 0$, for which $e^{-\sigma t} = 1$, expresses the Fourier transform as a subset of the Laplace transform, that is,

$$F(j\omega) = F(s)|_{s=j\omega}$$

A function $f(t)$ for which the integral 2.5 fails to converge, so that it does not possess a Fourier transform, may still have a Laplace transform, provided a value of σ exists for which the integral 3.2 converges. The real exponential $e^{-\sigma t}$ is therefore called the **convergence factor** of the Laplace transform, and the region of the s-plane for which the transform exists is its **region of convergence**, to be examined in Section 3.3.

The relationship 3.3 is used next to extend to the Laplace transform many concepts developed earlier for the Fourier transform, in particular the graphical imagery.

The definition 3.1 is sometimes called the **bilateral** or **two-sided** Laplace transform, to indicate integration over the entire time axis. This will be contrasted in Section 3.6 with the **unilateral** or **one-sided** Laplace transform, which only involves integration over the non-negative semi-axis.

Complex functions and complex variables

To avoid confusions regarding real and imaginary parts of functions and of variables, we briefly take stock. In this book time domain functions $f(t)$ are restricted to **complex functions** of the **real time variable** t, typically of the form

$$f(t) = f_1(t) + jf_2(t) = \mathrm{Re}\, f(t) + j\,\mathrm{Im}\, f(t)$$

whose real part $f_1(t) = \mathrm{Re}\, f(t)$ and imaginary part $f_2(t) = \mathrm{Im}\, f(t)$ are both **real functions**. A further restriction applies to signals of the physical world, where $f(t)$ is usually a **real function**, with zero imaginary part.

By contrast, frequency domain functions $F(s)$ are typically **complex functions** of the **complex frequency variable** s, of the form

$$F(s) = \mathrm{Re}\, F(s) + j\,\mathrm{Im}\, F(s)$$

where $s = \sigma + j\omega$ itself has a real part σ and an imaginary part ω, both of which are **real variables**.

The distinction extends to **complex function planes**, which contain the real and imaginary parts of the functions $f(t)$ or $F(s)$ on the one hand, and the **complex variable s-plane**, which represents the real and imaginary parts of points of the variable s, on the other.

3.1.2 Graphical interpretations

We now make use of the relationship 3.3 to extend the Fourier transform graphics of Section 2.2 to the Laplace transform. Besides stressing the underlying unity between these two, we are also preparing for Chapter 7, where the graphics of the Laplace transform are used to introduce the discrete-time Laplace transform and its variant, the z-transform.

Having visualized the Fourier transform as a superposition of exponentials of the form $e^{j\omega t}$, it is a small conceptual step to visualize the Laplace transform as a superposition of the more general exponentials e^{st}. But we need some familiarity with these functions.

Generalized complex exponential

The complex exponential $e^{j\omega t}$, of unit magnitude and argument $\theta = \omega t$, was introduced in Section 1.3 for the immediate Fourier requirements. The **complex frequency variable** $s = \sigma + j\omega$ simply attaches magnitude information to this exponential. Interpreting e^{st} as a function of time,

$$f(t) = e^{st} = e^{\sigma t}e^{j\omega t} = r(t)e^{j\omega t} \tag{3.4}$$

where the real exponential $r(t) = e^{\sigma t}$ represents the magnitude of e^{st}, has initial value $r(0) = 1$ and grows exponentially at a rate set by σ.

The variable s thus extends the concept of frequency, in that ω retains the usual meaning of angular frequency, or rate of change of angle, and σ represents the rate of change of magnitude. Since the argument of e^{st} is a dimensionless number, the units of s must be those of $1/t$, that is, of frequency.

Complex plane representations of two typical cases of the function $f(t)$ are given in Figure 3.2, together with a graphical meaning of the growth parameters σ and ω. Borrowing the imagery of kinematics, the logarithmic spiral $f(t)$ is a parametric representation in the complex plane of the trajectory of a point, with time t as the independent parameter. The velocity of the point is the time derivative of position,

$$v(t) = \frac{df(t)}{dt} = se^{st} = \sigma f(t) + j\omega f(t)$$

The term $\sigma f(t)$ is a radial velocity component, proportional to the magnitude of $f(t)$ and in line with it, while $\omega f(t)$ is a lateral velocity

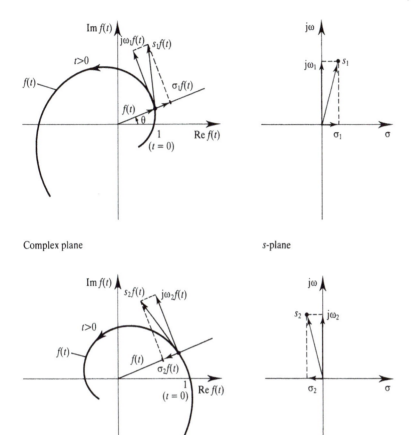

Complex plane

s-plane

Figure 3.2 Complex plane of $f(t) = e^{st}$ and associated s-plane for $\sigma_1 > 0$ (upper half) and $\sigma_2 < 0$ (lower half).

component, at right angles to $f(t)$, as expressed by the factor j. Their sum, the total velocity $v(t) = sf(t)$ is tangent to the spiral.

Positive values of σ indicate radial growth and an increasing spiral, negative values a decreasing spiral. Examples of both cases are illustrated in Figure 3.2, both shown with positive angular frequency $\omega > 0$. The orthogonal components σ and ω are parameters of the spiral. These are associated with the real and imaginary axes of the **complex s-plane**, each point of which identifies a specific spiral, as illustrated for the two cases of Figure 3.2.

For a full three-dimensional representation we interpret the real and imaginary parts of e^{st} by writing Equation 3.4 as

$$f(t) = e^{\sigma t} \cos \omega t + j e^{\sigma t} \sin \omega t = \operatorname{Re} f(t) + j \operatorname{Im} f(t)$$

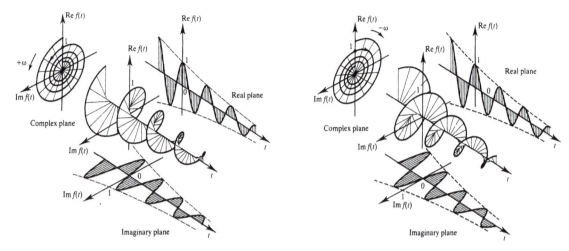

Figure 3.3 Full representation of e^{st}, $\sigma < 0$, $\omega > 0$ (left), $\omega < 0$ (right).

and place them on the real and imaginary planes of Figure 3.3, which illustrates cases of positive and negative ω, both with negative σ.

We can interpret the real and imaginary planes of Figure 3.3 in terms of **amplitude modulation**, considering the cosine or sine functions as carriers, with their amplitudes modulated by the envelope $e^{\sigma t}$. This concept can be extended to the three-dimensional representations, by interpreting the helices of Figures 1.9 and 1.10 as carriers, whose amplitudes are modulated by an exponentially varying surface of revolution of amplitude $e^{\sigma t}$.

Changing the parameter value σ modifies the rate of growth or decay, while changes of ω affect the frequency and sense of the helix. The results of various combinations are illustrated in Figure 3.4, arranged on an s-plane grid representing the corresponding parameter values. Real exponentials $e^{\sigma t}$ appear as special cases on the real axis, while complex exponentials $e^{j\omega t}$, of unit magnitude, appear on the imaginary axis.

These functions are to the Laplace transform what the exponentials $e^{j\omega t}$ are to the Fourier transform.

Laplace transform, slice of constant σ

We first investigate the representation of a single slice of the Laplace transform. For this we treat σ as a parameter, assign it an arbitrary value, say $\sigma = \sigma_0$, and form the product

$$f_{\sigma_0}(t) = f(t)e^{\sigma_0 t}$$

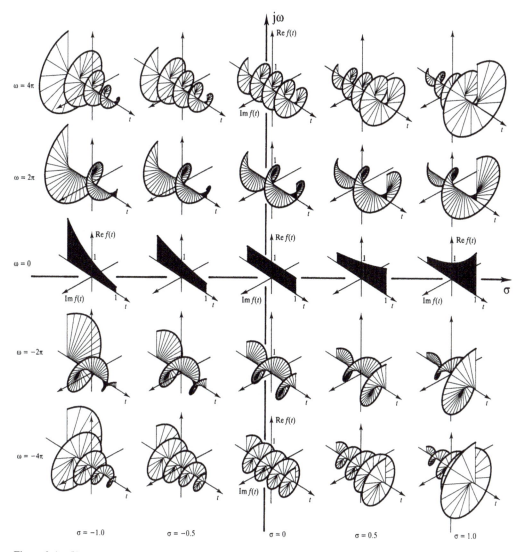

Figure 3.4   Chart of general complex exponentials e^{st} arranged on s-plane grid.

We then take the Fourier transform of this modified function, according to the relationship 3.3, that is,

$$F(\sigma_0 + j\omega) = \mathcal{F}\{f_{\sigma_0}(t)\}$$

This process is represented in Figure 3.5, based on the concepts outlined earlier in Figures 2.15 and 2.17. Both, the original $f(t)$ and the modified $f_{\sigma_0}(t)$ are shown in the time domain of Figure 3.5. Although we chose a real even-symmetrical function $f(t)$, this symmetry is destroyed by the convergence factor. The Fourier transform of $f_{\sigma_0}(t)$ thus contains both a real and an imaginary part, for the reasons given earlier with Figure 2.17.

The pair of real and imaginary parts thus obtained give one possible Laplace representation of $f(t)$, that corresponding to one particular value of σ. These are combined in the forefront of Figure 3.5 to reveal magnitude and phase implications.

To construct Figure 3.5, and anticipating the full representation that follows, the frequency domain of earlier figures was replaced by the complex s-plane, by adding the real σ-axis to the existing imaginary $j\omega$-axis. The orientations of the time and frequency axes were also changed to give the s-plane a more convenient orientation. A σ-axis was also grafted onto the time axis, to form an auxiliary plane for the display of $f_\sigma(t)$.

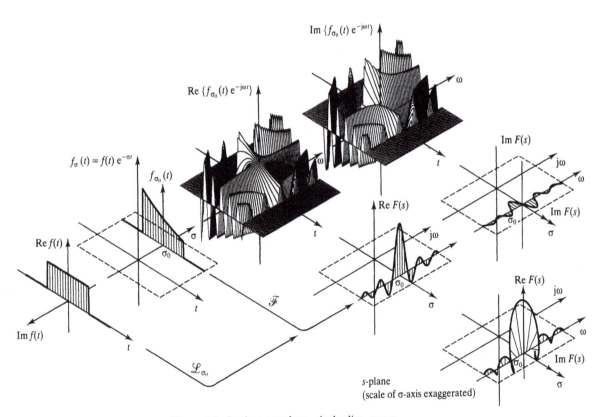

Figure 3.5 Laplace transform, single slice, $\sigma = \sigma_0$.

Laplace transform, full representation

Each value of σ modifies the function $f(t)$ by a different convergence factor. To show the overall shape of the Laplace transform, $F(s)$ requires the computation of its real and imaginary parts for all the values of σ for which the transform exists.

This is illustrated in Figure 3.6. The original time signal $f(t)$ is shown in the foreground of the time domain. Different values of σ give rise to modified functions $f_\sigma(t)$, displayed as slices of constant σ on the auxiliary $t\sigma$-plane. The individual slices are Fourier transformed, each yielding a corresponding slice $F(\sigma + j\omega)$ of the Laplace transform. Since each slice produces its own pair of integrand surfaces, we indicate these generically by the cosine and sine carriers.

The particular case of Figure 3.5 is highlighted as a slice of Figure 3.6. Another familiar example is the case $\sigma = 0$. It identifies the Fourier transform of $f(t)$ with the slice on the s-plane's imaginary axis, with the expected $\sin x/x$ function for its real part and zero for its imaginary part. As the absolute value of σ increases from zero, we clearly see the emergence of Im $F(s)$ from nothing.

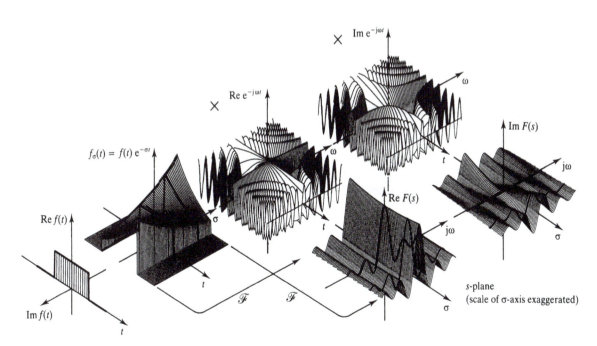

Figure 3.6 Laplace transform, full representation.

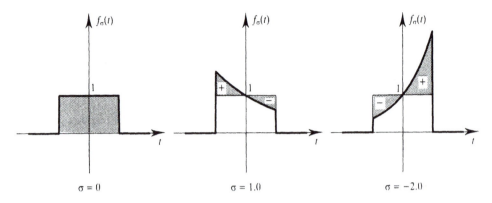

Figure 3.7   Growth of d.c. magnitude with increasing $|\sigma|$.

Less obvious is the amplitude increase of the real part. It is the direct result of an asymmetry introduced by the real exponential $e^{\sigma t}$ as it modulates the cosine carrier surface. This can be simply visualized by considering how the integral 3.3 varies with σ for a fixed frequency value, for example the d.c. component $\omega = 0$ shown in Figure 3.7 where shaded areas indicate differences with respect to the Fourier case $\sigma = 0$. There is a positive surplus, regardless of whether σ is positive or negative.

3.2   Inverse transform

The relationship 3.3 interprets the Laplace transform as an extension of the Fourier transform. It also leads to the inversion formula of the Laplace transform. For this we take the inverse Fourier transform of both sides of 3.3, written symbolically as

$$\mathcal{F}^{-1}\{F(\sigma + j\omega)\} = \mathcal{F}^{-1}\{\mathcal{F}\{f(t)e^{-\sigma t}\}\}$$

But the two operators on the right cancel, yielding $f(t)$ as

$$f(t) = e^{\sigma t}\mathcal{F}^{-1}\{F(\sigma + j\omega)\} \tag{3.5}$$

which is formally the inverse of Equation 3.3.

3.2.1   Inversion formula

Applying the inverse Fourier transform formula 2.6 to $F(\sigma + j\omega)$ of 3.5 and taking $e^{\sigma t}$ inside the integral yields

$$f(t) = \frac{1}{2\pi} \int_{-\infty}^{\infty} F(\sigma + j\omega)\, e^{(\sigma + j\omega)t}\, d\omega$$

A change of integration variable from ω to $s = \sigma + j\omega$, so that $ds = jd\omega$, and a corresponding change of integration limits from $-\infty < \omega < +\infty$ to $\sigma - j\infty < \sigma + j\omega < \sigma + j\infty$ yields the **Laplace inversion formula**

$$f(t) = \frac{1}{2\pi j} \int_{\sigma - j\infty}^{\sigma + j\infty} F(s)\, e^{st}\, ds \tag{3.6}$$

This involves integration along a contour of the complex s-plane, as explained in the graphical interpretation.

3.2.2   Graphical interpretation

The relationship 3.5 suggests that the original function $f(t)$ can be recovered from any slice of the Laplace transform, by an inverse process to that described in Figure 3.5.

Indeed, the slice $\sigma = \sigma_0$ of the function $F(\sigma + j\omega)$, with its real and imaginary parts, is transformed back by the relationship

$$f(t) = e^{\sigma_0 t} \mathcal{F}^{-1}\{\mathrm{Re}\, F(\sigma_0 + j\omega) + j\,\mathrm{Im}\, F(\sigma_0 + j\omega)\}$$

This is interpreted in Figure 3.8, which exactly reverses the operations visualized earlier in Figure 3.5. The real even function $f_1(t)$ represents the inverse transform of the real even part of $F(\sigma_0 + j\omega)$, the real odd function $f_2(t)$ that of the imaginary odd part. The sum of these two components reconstructs the modified function $f_{\sigma_0}(t) = f_1(t) + f_2(t)$, and this multiplied by $e^{\sigma_0 t}$, the inverse of the convergence factor, leads back to the original function $f(t)$.

This interpretation is valid for any slice σ for which the Laplace transform exists. It shows that each slice constitutes a complete and self-sufficient alternative description of $f(t)$, and that it contains enough information to recover the original signal. In this context, the Fourier transform $F(j\omega)$ simply becomes one of the possible Laplace descriptions, $F(j\omega) = F(\sigma + j\omega)|_{\sigma=0} = F(s)|_{s=j\omega}$

The frequency domain of Figure 3.8 also interprets the integration

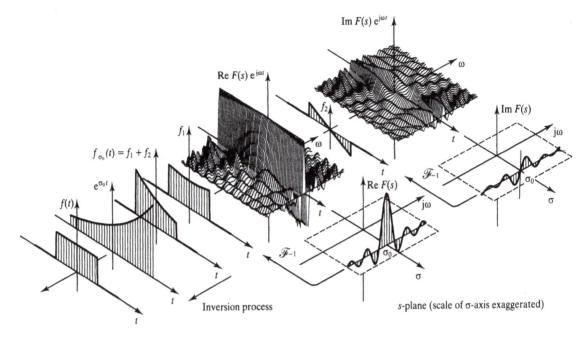

Figure 3.8 Inverse Laplace transform, single slice σ_0.

contour of Equation 3.6 as a line of constant σ, that is, the line $s = \sigma_0 + j\omega$, parallel to the imaginary axis. The integration limits are the points at infinity of that line, much as the Fourier limits $-\infty < \omega < +\infty$ are the points at infinity of the imaginary axis.

3.3 Poles, zeros and region of convergence

The rectangular pulse used for illustration so far provided the desired continuity with the Fourier transform, but failed to reveal some important features of the Laplace transform. We will now introduce the concepts of poles, zeros and region of convergence, and in the process build up a small table of Laplace pairs, which we collect in Figure 3.9 for later use.

In this section we will make repeated use of the standard integral

$$\int_{t_1}^{t_2} e^{pt}\, dt = \left.\frac{e^{pt}}{p}\right|_{t_1}^{t_2} = \frac{e^{pt_2} - e^{pt_1}}{p}$$

where p can be either real or a complex number $p = \alpha + j\beta$. The special case $t_1 = 0$, $t_2 = \infty$ yields the important result

$$\int_0^\infty e^{pt} \, dt = \frac{1}{-p} \qquad \text{for Re } p < 0 \tag{3.7}$$

The condition Re $p < 0$ can be verified by writing $e^{pt} = e^{\alpha t} e^{j\beta t}$, whose amplitude $e^{\alpha t}$ decays with increasing t, provided $\alpha < 0$, as illustrated earlier in Figure 3.4.

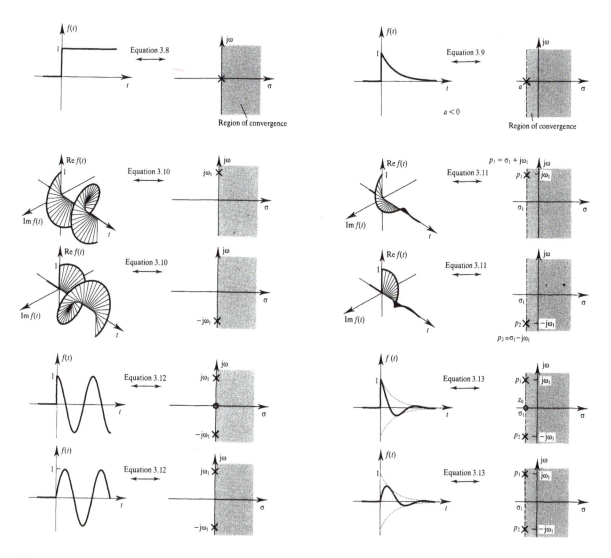

Figure 3.9   Related transform pairs and their regions of convergence.

3.3.1   Examples

Unit step

Consider the unit step function $f(t) = u(t)$. Applying the Laplace transform definition 3.1 we have

$$F(s) = \int_{-\infty}^{\infty} u(t)e^{-st}\, dt = \int_{0}^{\infty} e^{-st}\, dt$$

because $u(t)$ is zero for $t < 0$ and unity elsewhere. Making $p = -s$ in Equation 3.7 yields the Laplace pair

$$u(t) \quad \overset{\mathscr{L}}{\longleftrightarrow} \quad \frac{1}{s} \qquad \mathrm{Re}\, s > 0 \tag{3.8}$$

As the complex frequency variable s approaches zero, the magnitude of $F(s) = 1/s$ grows to infinity. A value of s for which such growth occurs is a **pole of the function**, and the corresponding point of the s-plane is identified by a cross.

The unit step $u(t)$ has a pole at the s-plane origin $s = 0$ as indicated at the top left of Figure 3.9. This pole also marks the lower boundary of the shaded half-plane to the right of the pole, expressed by the condition $\mathrm{Re}\, s > 0$, which represents the **region of convergence** of this function, to be formalized in Section 3.3.3.

Real exponential

We next consider the causal real exponential

$$f(t) = e^{at} u(t)$$

where a is a real constant, illustrated for a value $a < 0$ at the top right of Figure 3.9. The Laplace definition 3.1 yields

$$F(s) = \int_{-\infty}^{\infty} [e^{at} u(t)]e^{-st}\, dt = \int_{0}^{\infty} e^{-(s-a)t}\, dt$$

and, making $-(s - a) = p$, the result 3.7 leads to the pair

$$e^{at} u(t) \quad \overset{\mathscr{L}}{\longleftrightarrow} \quad \frac{1}{s-a} \qquad \mathrm{Re}\, s > a \tag{3.9}$$

When $s - a = 0$ the frequency domain denominator vanishes, thus revealing a pole at $s = a$, as well as the boundary of the region of convergence, as indicated in Figure 3.9. Note that multiplying the unit step by the real exponential e^{at} has the net effect of shifting the pole from the origin to the location $s = a$.

The shape of the Laplace transform and the concept of region of convergence are clarified further in Figure 3.10. The upper half represents one particular slice $\sigma_0 > 0$, and shows that the originally decaying exponential e^{at} decays even faster when multiplied by the diminishing factor $e^{-\sigma_0 t}$, thus assisting convergence.

The lower half of Figure 3.10 gives a full representation, and includes the above slice and the slice at $\sigma = 0$, that is, the Fourier transform

$$F(j\omega) = F(s)|_{s=j\omega} = \frac{1}{j\omega - a}$$

As σ becomes negative, in the range $a < \sigma < 0$ the modified function $f_\sigma(t)$ decays less rapidly than $f(t)$, until the limiting case $\sigma = a$ is reached, where $f_\sigma(t)$ becomes the unit step function $u(t)$, beyond which the Fourier integral of $f_\sigma(t)$ fails to converge and the Laplace transform does not exist.

This progression is reflected in the frequency domain, where the function $|F(s)|$ takes the shape of a volcano, located at the point $s = a$ and truncated at the boundary $\sigma = a$. Outside the region of convergence, the **algebraic function** $F(s) = 1/(s - a)$ still exists and can be evaluated for arbitrary points s, but the computed values do not represent the Laplace transform of the original function $f(t) = e^{at}u(t)$, as will be seen in the result 3.14.

Complex exponentials

The Laplace pair 3.9 is also valid when the parameter a takes complex values. The cases $a = j\omega_1$ and $a = -j\omega_1$ yield the transforms of the causal exponentials

$$e^{j\omega_1 t}u(t) \quad \overset{\mathcal{L}}{\longleftrightarrow} \quad \frac{1}{s - j\omega_1} \qquad \mathrm{Re}\, s > 0$$

$$e^{-j\omega_1 t}u(t) \quad \overset{\mathcal{L}}{\longleftrightarrow} \quad \frac{1}{s + j\omega_1} \qquad \mathrm{Re}\, s > 0 \tag{3.10}$$

with poles at $s = j\omega_1$ and $s = -j\omega_1$, respectively, and regions of convergence to their right.

Giving a the complex values $p_1 = \sigma_1 + j\omega_1$ and $p_2 = \sigma_1 - j\omega_1$, we have

$$e^{p_1 t}u(t) \quad \overset{\mathcal{L}}{\longleftrightarrow} \quad \frac{1}{s - p_1} \qquad \mathrm{Re}\, s > \sigma_1$$

$$e^{p_2 t}u(t) \quad \overset{\mathcal{L}}{\longleftrightarrow} \quad \frac{1}{s - p_2} \qquad \mathrm{Re}\, s > \sigma_1 \tag{3.11}$$

where p_1 and p_2 represent poles.

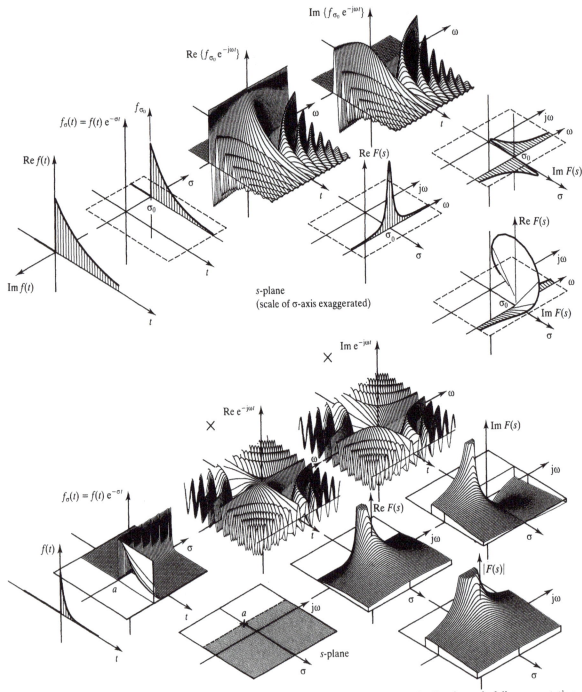

Figure 3.10 Laplace transform of real exponential, single slice (upper), full representation (lower).

These functions are also included in the table of Figure 3.9. The corresponding three-dimensional Laplace representations would simply shift the 'volcano' of Figure 3.10 to one of these new pole locations.

Sinusoidal functions

The semi-sum of both sides of the pairs 3.10 gives

$$\tfrac{1}{2}(e^{j\omega_1 t} + e^{-j\omega_1 t})u(t) = \cos \omega_1 t\, u(t)$$

$$\xleftrightarrow{\ \mathscr{L}\ } \quad \frac{\tfrac{1}{2}}{s - j\omega_1} + \frac{\tfrac{1}{2}}{s + j\omega_1} = \frac{s}{s^2 + \omega_1^2}$$

and a similar result holds for the semi-difference. These yield the pairs

$$\cos \omega_1 t\, u(t) \quad \xleftrightarrow{\ \mathscr{L}\ } \quad \frac{s}{s^2 + \omega_1^2} \qquad \mathrm{Re}\, s > 0$$

$$\tag{3.12}$$

$$\sin \omega_1 t\, u(t) \quad \xleftrightarrow{\ \mathscr{L}\ } \quad \frac{\omega_1}{s^2 + \omega_1^2} \qquad \mathrm{Re}\, s > 0$$

Both functions have the same second-order denominator, whose roots are those of $(s - j\omega_1)$ and $(s + j\omega_1)$. Their poles $p_1 = j\omega_1$ and $p_2 = -j\omega_1$ are therefore those of the exponentials 3.10, as indicated in Figure 3.9.

The sum leading to the cosine function also introduces the variable s into the numerator of the transform. Values of s for which the numerator vanishes are **zeros of the function**, indicated in the s-plane by a small circle. The cosine function has such a zero at the origin $s = 0$ as shown in the corresponding s-plane of Figure 3.9, but not the sine function.

Similar semi-sums of the functions 3.11 can be arranged as

$$e^{\sigma_1 t} \cos \omega_1 t\, u(t) \quad \xleftrightarrow{\ \mathscr{L}\ } \quad \frac{(s - \sigma_1)}{(s - \sigma_1)^2 + \omega_1^2} \qquad \mathrm{Re}\, s > \sigma_1$$

$$\tag{3.13}$$

$$e^{\sigma_1 t} \sin \omega_1 t\, u(t) \quad \xleftrightarrow{\ \mathscr{L}\ } \quad \frac{\omega_1}{(s - \sigma_1)^2 + \omega_1^2} \qquad \mathrm{Re}\, s > \sigma_1$$

Compared with expressions 3.12, the time domains were multiplied by the real exponential $e^{\sigma_1 t}$, with the effect of shifting all s-plane poles and zeros by the real value σ_1, as shown in Figure 3.9. The full Laplace transform representation of the sine case, Figure 3.11, shows a 'double volcano' in correspondence with the poles.

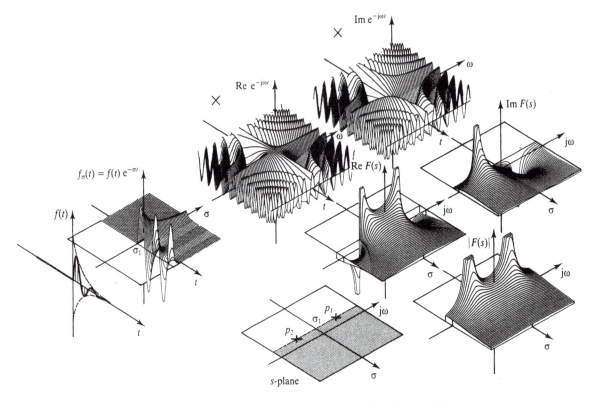

Figure 3.11 Full Laplace representation of $e^{\sigma_1 t}\sin\omega_1 t\, u(t)$.

3.3.2 Left-sided and two-sided functions

Consider the left-sided function $f(t) = -e^{bt}u(-t)$ of Figure 3.12. The Laplace definition 3.1 gives

$$F(s) = \int_{-\infty}^{\infty} [-e^{bt}u(-t)]e^{-st}\,\mathrm{d}t - \int_{-\infty}^{0} e^{-(s-b)t}\,\mathrm{d}t$$

Writing $(s - b) = p$, this integral is of the form

$$-\int_{-\infty}^{0} e^{-pt}\,\mathrm{d}t = -\int_{0}^{\infty} e^{pt}\,\mathrm{d}t = \frac{1}{p} \qquad \mathrm{Re}\, p < 0$$

which yields the transform pair

$$-e^{bt}u(-t) \quad \overset{\mathscr{L}}{\longleftrightarrow} \quad \frac{1}{s-b} \qquad \mathrm{Re}\, s < b \tag{3.14}$$

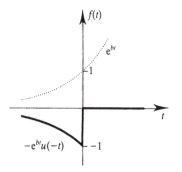
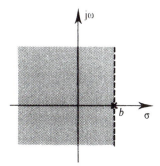

Figure 3.12   Left-sided exponential.

where the convergence condition is an alternative statement for $\text{Re}(s - b) < 0$. The frequency representation of this function is algebraically identical to that of expression 3.9, but its region of convergence is the half-plane **to the left of the pole** at $s = b$, as shown in Figure 3.12.

We alluded to this situation in the context of Figure 3.10. If b were to be given the same negative value, $b = a$, the full representation would only show 'the other side of the volcano'. This demonstrates the need for associating a region of convergence with the algebraic equation, to resolve any ambiguity.

A function that extends over the entire time axis is double-sided and can be interpreted as the sum to two single-sided functions. The top three examples of Figure 3.13 illustrate this, where the third is the sum of the other two,

$$e^{at}u(t) + e^{bt}u(-t) \xleftrightarrow{\mathscr{L}} \frac{a - b}{(s - a)(s - b)} \qquad a < \text{Re}\,s < b$$

$$(3.15)$$

The region of convergence is the intersection, or overlap, of the two component regions.

The left-sided function

$$e^{bt}u(-t) \xleftrightarrow{\mathscr{L}} \frac{-1}{s - b} \qquad \text{Re}\,s < b$$

is simply the pair 3.14 amplitude scaled by the factor -1. It has the same s-plane representation because the pole is fully capable of defining the shape of $F(s)$, but not the associated absolute values.

The remainder of Figure 3.13 gives a selection of real single-sided and double-sided exponentials with their transforms and s-plane representations. In each case, the right-sided function defines a right-sided region of convergence, while the left-sided function defines a left-sided region. The Laplace transform only exists in the intersection of the two regions. If they do not intersect, the Laplace transforms does not exist.

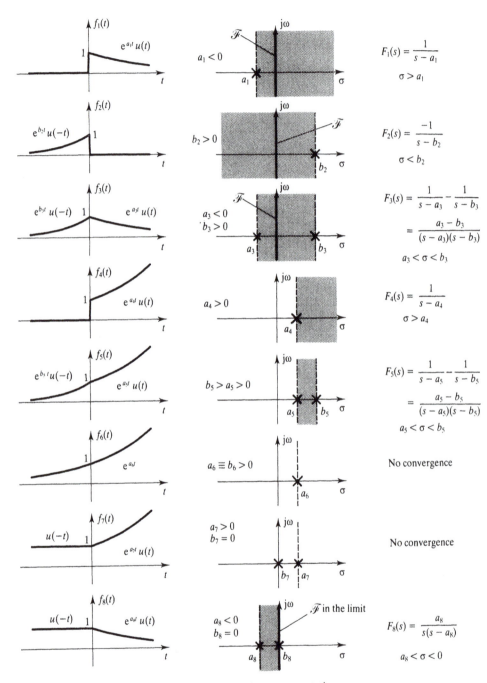

Figure 3.13   Exponentials and their s-plane representations.

If the region of convergence contains the imaginary axis, then the Fourier transform exists and can be obtained by evaluating $F(s)$ for $s = j\omega$,

$$F(j\omega) = F(s)|_{s=j\omega}$$

Laplace convergence conditions are inequalities, such as $\operatorname{Re} s > a$, which exclude the boundary $\operatorname{Re} s = a$ itself. This is called an **open boundary**, indicated by a broken line in the s-plane representation, which implies that the Laplace transform does not exist on the boundary itself.

If a boundary coincides with the imaginary axis, the Fourier transform can only exist as an **integral in the limit**, and can not be obtained from $F(s)$. The unit step $u(t)$ provides a typical example. The region of covergence of expression 3.8 does not include the imaginary axis but the Fourier transform was found in Example 2.9 with the help of the function sgn(t) and the impulse $\delta(\omega)$.

3.3.3   Convergence properties

We now re-examine the functions collected in Figure 3.9 for signs of convergence. All the time functions of the right half of the figure are contained within a surface of revolution of magnitude $e^{at}u(t)$, whose rate of decay a sets the common boundary of their regions of convergence. The left half of the figure represents the special case $a = 0$.

We conclude that the Laplace transform of a right-sided function $f(t)u(t)$ exists for $\operatorname{Re} s > \alpha$, provided a positive scaling factor M can be found, such that the function is contained within an exponential envelope of amplitude $Me^{\alpha t}u(t)$. This condition is expressed as

$$|f(t)u(t)| < Me^{\alpha t}u(t)$$

and confirmed when inserted in the inequality

$$|F(s)| \leq \int_{-\infty}^{\infty} |f(t)u(t)e^{-st}|\,dt < \int_{-\infty}^{\infty} |f(t)u(t)|e^{-\sigma t}\,dt$$

in which $e^{-\sigma t}$ represents the magnitude of e^{-st}, to yield

$$|F(s)| < \int_{0}^{\infty} Me^{(\alpha-\sigma)t}\,dt = \frac{M}{\sigma - \alpha} \qquad \sigma > \alpha$$

where we used the result 3.7 to evaluate the integral and its condition of validity $\alpha - \sigma < 0$.

This result is also seen as the limit case of the full representation of the lower half of Figure 3.10, which illustrates the gradual transformation of the exponential envelope into the unit step function, as the convergence boundary $\sigma = a$ is approached.

A similar argument applies to a left-sided function $f(t)u(-t)$ contained within an envelope $Me^{\beta t}u(-t)$, whose transform exists for $\operatorname{Re} s < \beta$.

The combination of the two cases leads to the general condition of convergence for arbitrary two-sided functions, real or complex, contained within an envelope of the form of expression 3.15, that is,

$$|f(t)| < \begin{cases} Me^{\alpha t} & t > 0 \\ Me^{\beta t} & t < 0 \end{cases} \tag{3.16}$$

Such a function is said to be **of exponential order**, and its Laplace transform converges in the strip $\alpha < \operatorname{Re} s < \beta$. An arbitrary example is shown in Figure 3.14.

The real exponential functions of Figure 3.13 can be interpreted as the magnitudes of envelopes of revolution, and their regions of convergence are also those of all the possible functions contained within such envelopes.

It is clear that the rate of decay at infinity of the right-sided function controls the lower boundary α, while that of the left-sided function controls the upper boundary β. If the two regions do not intersect, there is no Laplace transform.

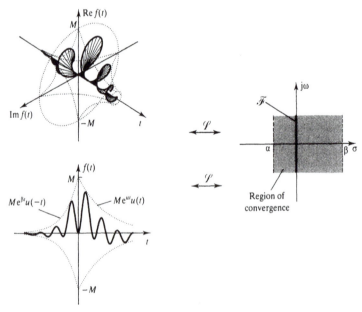

Figure 3.14 Region of convergence of arbitrary function contained within exponential envelope.

Convergence properties are mainly concerned with the functions' behaviour at infinite time. For functions defined between finite lower and upper time limits, such as those of Figure 3.6 and of the later Figure 3.15, the question of convergence does not arise, their regions of convergence extend over the entire s-plane.

3.4   Function symmetries, real $f(t)$

In Section 2.3 we examined the behaviour of the Fourier transform towards function symmetries, summarizing the general results in scheme 2.16 and those for real $f(t)$ in scheme 2.17. Both cases are readily extended to the Laplace transform, but because they involve lengthy algebraic expressions, we relegate the derivations to the appendix at the end of this chapter. In this section we interpret the results for the most common case of real $f(t)$.

In the earlier Figure 3.6 we saw that the Laplace transform of a typical real even function has real and imaginary parts that are respectively even and odd in the frequency variable ω. The transform of a typical real odd function is shown in Figure 3.15, and it too has real even and imaginary odd frequency components.

We conclude that the Laplace transform $F(s)$ of a real function $f(t)$ with even and odd components, will have real even and imaginary odd parts, which makes it a Hermitian function. Conversely, both the even and the odd components of $f(t)$ contribute to each of the Laplace transform components, as expressed in the first and last of Equations 3.47 in the appendix, namely

$$\mathrm{Re}\, F_e(\sigma + j\omega) = \int_{-\infty}^{\infty} [f_e(t)\cosh \sigma t - f_o(t)\sinh \sigma t]\cos \omega t\, dt$$

$$(3.17)$$

$$\mathrm{Im}\, F_o(\sigma + j\omega) = \int_{-\infty}^{\infty} [f_e(t)\sinh \sigma t - f_o(t)\cosh \sigma t]\sin \omega t\, dt$$

The cited Figures 3.6 and 3.15 represent terms of these expressions.

The Fourier transform of a real function with even and odd components was illustrated in the sequence of Figures 2.15 to 2.17. That sequence can also be interpreted as the illustration for one particular slice, $\sigma = 0$, of the corresponding Laplace transform.

The inverse Laplace transform corresponding to a real even $f(t)$ was shown in Figure 3.8. It represented a special case of the more common situation where $f(t)$ is real and has even and odd components. These components are obtained by the expressions

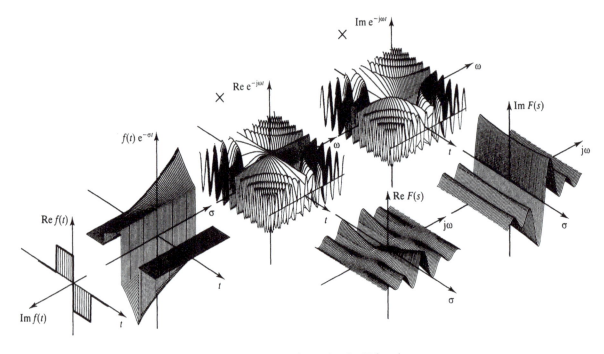

Figure 3.15   Laplace transform of real odd function.

$$f_e(t) = \cosh \sigma t \, \frac{1}{2\pi} \int_{-\infty}^{\infty} \operatorname{Re} F_e(\sigma + j\omega) \cos \omega t \, d\omega$$

$$- \sinh \sigma t \, \frac{1}{2\pi} \int_{-\infty}^{\infty} \operatorname{Im} F_o(\sigma + j\omega) \sin \omega t \, d\omega \tag{3.18}$$

$$f_o(t) = \sinh \sigma t \, \frac{1}{2\pi} \int_{-\infty}^{\infty} \operatorname{Re} F_e(\sigma + j\omega) \cos \omega t \, d\omega$$

$$- \cosh \sigma t \, \frac{1}{2\pi} \int_{-\infty}^{\infty} \operatorname{Im} F_o(\sigma + j\omega) \sin \omega t \, d\omega$$

which were extracted from the general expressions 3.49 of the appendix. The two sets of Equations 3.17 and 3.18 are expressed schematically as

$$\overbrace{
\begin{array}{ccc}
f_e(t) & + & f_o(t) \\
\updownarrow\!\!\!\searrow & & \nearrow\!\!\!\updownarrow \\
\operatorname{Re} F_e(s) & + & j \operatorname{Im} F_o(s)
\end{array}
}^{f(t)} \tag{3.19}$$

$$\underbrace{\phantom{\operatorname{Re} F_e(s) + j \operatorname{Im} F_o(s)}}_{F(s)}$$

3.5   Transform properties

The formal correspondence of Equation 3.3 between Laplace and Fourier transforms suggests a similar correspondence of properties. Laplace properties are derived by formally identical means to those presented in Section 2.4 for the Fourier transform, except that the region of convergence is also affected.

3.5.1   Linearity

The Laplace transform is a linear operator, and this can be verified from its definition. Two functions, with known transforms,

$$f_1(t) \overset{\mathcal{L}}{\longleftrightarrow} F_1(s) \qquad \alpha_1 < \mathrm{Re}\, s < \beta_1$$

$$f_2(t) \overset{\mathcal{L}}{\longleftrightarrow} F_2(s) \qquad \alpha_2 < \mathrm{Re}\, s < \beta_2$$

can be linearly combined, to give a result equivalent to expression 2.24

$$af_1(t) + bf_2(t) \overset{\mathcal{L}}{\longleftrightarrow} aF_1(s) + bF_2(s)$$

$$\max(\alpha_1, \alpha_2) < \mathrm{Re}\, s < \min(\beta_1, \beta_2)$$

$$(3.20)$$

The region of convergence of the sum is the intersection of the component regions, and is usually smaller than these. The pair 3.15, combined with the top three functions of Figure 3.13, provides an example.

3.5.2   Time and frequency scaling

These properties are formally similar to expressions 2.26 and 2.27, namely

$$f(at) \overset{\mathcal{L}}{\longleftrightarrow} \frac{1}{|a|}F\left(\frac{s}{a}\right) \qquad a\alpha < \mathrm{Re}\, s < a\beta$$

$$\frac{1}{|b|}f\left(\frac{t}{b}\right) \overset{\mathcal{L}}{\longleftrightarrow} F(bs) \qquad \frac{\alpha}{b} < \mathrm{Re}\, s < \frac{\beta}{b}$$

$$(3.21)$$

An expansion of the time variable t leads to a proportional compression of the frequency variable s, and vice versa. This applies to both the real and the imaginary parts of the frequency variable s and causes a corresponding compression of the region of convergence.

This is illustrated in Figure 3.16 for two cases, $a = 2$ and $a = -1$. The latter example also yields the result

$$h(t) = f(-t) \quad \overset{\mathscr{L}}{\longleftrightarrow} \quad H(s) = F(-s) \qquad -\alpha > \operatorname{Re} s > -\beta \quad (3.22)$$

3.5.3 Time and frequency shifting

The Laplace counterparts to expressions 2.31 and 2.33 are

$$f(t - t_0) \quad \overset{\mathscr{L}}{\longleftrightarrow} \quad e^{-st_0} F(s) \qquad \alpha < \operatorname{Re} s < \beta \qquad (3.23)$$

$$e^{at} f(t) \quad \overset{\mathscr{L}}{\longleftrightarrow} \quad F(s - a) \qquad \alpha - \operatorname{Re} a < \operatorname{Re} s < \beta - \operatorname{Re} a \qquad (3.24)$$

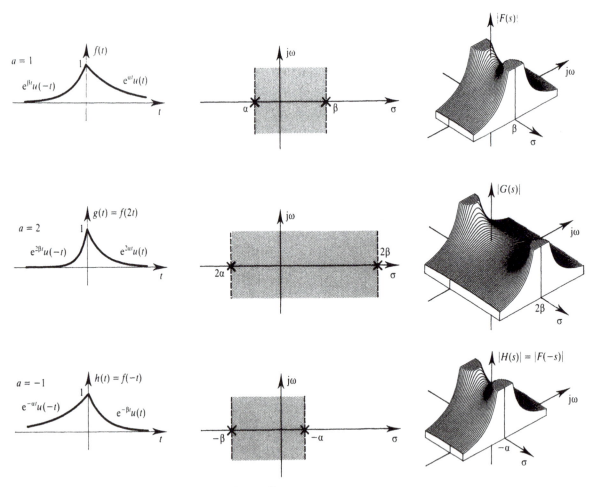

Figure 3.16 Time scaling.

Examples are found in Figure 3.9, where all the functions in the right half are obtained by modulating the amplitudes of the corresponding functions of the left half by the real exponential e^{at}, $a < 0$. This is equivalent to shifting all features of the s-plane by the real amount a. The frequency results could have been obtained directly by substituting $(s - a)$ for s in the appropriate functions $F(s)$.

Further examples, involving shifted unit steps, are given in Figure 3.17.

3.5.4   Convolutions

The Fourier convolution properties 2.35 and 2.36 become

$$y(t) = x(t) * h(t) \quad \overset{\mathscr{L}}{\longleftrightarrow} \quad Y(s) = X(s) \cdot H(s)$$

$$\max(\alpha_1, \alpha_2) < \mathrm{Re}\,s < \min(\beta_1, \beta_2)$$

$$(3.25)$$

$$g(t) = x(t) \cdot h(t) \quad \overset{\mathscr{L}}{\longleftrightarrow} \quad G(s) = \frac{1}{2\pi\mathrm{j}} X(s) * H(s)$$

$$\alpha_1 + \alpha_2 < \mathrm{Re}\,s < \beta_1 + \beta_2$$
$$(3.26)$$

Two examples of expression 3.25 are given in Figure 3.18. The first derives the Laplace transform of the unit ramp $r(t)$ as a convolution of two unit steps,

$$r(t) = u(t) * u(t) \quad \overset{\mathscr{L}}{\longleftrightarrow} \quad R(\mathrm{j}\omega) = \frac{1}{s^2} \quad \mathrm{Re}\,s > 0$$

which has two poles at the s-plane origin. The second derives the transform of a triangular pulse.

3.5.5   Differentiation and integration

The counterparts to expressions 2.37 and 2.38 are

$$\frac{\mathrm{d}f(t)}{\mathrm{d}t} \quad \overset{\mathscr{L}}{\longleftrightarrow} \quad sF(s) \quad \alpha < \mathrm{Re}\,s < \beta$$

$$(3.27)$$

$$\frac{\mathrm{d}^n f(t)}{\mathrm{d}t^n} \quad \overset{\mathscr{L}}{\longleftrightarrow} \quad s^n F(s) \quad \alpha < \mathrm{Re}\,s < \beta$$

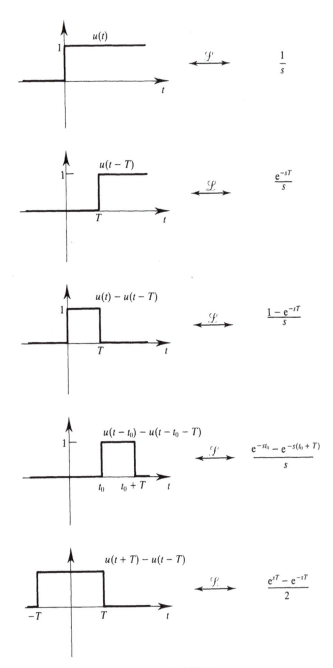

Figure 3.17 Functions as sums of shifted steps.

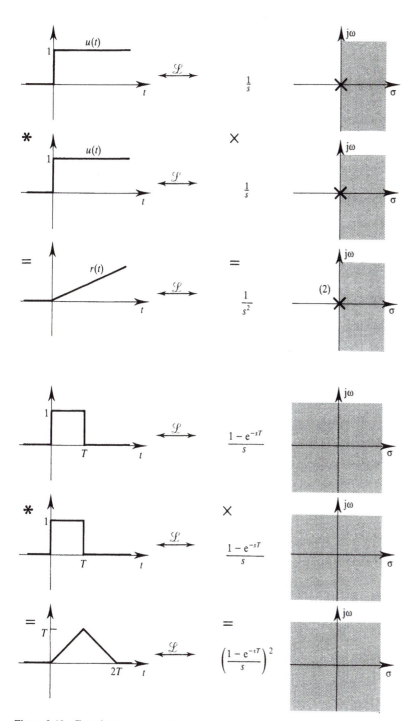

Figure 3.18 Functions as convolutions.

An example is given in Figure 3.19, where we transform the derivative of the function $f(t) = e^{at}u(t)$. The time domain of the derivative has two terms

$$g(t) = \frac{df(t)}{dt} = ae^{at}u(t) + e^{at}\delta(t) = ae^{at}u(t) + \delta(t) \tag{3.28}$$

where the impulse represents the derivative of the discontinuity and we used the impulse property $f(t)\delta(t) = f(0)\delta(t)$. The transform of $f(t)$ itself was given in the pair 3.9, which, with the property 3.27 yields the pair

$$ae^{at}u(t) + \delta(t) \overset{\mathcal{L}}{\longleftrightarrow} \frac{s}{s-a} \tag{3.29}$$

The Laplace counterpart of 2.39 for the derivative in frequency is

$$(-t)^n f(t) \overset{\mathcal{L}}{\longleftrightarrow} \frac{d^n F(s)}{ds^n} \tag{3.30}$$

and the special case of the first derivative $n = 1$ gives the useful pair

$$tf(t) \overset{\mathcal{L}}{\longleftrightarrow} -\frac{dF(s)}{ds} \tag{3.31}$$

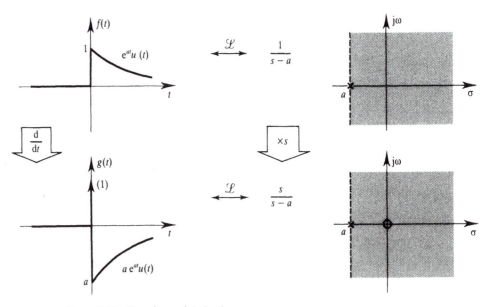

Figure 3.19 Transform of derivative.

Finally, interpreting integration as convolution with the unit step, as in Equation 1.9, the time convolution property 3.25 yields

$$g(t) = \int_{-\infty}^{t} f(\tau)\,d\tau \quad \xleftrightarrow{\mathscr{L}} \quad \frac{F(s)}{s} \qquad \max(\alpha, 0) < \mathrm{Re}\,s < \beta \qquad (3.32)$$

where the lower limit of the region of convergence reflects the pole at the s-plane origin associated with that step. The earlier example of the unit ramp $r(t)$ as convolution of two unit steps, see Figure 3.18, can thus be reinterpreted as the integral of one of those steps.

3.6 Unilateral Laplace transform

The Laplace transform definition 3.1 involves integration over the entire time range, $-\infty < t < \infty$, for which it is also called the **two-sided** or **bilateral Laplace transform**. But most physical systems are causal and their time response is described by a one-sided function. If the input signal is causal too, such systems can be solved using the **one-sided** or **unilateral Laplace transform**. Furthermore, if the system or the signal or both have non-zero initial conditions, the unilateral Laplace transform offers distinct advantages.

3.6.1 Definition

The unilateral Laplace transform, indicated by the pair

$$f(t) \quad \xleftrightarrow{\mathscr{L}_\mathrm{I}} \quad F_\mathrm{I}(s)$$

is defined by the integral

$$F_\mathrm{I}(s) = \mathscr{L}_\mathrm{I}\{f(t)\} = \int_{0}^{\infty} f(t)\mathrm{e}^{-st}\,dt \qquad (3.33)$$

which only differs from Equation 3.1 in the lower integration limit. The time domain functions of the bilateral transform examples of Section 3.3.1 are causal, and for them the unilateral definition 3.33 gives the same results.

In Section 3.3.2 we concluded that the region of convergence of a right-sided function is a right-sided half-plane. This is always the case with the unilateral Laplace transform, which obviates the need to specify such region, as it always lies to the right of the rightmost pole.

The inversion formula remains formally identical to Equation 3.6, namely

$$f(t) = \frac{1}{2\pi j} \int_{\sigma - j\infty}^{\sigma + j\infty} F_{\mathrm{I}}(s) e^{st}\, ds \qquad (3.34)$$

For a causal function $f(t)$ the unilateral transform $F_{\mathrm{I}}(s)$ is identical to the bilateral transform $F(s)$ and either form of the inversion formula would recreate the original causal function. In contrast, for an acausal time function the unilateral transform 3.33 would simply ignore all function values for $t < 0$. A subsequent inversion according to Equation 3.34 would return a causal function $f(t)u(t)$, that is, a truncated version of the original function, as expressed in

$$f(t) \xrightarrow{\;\mathscr{L}_{\mathrm{I}}\;} F_{\mathrm{I}}(s) \xrightarrow{\;\mathscr{L}_{\mathrm{I}}^{-1}\;} f(t)u(t)$$

3.6.2   Bilateral transform as sum of two unilateral transforms

Most tables of Laplace transform pairs give the unilateral form. These can be used to find the transform of a bilateral function, as well as the inverse transform, by considering the acausal time function $f(t)$ as a sum of a causal and an anticausal function, that is,

$$f(t) = g(t)u(t) + h(t)u(-t) \xrightarrow{\;\mathscr{L}\;} F(s) = G(s) + H(s)$$

Such sum was implied in the case of the pair 3.15. The first term, $g(t)u(t)$, is causal and its unilateral transform $G_{\mathrm{I}}(s)$ is identical to $G(s)$, as indicated in the upper loop of Figure 3.20.

To transform the second term, $h(t)u(-t)$, which is anticausal, we change the sign of the time variable, thus forming the causal function $h(-t)u(t)$, whose bilateral transform, according to the scaling property 3.22, is

$$h(-t)u(t) \xrightarrow{\;\mathscr{L}\;} H(-s) \qquad -\alpha > \mathrm{Re}\, s > -\beta$$

Being causal, this function also has a unilateral transform $H_{\mathrm{I}}(s)$ that is identical to $H(-s)$. Thus, to obtain the term $H(s)$ we only need to change the sign of the frequency variable s in $H_{\mathrm{I}}(s)$, as indicated in the lower loop of Figure 3.20, with the result

$$F(s) = G(s) + H_{\mathrm{I}}(-s)$$

Example 3.1

We illustrate the procedure by reinterpreting the pair 3.15. With

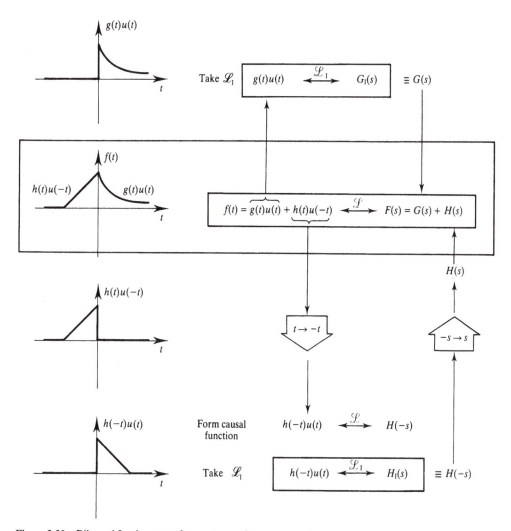

Figure 3.20 Bilateral Laplace transform as sum of two unilateral transforms.

$$g(t)u(t) = e^{at}u(t) \xleftrightarrow{\mathscr{L}_I} G_I(s) = \frac{1}{s-a} \qquad \text{Re}\, s > a$$

and $h(t)u(-t) = e^{bt}u(-t)$, so that

$$h(-t)u(t) = e^{-bt}u(t) \xleftrightarrow{\mathscr{L}_I} H_I(s) = \frac{1}{s+b} \qquad \text{Re}\, s > -b$$

we have

$$F(s) = G_I(s) + H_I(-s) = \frac{1}{s-a} + \frac{1}{-s+b}$$

$$= \frac{a-b}{(s-a)(s-b)} \qquad a < \operatorname{Re} s < b$$

which confirms expression 3.15.

___

3.6.3 Properties

Most of the properties presented in Section 3.5 for the bilateral Laplace transform are valid for the unilateral case. Notable exceptions are the transforms of the first and higher order derivatives of a function, which incorporate initial values of the time function and of its derivatives, thus giving the unilateral Laplace transform its special qualities.

Time shifting

When time shifting a function it must be remembered that the transform of the original function $f(t)$ ignores any values of negative time. With

$$f(t) \quad \overset{\mathscr{L}_I}{\longleftrightarrow} \quad F_I(s) \qquad \operatorname{Re} s > \alpha$$

when $f(t)$ is causal, the expression 3.23 is valid, in the form

$$f(t - t_0) \quad \overset{\mathscr{L}_I}{\longleftrightarrow} \quad e^{-st_0} F_I(s) \qquad \operatorname{Re} s > \alpha$$

When $f(t)$ is acausal, to make use of $F_I(s)$ the truncated function $f(t)u(t)$ must be used, or implied, giving

$$f(t - t_0)u(t - t_0) \quad \overset{\mathscr{L}_I}{\longleftrightarrow} \quad e^{-st_0} F_I(s) \qquad \operatorname{Re} s > \alpha \tag{3.35}$$

as is clear from Figure 3.21.

Transform of a derivative

Given the unilateral transform pair

$$f(t) \quad \overset{\mathscr{L}_I}{\longleftrightarrow} \quad F_I(s)$$

the unilateral definition 3.33 expresses the transform of the derivative $f'(t)$ as

$$\mathscr{L}_I\{f'(t)\} = \int_0^\infty f'(t)e^{-st}\, dt$$

We use the standard technique of integration by parts,

$$\int_a^b u'v\, dt = uv \Big|_a^b - \int_a^b uv'\, dt$$

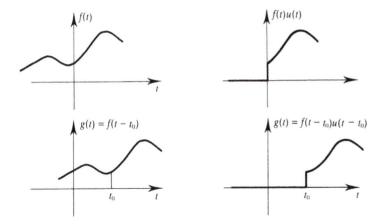

Figure 3.21   Time shifting of acausal and causal functions.

which, with the substitutions

$$f'(t) = u' \qquad \text{hence} \qquad u = f(t)$$

$$e^{-st} = v \qquad \text{hence} \qquad v' = -se^{-st}$$

yields the expression

$$\mathcal{L}_1\{f'(t)\} = f(\infty)e^{-s\infty} - f(0)e^{-s0} + s\int_0^\infty f(t)e^{-st}\,dt \qquad (3.36)$$

The integral in the third term represents the transform $F_1(s)$ of the original function, the factor e^0 is one, which leaves the transform pair

$$\boxed{f'(t) \quad \overset{\mathcal{J}_1}{\longleftrightarrow} \quad sF_1(s) - f(0)} \qquad\qquad (3.37)$$

because the first term on the right of Equation 3.36 is zero, as it represents the limit

$$\lim_{t \to \infty} f(t)e^{-st} = 0$$

which is a prerequisite for the existence of the original transform $F_1(s)$.

The great importance of this property lies in its ability to introduce the initial time domain value $f(0)$ into the frequency domain.

A recursive application of the result 3.37 yields the transforms of higher derivatives. For instance, the second derivative gives

$$\mathcal{L}_1\{f''(t)\} = s\mathcal{L}_1\{f'(t)\} - f'(0) = s[sF_1(s) - f(0)] - f'(0)$$
$$= s^2 F_1(s) - sf(0) - f'(0)$$

Similarly, the unilateral transform of the nth-order derivative becomes

$$f^{(n)}(t) \xleftrightarrow{\mathscr{L}_I} s^n F_I(s) - P(s) \tag{3.38}$$

where $P(s)$ is a polynomial in s, whose coefficients are the initial values of $f(t)$ and of its $n-1$ derivatives,

$$P(s) = s^{n-1} f(0) + s^{n-2} f'(0) + \ldots + s f^{(n-2)}(0) + f^{(n-1)}(0)$$

$$= \sum_{k=1}^{n} s^{n-k} f^{(k-1)}(0)$$

Initial value

The initial value of the function $f(t)$ is determined by the behaviour at infinity of its transform $F(s)$. The expression 3.37 permits finding the value $f(0)$ directly, without needing to invert the transform.

The unilateral transform of a function $g(t)$ is the integral

$$G_I(s) = \int_0^\infty g(t) e^{-st} \, dt$$

For $t > 0$, the exponential e^{-st} tends to zero as s tends to infinity. If $g(t)$ does not contain impulses at the origin, then

$$\lim_{s \to \infty} G_I(s) = \lim_{s \to \infty} \int_0^\infty g(t) e^{-st} \, dt = 0$$

Applying this result to the function $g(t) = f'(t)$ and using the property 3.37, we have

$$\lim_{s \to \infty} \int_0^\infty f'(t) e^{-st} \, dt = \lim_{s \to \infty} s F_I(s) - f(0) = 0$$

so that

$$f(0) = \lim_{s \to \infty} s F_I(s) \tag{3.39}$$

For example, the initial value $f(0)$ associated with the frequency domain function

$$F_I(s) = \frac{1}{s - a} \qquad a < 0$$

is given by Equation 3.40 as

$$f(0) = \lim_{s \to \infty} \frac{s}{s - a} = 1$$

This result was obtained without inverting the function $F_I(s)$ and without regard to the shape of $f(t)$, which incidentally is the function $f(t) = e^{at} u(t)$ of Example 3.1.

Final value

In contrast, the final value of the function $f(t)$, for t tending to infinity, is determined by the behaviour of its transform $F_1(s)$ at the s-plane origin. Again, the value $f(\infty)$ is found directly with the aid of expression 3.37.

As $s \to 0$ a function $G_1(s)$ approaches the limit

$$\lim_{s \to 0} G_1(s) = \lim_{s \to 0} \int_0^\infty g(t)e^{-st}\, dt = \int_0^\infty g(t)\, dt$$

which exists, provided the poles of the function $G_1(s)$ are all in the stable half of the s-plane. Applied to the function $g(t) = f'(t)$ and the result 3.37, this limit yields

$$\int_0^\infty f'(t)\, dt = \lim_{s \to 0} [sF_1(s) - f(0)]$$

that is,

$$\lim_{t \to \infty} f(t) - f(0) = \lim_{s \to 0} sF_1(s) - f(0)$$

so that

$$\lim_{t \to \infty} f(t) = \lim_{s \to 0} sF_1(s) \qquad\qquad (3.40)$$

For example, the final value of the time domain $f(t)$ of the function

$$F_1(s) = \frac{1}{s - a} \qquad a < 0$$

is given by Equation 3.40 as

$$\lim_{t \to \infty} f(t) = \lim_{s \to 0} \frac{s}{s - a} = 0$$

This result too was obtained without the knowledge of $f(t)$.

3.6.4   Interpretation of lower integration limit

The ability of the unilateral Laplace transform to deal with discontinuities at the origin of time, and with the associated impulses, gives it its strength but can also lead to confusion. This is because the impulse and its derivatives are centred about the origin, taking significant values in the infinitesimal interval $(-\varepsilon,\ \varepsilon)$ and making it necessary to choose between including or excluding them from the unilateral Laplace definition 3.33.

If the transform is defined by the limit

$$F_1^-(s) = \lim_{\varepsilon \to 0} \int_{-\varepsilon}^\infty f(t)e^{-st}\, dt = \int_{0^-}^\infty f(t)e^{-st}\, dt \qquad\qquad (3.41)$$

any impulses at the origin are included with the function $f(t)$. With this definition the transform of the derivative 3.37 becomes

$$\mathscr{L}_I^-\{f'(t)\} = sF_I^-(s) - f(0^-) \tag{3.42}$$

For an arbitrary function, $f(0^-)$ represents the value of the function when the origin is approached from negative time. For a causal function this value is $f(0^-) = 0$.

Alternatively, if we wish to exclude impulses at the origin from the transform definition, we must interpret Equation 3.33 as the limit

$$F_I^+(s) = \lim_{\varepsilon \to 0} \int_\varepsilon^\infty f(t)e^{-st}\,dt = \int_{0^+}^\infty f(t)e^{-st}\,dt \tag{3.43}$$

so that

$$\mathscr{L}_I^+\{f'(t)\} = sF_I^+(s) - f(0^+) \tag{3.44}$$

With this definition any impulses at the origin must be processed separately.

The choice is therefore whether to include impulses with the transform or with initial values. Either approach is valid as both lead to the same final results. The requirement is for consistency, to avoid including the same impulses twice or missing them altogether.

Example 3.2

We apply the unilateral definitions 3.41 and 3.43 to the impulse function $\delta(t)$.

The definition 3.41 captures the entire region in which the impulse function is non-zero, yielding

$$\mathscr{L}_I^-\{\delta(t)\} = \int_{0^-}^\infty \delta(t)e^{-st}\,dt = \int_{-\infty}^\infty \delta(t)e^{-st}\,dt = 1$$

while the definition 3.43 misses that region entirely, so that

$$\mathscr{L}_I^+\{\delta(t)\} = \int_{0^+}^\infty \delta(t)e^{-st}\,dt = 0$$

We conclude that, for causal functions not containing impulses at the origin, all definitions of the Laplace transform give the same results. In contrast, when an impulse is present at the origin the definition 3.41 gives the same result as the bilateral transform, while the definition 3.43 ignores the impulse, gives a different result and the impulse has to be dealt with separately.

Appendix: Signal symmetries and the Laplace transform

We follow the process of Section 2.3 to extend to the Laplace transform the Fourier properties relating to function symmetry.

Forward transform

Using the Fourier interpretation (3.3) of the Laplace transform, we apply the general expression 2.12 to the modified function $f_\sigma(t) = [f(t)e^{-\sigma t}]$

$$\mathcal{F}\{f(t)e^{-\sigma t}\} = \int_{-\infty}^{\infty} [f(t)e^{-\sigma t}]_e \cos \omega t \, dt - j \int_{-\infty}^{\infty} [f(t)e^{-\sigma t}]_o \sin \omega t \, dt$$

$$(3.45)$$

The left side is the Laplace transform $F(s) = \mathcal{L}\{f(t)\}$, with elemental components

$$F(\sigma + j\omega) = \operatorname{Re} F_e(\sigma + j\omega) + j \operatorname{Im} F_e(\sigma + j\omega)$$
$$+ \operatorname{Re} F_o(\sigma + j\omega) + j \operatorname{Im} F_o(\sigma + j\omega)$$

where even or odd symmetry still refers to the frequency variable ω. Equating terms with Equation 3.45 yields formally similar expressions to Equations 2.14,

$$\operatorname{Re} F_e(\sigma + j\omega) = \int_{-\infty}^{\infty} \operatorname{Re}\{f(t)e^{-\sigma t}\}_e \cos \omega t \, dt$$

$$\operatorname{Im} F_e(\sigma + j\omega) = \int_{-\infty}^{\infty} \operatorname{Im}\{f(t)e^{-\sigma t}\}_e \cos \omega t \, dt$$

$$(3.46)$$

$$\operatorname{Re} F_o(\sigma + j\omega) = \int_{-\infty}^{\infty} \operatorname{Im}\{f(t)e^{-\sigma t}\}_o \sin \omega t \, dt$$

$$\operatorname{Im} F_o(\sigma + j\omega) = -\int_{-\infty}^{\infty} \operatorname{Re}\{f(t)e^{-\sigma t}\}_o \sin \omega t \, dt$$

We next establish which parts of the function $\{f(t)e^{-\sigma t}\}$ lead, after integration, to even and odd parts of $F(s)$. For this we expand both $f(t)$ and the exponential, into even and odd parts. Since σ and t are real,

$$e^{-\sigma t} = \cosh \sigma t - \sinh \sigma t$$

where $\cosh \sigma t$ is an even function of time, and $\sinh \sigma t$ is odd. Hence

$$f(t)e^{-\sigma t} = [f_e(t) + f_o(t)](\cosh \sigma t - \sinh \sigma t)$$

From the rules of multiplication of even and odd functions, it follows that

$$\{f(t)e^{-\sigma t}\}_e = f_e(t)\cosh \sigma t - f_o(t)\sinh \sigma t$$

$$\{f(t)e^{-\sigma t}\}_o = -f_e(t)\sinh \sigma t + f_o(t)\cosh \sigma t$$

Being real, these hyperbolic functions do not affect the real or imaginary character of the functions they multiply, so that

$$\mathrm{Re}\, F_e(\sigma + j\omega) = \int_{-\infty}^{\infty}[\mathrm{Re}\, f_e(t)\cosh \sigma t - \mathrm{Re}\, f_o(t)\sinh \sigma t]\cos \omega t\, dt$$

$$\mathrm{Im}\, F_e(\sigma + j\omega) = \int_{-\infty}^{\infty}[\mathrm{Im}\, f_e(t)\cosh \sigma t - \mathrm{Im}\, f_o(t)\sinh \sigma t]\cos \omega t\, dt$$

$$\mathrm{Re}\, F_o(\sigma + j\omega)$$
$$= \int_{-\infty}^{\infty}[-\mathrm{Im}\, f_e(t)\sinh \sigma t + \mathrm{Im}\, f_o(t)\cosh \sigma t]\sin \omega t\, dt$$

$$\mathrm{Im}\, F_o(\sigma + j\omega) = \int_{-\infty}^{\infty}[\mathrm{Re}\, f_e(t)\sinh \sigma t - \mathrm{Re}\, f_o(t)\cosh \sigma t]\sin \omega t\, dt$$

$$(3.47)$$

In conclusion, each element of the time function $f(t)$ contributes to two elements of the transformed function $F(s)$. Conversely, each element of $F(s)$ is the result of contributions from two elements of $f(t)$.

The simple conservation of even and odd parts of the Fourier transform, which represents the case $\sigma = 0$, where $\cosh \sigma t = 1$ and $\sinh \sigma t = 0$, does not apply to the general case.

Inverse transform

We could follow the process of Section 2.3.2 and derive the elemental relationships for the inverse transform by using the Laplace inversion formula 3.6. But it is simpler to use the inverse Fourier interpretation of Equation 3.5 and the results 2.15 of Section 2.3.2. Denoting the symbolic inverse Fourier transform of Equation 3.5 by $\bar{f}(t)$, we write

$$f(t) = e^{\sigma t}\mathcal{F}^{-1}\{F(\sigma + j\omega)\} = e^{\sigma t}\bar{f}(t)$$

Expanding this expression into even and odd elements as

$$f(t) = (\cosh \sigma t + \sinh \sigma t)$$
$$\times [\mathrm{Re}\,\bar{f}_e(t) + j\,\mathrm{Im}\,\bar{f}_e(t) + \mathrm{Re}\,\bar{f}_o(t) + j\,\mathrm{Im}\,\bar{f}_o(t)]$$

we then extract the elements of $f(t)$ as

$$\text{Re } f_e(t) = \cosh \sigma t \text{ Re } \bar{f}_e(t) + \sinh \sigma t \text{ Re } \bar{f}_o(t)$$
$$\text{Im } f_e(t) = \cosh \sigma t \text{ Im } \bar{f}_e(t) + \sinh \sigma t \text{ Im } \bar{f}_o(t)$$
$$\text{Re } f_o(t) = \sinh \sigma t \text{ Re } \bar{f}_e(t) + \cosh \sigma t \text{ Re } \bar{f}_o(t)$$
$$\text{Im } f_o(t) = \sinh \sigma t \text{ Im } \bar{f}_e(t) + \cosh \sigma t \text{ Im } \bar{f}_o(t)$$

(3.48)

These are finally inserted into the Fourier results 2.15 to yield

$$\text{Re } f_e(t) = \cosh \sigma t \frac{1}{2\pi} \int_{-\infty}^{\infty} \text{Re } F_e(\sigma + j\omega) \cos \omega t \, d\omega$$
$$\qquad - \sinh \sigma t \frac{1}{2\pi} \int_{-\infty}^{\infty} \text{Im } F_o(\sigma + j\omega) \sin \omega t \, d\omega$$

$$\text{Im } f_e(t) = \cosh \sigma t \frac{1}{2\pi} \int_{-\infty}^{\infty} \text{Im } F_e(\sigma + j\omega) \cos \omega t \, d\omega$$
$$\qquad + \sinh \sigma t \frac{1}{2\pi} \int_{-\infty}^{\infty} \text{Re} F_o(\sigma + j\omega) \sin \omega t \, d\omega$$

(3.49)

$$\text{Re } f_o(t) = \sinh \sigma t \frac{1}{2\pi} \int_{-\infty}^{\infty} \text{Re } F_e(\sigma + j\omega) \cos \omega t \, d\omega$$
$$\qquad - \cosh \sigma t \frac{1}{2\pi} \int_{-\infty}^{\infty} \text{Im } F_o(\sigma + j\omega) \sin \omega t \, d\omega$$

$$\text{Im } f_o(t) = \sinh \sigma t \frac{1}{2\pi} \int_{-\infty}^{\infty} \text{Im } F_e(\sigma + j\omega) \cos \omega t \, d\omega$$
$$\qquad + \cosh \sigma t \frac{1}{2\pi} \int_{-\infty}^{\infty} \text{Re } F_o(\sigma + j\omega) \sin \omega t \, d\omega$$

The relationships 3.47 and 3.49 between elements are also expressed schematically as

(3.50)

This is a generalization of the Fourier schematic 2.16 and incorporates the latter as a special case.

In practice, the time function $f(t)$ is usually real. In such case the number of elements of each domain is halved, and the number of relationships reduces accordingly, as interpreted in Section 3.4.

Exercises

3.1 Sketch the real and imaginary parts of two sequences taken from the generalized complex exponentials of Figure 3.4:
(a) the sequence of constant growth factor $\sigma = -1$;
(b) the sequence of constant angular frequency $\omega = 4\pi$.
Discuss the pattern of the resulting envelopes and of the zero crossings.

3.2 Discuss the Fourier interpretations 3.3 and 3.5 of the Laplace transform and its inversion formula. With the aid of Figures 3.5 and 3.8 show that, within the region of convergence, any one slice $F(\sigma + j\omega)$ of constant σ carries all the information contained in the time domain description $f(t)$.

3.3 With the aid of Figure 3.10 discuss the meaning of the region of convergence in terms of the existence of the Fourier transform of the modified function $f_o(t)$. What role does the function's pole play in establishing the boundary of that region?

3.4 Sketch the magnitude $|F(s)|$ of the algebraic expression $F(s) = 1/(s - a)$ in the half-plane $\sigma < a$, for $a < 0$. Show that it represents the missing half of the corresponding 'volcano' of Figure 3.10. Find and plot the function $f(t)$ of which it represents the Laplace transform.

3.5 Derive the Laplace transforms of the functions $f(t) = \sin(\omega_1 t + \theta)\, u(t)$ and $g(t) = \cos(\omega_1 t + \theta)\, u(t)$ by the process used to derive the transform pairs 3.12. Plot the functions $f(t)$ and $g(t)$ for $\theta = 0$, $\theta = \pi/4$ and $\theta = \pi/2$, as well as the locations of the associated s-plane poles and zeros.

3.6 Prove that the Laplace transform is a linear operator (Equation 3.20). Interpreting the function $f_1(t)$ of the top row of Figure 3.13 as the sum $f_1(t) = f_3(t) - f_2(t)$, when $a_3 = a_1$ and $b_3 = b_2$, show that as a result of a pole/zero cancellation the region of convergence of the sum can extend beyond the component regions.

3.7 Derive the shifting properties of the Laplace transform (Equations 3.23 and 3.24) by the process used for the corresponding Fourier properties.

3.8 Derive the Laplace transform properties of convolution in time (Equation 3.25) and of differentiation in time (Equation 3.27, first part).

3.9 Find the Laplace transform of the functions

$$tu(t) \qquad t^2 u(t) \qquad t^3 u(t) \qquad t'' u(t)$$
$$te^{-t} u(t) \qquad t^n e^{-\alpha t} u(t) \qquad 2e^{-t} u(t) + 3e^{-2t} u(t)$$

3.10 Derive the Laplace transform $F(s)$ of the function $f(t) = e^{-\alpha|t|}$, where α is a positive constant, first by applying the Laplace definition 3.1, and again by treating $f(t)$ as the sum of a right-sided and a left-sided exponential. Plot the region of convergence of the Laplace transform, establish whether the Fourier transform exists and, if so, write the expression for $F(j\omega)$.

3.11 Discuss the significance of the two boundaries of the region of convergence of Exercise 3.10, in the same context as Exercise 3.3.

3.12 Write the bilateral Laplace transform of the differential equation

$$a_2 \frac{d^2 y(t)}{dt^2} + a_1 \frac{dy(t)}{dt} + a_0 y(t) = x(t)$$

Express the transform of $y(t)$ in terms of the transform of $x(t)$ and confirm that the relationship is algebraic.

3.13 Using the unilateral Laplace transform of the differential equation of Exercise 3.12, with the initial conditions $y(0) = y_0$ and $y'(0) = y_1$, write $Y(s)$ explicitly as a function of $X(s)$ and the initial conditions.

3.14 Verify the following results of the unilateral Laplace transform,

$$\int_0^t f(\tau) \, d\tau \quad \xleftrightarrow{\mathcal{L}_1} \quad \frac{1}{s} F_I(s)$$

$$\int_{-\infty}^t f(\tau) \, d\tau \quad \xleftrightarrow{\mathcal{L}_1} \quad \frac{1}{s} F_I(s) + \frac{1}{s} \int_{-\infty}^0 f(\tau) \, d\tau$$

3.15 Find the initial and final values of the time domain representation $f(t)$ corresponding to the function

$$F_I(s) = \frac{2s + 1}{s(s + 1)} = \frac{1}{s} + \frac{1}{s + 1}$$

using the initial value and final value properties of the unilateral Laplace transform. Verify the results by inverting $F_I(s)$.

CHAPTER 4

System Description

In this chapter we characterize a known continuous-time system, with a view to determining the system's effects on arbitrary input signals. The problem is one of **system analysis** and Figure 4.1 provides a guide to the aspects covered in this chapter.

We assume that a **physical description** of the system is available in the form of a circuit diagram, or a body schematic, showing the system's physical elements and their interconnections. This description can take place in the time domain or in the frequency domain, where the latter uses the concept of Laplacian impedance, and both domains are related by the Laplace transform.

The physical description provides the basis for an **analytical description** of the system, a mathematical relationship between the input and output signals. This involves the fundamental laws and methods of the discipline associated with each particular system, such as Kirchhoff's laws for electrical networks, D'Alambert's principle for mechanical systems and suitable combinations of such laws for hybrid systems.

In the time domain this leads to an integro-differential equation, whose frequency domain counterpart is an algebraic equation. These too are related by the Laplace transform, as indicated in Figure 4.1. Both equations can be expressed visually by block diagrams containing integrator and differentiator elements.

In **signal processing** it is customary to reduce the analytical description to one single Nth-order differential equation, sometimes one integral equation. In **control theory** the current trend is to break the description down into an equivalent system of N first-order differential equations and to use matrix notation. In this text we will use the first approach.

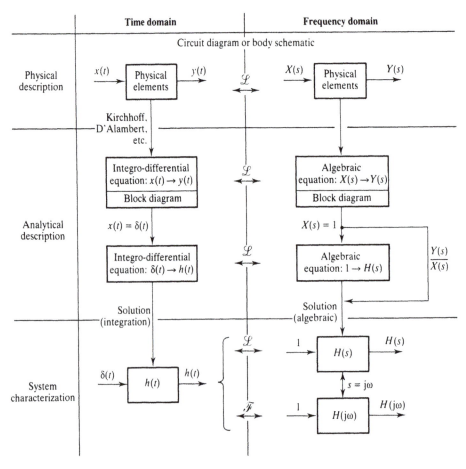

Figure 4.1 System descriptions and characterizations.

Such analytical descriptions provide a link, see Figure 4.1, between a physical description and a more abstract **system characterization**, such as the impulse response $h(t)$, system function $H(s)$ or frequency response $H(j\omega)$. These are the result of probing the system with an input signal that is an impulse $\delta(t)$ and solving the resulting differential or algebraic equations for the output signal. This is suggested in Figure 4.1, which stresses the origin and unity of all such descriptions and the roles played by the Laplace and Fourier transforms.

The chapter starts by describing the terminal properties of some typical system elements, with examples from electrical and mechanical engineering. Connecting such elements into first-order and second-order systems provides a vehicle for introducing fundamental system concepts and representations. These simple systems play an important role in the interpretation of general Nth-order systems, as the latter can be partitioned into series or parallel connections of first- and second-order sections. Being obviously realizable, these elementary systems provide the necessary building blocks for system synthesis, as will be seen in Chapter 10.

It is not our aim to develop any particular discipline, such as electrical networks or mechanical dynamic systems, rather to develop a global awareness of implications at the **systems level**. We thus classify system elements by their **fundamental properties**, such as their ability to store, exchange and dissipate energy.

Such properties provide the common ground for using the same mathematical model to represent physically different entities from different disciplines. This is virtually the only chapter of the book to deal with systems at physical component level. Once an analytical representation has been obtained, it is more expedient to proceed at the more abstract **systems level**. Systems from different disciplines exhibit similar behaviour patterns and it is often possible to interpret the response of one particular system in terms of that of a more familiar case.

The system characterizations derived in this chapter are independent of subsequent input signals and of the system's initial conditions. How these characterizations are used with arbitrary input signals is the subject of Chapter 5. Note that **the characterizations themselves are made with the system initially at rest**, that is, with all its energy storing elements discharged. This statement will be clarified in Section 5.4.

4.1 Idealized system elements

Consider the arbitrary system of Figure 4.2, which contains a finite number of idealized elements R, L, C, m, k, c interconnected in some

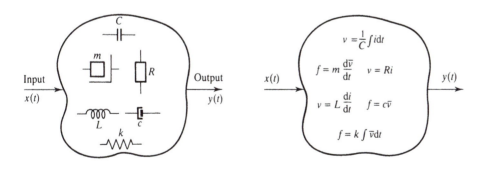

Physical   Analytical

Figure 4.2 Typical system elements and their analytical representations.

specified way. The overall system is described by a mathematical relationship between the input signal $x(t)$ and the output signal $y(t)$, in terms of the elementary components of the system.

These elements are themselves described in terms of their local variables, as suggested in the figure. The main aim in this section is to characterize some familiar element types according to their energy handling capability, a fundamental property of systems.

The resulting integro-differential relationships have to be linear and time-invariant, to produce linear and time-invariant systems. This implies that each element must hold its characteristics, regardless of signal level or time of application.

4.1.1 Two-terminal elements

Interpreting each component as an elementary system, we characterize it by a relationship between two associated variables. This may either involve a direct linear relationship between the two variables, or between one variable and the integral or derivative of the other. Which of these applies depends on the element type and on the variable chosen for input.

Taking a broad view of physical components, there are two different approaches for classifying the variables: either into **effort and flow variables**, associated with the so called **force–voltage analogy**, or, alternatively, into **through and across variables**, associated with the **force–current analogy**. The two approaches are equivalent, but not consistently so across all disciplines. We will use the first form.

For electrical elements the **applied voltage** is the **effort variable**, so that the **current flowing through the element** becomes the **flow variable**. For mechanical elements the **applied force** is the **effort variable**, and the **velocity difference** of its terminals is the **flow variable**. Similar associations are made in other disciplines.

According to this classification, energy supplied through the terminals of an element can be handled in one of three ways. It can be dissipated to the outside world in the form of heat, or stored as **potential energy** or as **kinetic energy**. When connected into systems, elements exchange energy through their terminals, causing oscillations between the two types of energy storage elements.

The conclusions of the following Sections 4.1.2 and 4.1.3 are collected in Figure 4.3, and build up a summary chart.

4.1.2 Dissipators, or loss elements

This element type is characterized by a direct linear relationship between effort and flow variables. It is typified by the electrical resistor,

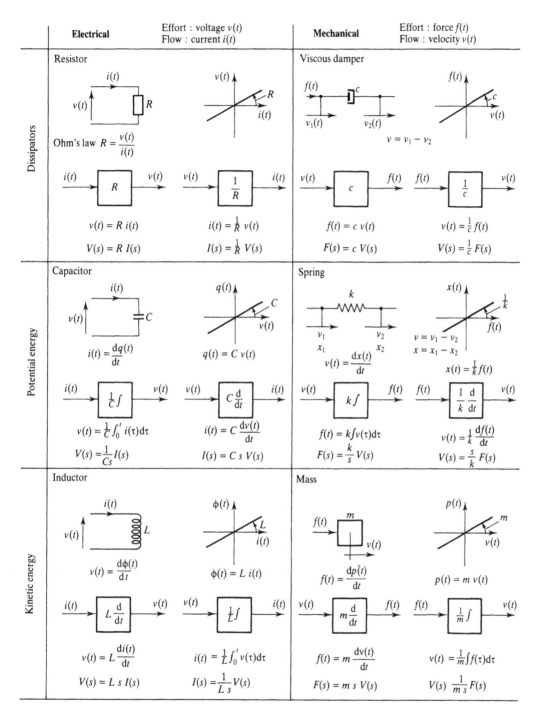

Figure 4.3   Chart of two-terminal elements.

whose proportionality constant is the resistance R, expressed by Ohm's law

$$R = \frac{v(t)}{i(t)}$$

The corresponding voltage–current graph in the top row of Figure 4.3 reflects the linearity of this relationship. Depending on which variable is chosen for input, the element gives two alternative system interpretations represented by the expressions

$$v(t) = Ri(t) \quad \text{and} \quad i(t) = \frac{1}{R}v(t) = Gv(t)$$

where G is the conductance of the element.

These are time domain relationships. Taking the Laplace transform of both sides gives the equivalent frequency domain representations

$$V(s) = RI(s) \quad \text{and} \quad I(s) = \frac{1}{R}V(s)$$

These lead to the concepts of **Laplacian impedance** $Z_R(s)$ and **Laplacian admittance** $Y_R(s)$

$$Z_R(s) = \frac{V(s)}{I(s)} = R \quad \text{and} \quad Y_R(s) = \frac{I(s)}{V(s)} = \frac{1}{R}$$

which are formally similar to the impedance and admittance concepts found in the phasor interpretation of electrical network theory.

Energy handling is reflected in the resistor's power dissipation capability

$$P(t) = v(t)i(t) = i^2(t)R = v^2(t)/R$$

The mechanical equivalent of the resistor is the velocity-dependent damper. The damping coefficient c provides the linear relationship between force and velocity, expressed by Coulomb's law

$$c = \frac{f(t)}{v(t)}$$

This gives rise to alternative system relationships

$$f(t) = cv(t) \quad \text{and} \quad v(t) = \frac{1}{c}f(t)$$

whose Laplace transforms lead, by analogy, to the concepts of **mechanical impedance** $Z_c(s)$ and **mechanical mobility** $Y_c(s)$,

$$Z_c(s) = \frac{F(s)}{V(s)} = c \quad \text{and} \quad Y_c(s) = \frac{V(s)}{F(s)} = \frac{1}{c}$$

A variant of the above translating mechanical element is the rotating viscous damper, which relates torque $T(t)$, the rotational equivalent of force $f(t)$, to angular velocity $\omega(t)$, the rotational equivalent of $v(t)$, as

$$c = \frac{T(t)}{\omega(t)}$$

so that

$$T(t) = c\omega(t) \quad \text{and} \quad \omega(t) = \frac{1}{c} T(t)$$

Related elements from other disciplines are the fluid resistor of fluid mechanics and the thermal resistor of thermodynamic systems.

4.1.3   Energy-storing elements

The product of the effort and flow variables defines power $P(t)$. The integral of power is energy $E(t)$, which, for electrical elements, takes the form

$$E(t) = \int v(t)i(t)\,dt$$

and for mechanical elements the form

$$E(t) = \int f(t)v(t)\,dt$$

Energy-storing elements are characterized by a linear relationship between one variable and the integral or derivative of the other. We distinguish two types, according to their energy-storing properties.

Potential energy storage

The electrical capacitor is characterized by the capacitance C, defined by the linear relationship between electrical charge $q(t)$ and voltage,

$$q(t) = Cv(t)$$

while current is defined as the rate of change of charge,

$$i(t) = \frac{dq(t)}{dt}$$

These lead to equivalent integral or differential relationships between $v(t)$ and $i(t)$ and their Laplace transforms, to be found in the middle row of Figure 4.3.

The energy stored by the capacitor is the integral

$$E(t) = \int Cv(t) \frac{dv(t)}{dt} dt = \tfrac{1}{2} Cv^2(t)$$

The relevant mechanical element is the spring characterized by the stiffness k defined by Hooke's law as a linear relationship between force $f(t)$ and displacement $x(t)$, the integral of velocity $v(t)$. This leads to the relationships shown in Figure 4.3. The potential energy stored in the spring is

$$E(t) = \int kx(t) \frac{dx(t)}{dt} dt = \tfrac{1}{2} kx^2(t)$$

Kinetic energy storage

These are represented by the electrical inductor and the mechanical mass. The inductor is characterized by the linear relationship $L = \phi(t)/i(t)$, where L is inductance and $\phi(t)$ represents flux linkage, which in turn is related to voltage by Faraday's law, $v(t) = d\phi(t)/dt$. The energy stored in the inductor is $E(t) = \tfrac{1}{2} Li^2$.

The mass m is characterized by its momentum $p(t) = mv(t)$, force is the rate of change of momentum $f(t) = dp(t)/dt$ and the kinetic energy is expressed as $E(t) = \tfrac{1}{2} mv^2$. These results are summarized in the bottom row of Figure 4.3.

4.1.4 Four-terminal elements

The preceding elements are sufficient to describe the systems covered in this book. We list other element types that may be required to model practical applications.

The most common additional types relate two pairs of terminals, one pair representing the input, the other the output. Each pair has an effort variable and a flow variable. One variable from the input side becomes the input to the elementary system, while one variable from the output side becomes the system's output. The resulting system is characterized by the relationship between these input/output variables. We distinguish two main classes, namely passive and active elements.

A **passive** element is characterized by energy conservation, that is, all energy is exchanged through the terminals, no additional energy is gained from or dissipated to the external world. The electrical transformer and the mechanical gear train are typical examples.

Transducers belong to this class, they convert signals from two different disciplines. The ideal ironless d.c. electric motor provides the typical electro-mechanical example. It relates current to torque and voltage to angular velocity, so that all electrical energy is converted to mechanical energy, and vice versa.

In contrast, **active** four-terminal elements do not conserve energy, they control the transfer of energy to the outside world. Amplifiers and instruments belong to this class. An amplifier provides a power output signal in response to an input signal of nominally zero power, thus amplifying power. Typical examples are the operational amplifier and the voltage-controlled current source with its input/output permutations. An instrument behaves in the opposite way, it provides an output signal of nominally zero power in response to a power input signal.

4.2   First-order systems

We now examine first-order systems in some depth, introducing many of the concepts and representations to be found later in more general systems.

4.2.1   Time domain representations

Consider the series RC circuit of Figure 4.4, where the variable voltage source $e(t)$ is the system's input and the capacitor voltage $v(t)$ is its

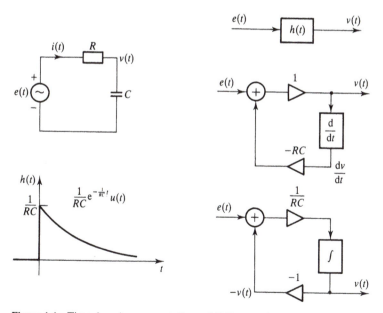

Figure 4.4   Time domain representations of RC network.

output. We wish to establish an analytical relationship between $v(t)$ and $e(t)$. Kirchhoff's voltage law relates the voltage drops round the mesh as

$$v_R(t) + v(t) = e(t)$$

while the elementary expressions for the resistor and the capacitor,

$$v_R(t) = Ri(t) \quad \text{and} \quad i(t) = C \frac{dv(t)}{dt}$$

provide the means of expressing the mesh equation in terms of input and output variables only as

$$RC \frac{dv(t)}{dt} + v(t) = e(t) \tag{4.1}$$

This is a first-order differential equation, which can be solved by routine analysis methods, or by approximate numerical methods. Note that Equation 4.1 represents an alternative way of describing the information contained in the circuit diagram.

Block diagram

The expression 4.1 contains one differentiation, one multiplication and one sum, each of which may be represented by a graphical symbol. Suitably linked, these symbols give a block diagram representation of Equation 4.1, also shown in Figure 4.4.

This can be verified by rearranging Equation 4.1 in the form

$$v(t) = e(t) - RC \frac{dv(t)}{dt}$$

and observing the block diagram output $v(t)$. It is the sum of the input $e(t)$ and of its own derivative dv/dt, scaled by the factor $-RC$.

This block diagram, being an alternative representation of the differential equation (4.1), provides another complete and self-contained description of the system given in the original electrical diagram.

Integral equation and block diagram

By integrating both sides of Equation 4.1, we can give it the form

$$RCv(t) + \int_{-\infty}^{t} v(\tau)\, d\tau = \int_{-\infty}^{t} e(\tau)\, d\tau \tag{4.2}$$

This is also visualized in Figure 4.4 by a block diagram containing an integrator element. Its output $v(t)$ is interpreted as

$$v(t) = \frac{1}{RC} \int_{-\infty}^{t} e(\tau)\, d\tau - \frac{1}{RC} \int_{-\infty}^{t} v(\tau)\, d\tau$$

where the last term represents the feedback path of the diagram. An alternative view of the output is

$$v(t) = \frac{1}{RC} \int_{-\infty}^{t} [e(\tau) - v(\tau)] \, d\tau$$

The expression 4.2 and its block diagram provide yet more equivalent descriptions of the original circuit.

Impulse response h(t)

Any of the system descriptions presented so far would provide a particular solution for a given input function $e(t)$. But every new input would require a repetition of the solution process. A different type of system characterization is required, which is not dependent on the input function.

The most common time domain characterization of this type is the impulse response $h(t)$. As the terminology implies, it represents the system by means of the output signal $y(t) = h(t)$, that results when the input signal $x(t)$ is the impulse $\delta(t)$.

In the case of our RC example, if $e(t) = \delta(t)$, then the system's impulse response is given by $v(t) = h(t)$. Under these circumstances, the system's differential equation (4.1) becomes

$$RC \frac{dh(t)}{dt} + h(t) = \delta(t) \tag{4.3}$$

which can be solved by standard integration methods to yield $h(t)$.

Our aim is to show that, although the time domain expression 4.3 can be solved by integration, it is more expedient and enlightening to derive the result $h(t)$ by the transform methods of the coming Section 4.2.2.

Solving Equation 4.3 for $h(t)$ involves the integration of both sides between the limits $(-\infty, t)$. Because the impulse function values are non-zero only in the infinitesimal neighbourhood of the time origin, it is expedient to partition the integration interval accordingly, that is,

$$\int_{-\infty}^{t} = \int_{-\infty}^{0^-} + \int_{0^-}^{0^+} + \int_{0^+}^{t}$$

In the interval $-\infty < t < 0^-$ both sides of Equation 4.3 are zero, because the impulse function is zero-valued for $t < 0$ and causality requires that both $h(t)$ and its derivative are zero.

Jumping to the last integral, in the interval $(0^+, t)$ the impulse function is again zero-valued for $t > 0$, so that Equation 4.3 reduces to

$$RC \frac{dh(t)}{dt} + h(t) = 0$$

This is a first-order ordinary differential equation, with general solution

$$h(t) = h(0^+)e^{-\frac{1}{RC}t} \tag{4.4}$$

where $h(0^+)$ represents its yet unknown initial value.

The initial value is determined by the integral in the infinitesimal region surrounding the time origin,

$$\int_{0^-}^{0^+} RC \frac{dh(t)}{dt}\, dt + \int_{0^-}^{0^+} h(t)\, dt = \int_{0^-}^{0^+} \delta(t)\, dt$$

By the definition of the impulse function, the right side integral equals unity. Assuming $h(t)$ has no impulses at the origin, the second term on the left vanishes, as it represents the integral of a finite function over an infinitesimal interval. The first term reduces to

$$RC \int_{0^-}^{0^+} dh(t) = RC[h(0^+) - h(0^-)] = RCh(0^+)$$

because $h(0^-) = 0$ for causality. This leaves the initial value as

$$h(0^+) = \frac{1}{RC}$$

Inserting this in Equation 4.4 gives the impulse response of the system as

$$h(t) = \frac{1}{RC}e^{-\frac{1}{RC}t}\, u(t) \tag{4.5}$$

This is a causal exponential, with initial value $1/RC$, decaying at a rate $-1/RC$, as shown in Figure 4.4. The factor $u(t)$ represents causality.

This derivation illustrates the lengthy process involved in the time domain solution of even the simplest first-order system.

4.2.2 Frequency domain representations

Each of the time domain concepts of Section 4.2.1 has a frequency domain counterpart, and these are most generally related by the Laplace transform. We will show that frequency domain manipulations are simpler, making it expedient to derive time domain representations by using transforms.

To stress the close correspondence between representations, we use the same RC example, rearranging the time domain results of Figure 4.4 in the left half of Figure 4.5, and placing their frequency counterparts on the right. The Laplace equivalence of the circuit diagram at the top of Figure 4.5 will become clear in Section 4.2.5.

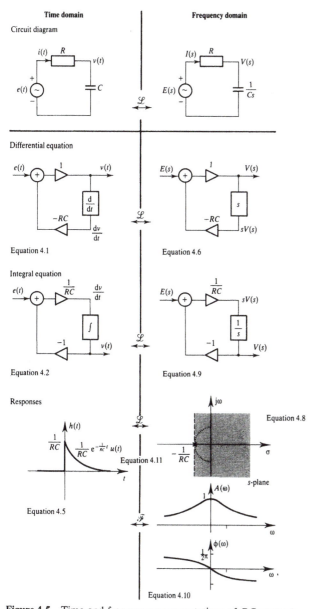

Figure 4.5   Time and frequency representations of *RC* network.

System function *H*(s)

The system's differential equation (4.1) provides a starting point. Taking the Laplace transform of both sides yields

$$RCsV(s) + V(s) = E(s) \tag{4.6}$$

which **transforms a differential equation into an algebraic equation** that is much easier to solve.

The expression 4.6 can also be represented in block diagram form, as shown in Figure 4.5 next to its time domain counterpart. The diagram structure and the coefficient values are the same, only the symbols for the operator and the variables have changed.

We define the **system function** $H(s)$ by the ratio between the Laplace transform $Y(s)$ of the output signal and the Laplace transform $X(s)$ of the input signal,

$$H(s) = \frac{Y(s)}{X(s)} \tag{4.7}$$

Applied to Equation 4.6 it yields

$$H(s) = \frac{V(s)}{E(s)} = \frac{1}{RCs + 1} = \frac{1/RC}{s + 1/RC} \tag{4.8}$$

which is a complex function of the complex variable s.

The Laplace transform of the integral equation (4.2) would have led to the same system function, since the result

$$RCV(s) + \frac{1}{s} V(s) = \frac{1}{s} E(s) \tag{4.9}$$

multiplied through by s leads back to Equation 4.6, hence to Equation 4.8. The block diagram of Equation 4.9 is simply a variant of that of Equation 4.6, as will be explained in the generalization of Section 4.3.4.

Frequency response $H(j\omega)$

The frequency response $H(j\omega)$ of the RC system could be similarly derived by taking the Fourier transform of the differential equation (4.1) and defining $H(j\omega)$ as the ratio of the Fourier transforms of output and input,

$$H(j\omega) = \frac{Y(j\omega)}{X(j\omega)}$$

But having earlier interpreted the Fourier transform as a subset of the Laplace transform, we can write the frequency response expression by simply substituting $j\omega$ for s in the system function 4.8, that is

$$H(j\omega) = H(s)|_{s=j\omega} = \frac{1/RC}{j\omega + 1/RC} \tag{4.10}$$

4.2.3 Links between responses

Note that the system response expression 4.8 does not depend on the input signal $E(s)$, and therefore represents the system for any input.

This is a consequence of having defined $H(s)$ as the **output-to-input ratio**, so that any other input function would give the same result.

If we use the impulse function as an input, $e(t) = \delta(t)$, it gives the impulse response $h(t)$ as output, and these are related by the differential equation 4.3. Taking the Laplace transform of both sides, and denoting the transform of $h(t)$ by $H(s)$, we have

$$RCsH(s) + H(s) = 1$$

which, solved for $H(s)$, leads back to Equation 4.8, thus confirming independence from the input signal.

But, more importantly, since the current function $H(s)$ is the Laplace transform of the impulse response $h(t)$, and $H(s)$ of Equation 4.6 is the system function, we conclude that the latter is the Laplace transform of the system's impulse response $h(t)$.

We can therefore derive $h(t)$ indirectly, by first finding $H(s)$ and then its inverse Laplace transform. Using standard input/output notation, this follows the scheme

$$a_1 y'(t) + a_0 y(t) = b_0 x(t) \quad \xrightarrow{\mathscr{L}} \quad a_1 s Y(s) + a_0 Y(s) = b_0 X(s)$$

$$h(t) = \frac{b_0}{a_1} e^{-\frac{a_0}{a_1} t} u(t) \left\{ \begin{array}{l} \xleftarrow{\mathscr{L}^{-1}} \quad H(s) = \dfrac{Y(s)}{X(s)} = \dfrac{b_0}{a_1 s + a_0} \\[2mm] \qquad\qquad \downarrow \; s = \mathrm{j}\omega \qquad\qquad (4.11) \\[2mm] \xleftarrow{\mathscr{F}^{-1}} \quad H(\mathrm{j}\omega) = \dfrac{b_0}{a_1 \mathrm{j}\omega + a_0} \end{array} \right.$$

where the inverse Laplace transform is of the form of expression 3.9,

$$e^{p_1 t} u(t) \quad \xleftrightarrow{\mathscr{L}} \quad \frac{1}{s - p_1}$$

with $p_1 = -a_0/a_1$. The alternative Fourier path (expressions 4.11) is obvious.

All the relevant results of our RC example can be related to the scheme 4.11 by setting the coefficients to $a_1 = RC$, $a_0 = 1$ and $b_0 = 1$. It is clear that this seemingly lengthy process leads to the impulse response $h(t)$ with much less effort than the direct time domain method.

Graphical representation

Having thus established the Laplace transform link between the various time domain and frequency domain responses, all the graphical relationships developed in Chapter 3 are applicable. For instance, scaling amplitudes by the common factor $1/RC$ and locating the pole at $a = p_1 = -1/RC$, the diagrams of Figure 3.10 can be taken to represent the responses of the RC system of Figure 4.5.

The system function $H(s)$ was not shown in full in Figure 4.5, it was merely suggested by the s-plane locations of the system's poles and zeros, a single pole in this case. This can be considered to be a shorthand version of the full Laplace representation, and is the most frequently used graphical aid. Despite its simplicity, it conveys all the information contained in Equation 4.8, apart from an overall amplitude scaling factor.

The frequency response $H(j\omega)$, which is the slice of the Laplace transform in correspondence with the imaginary axis, was shown in Figure 4.5 in the most familiar form, that is, in terms of magnitude $A(\omega)$ and phase $\phi(\omega)$,

$$H(j\omega) = A(\omega)e^{j\phi(\omega)}$$

A full representation is given in Figure 4.6, including the components

$$H(j\omega) = \operatorname{Re} H(j\omega) + j\operatorname{Im} H(j\omega) = \frac{b_0}{a_1}\left[\frac{-p_1}{\omega^2 + p_1^2} - j\frac{\omega}{\omega^2 + p_1^2}\right]$$

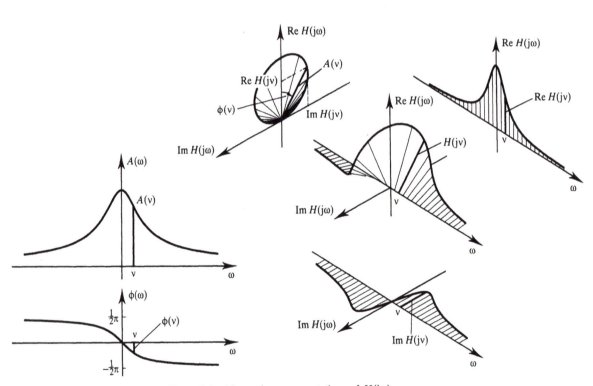

Figure 4.6   Alternative representations of $H(j\omega)$.

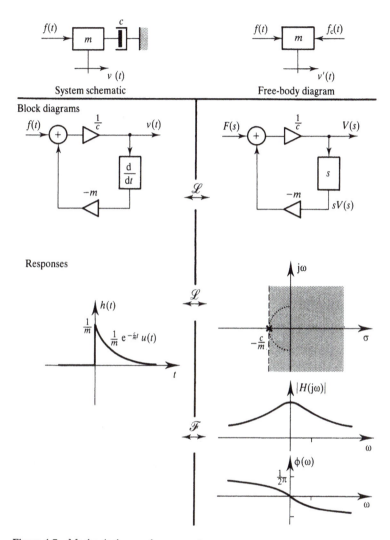

Figure 4.7   Mechanical mass-damper system.

Example 4.1

We characterize the mechanical mass-damper system described by the schematic diagram of Figure 4.7, where the applied force $f(t)$ represents the input signal and the velocity $v(t)$ the output signal.

To write the system's differential equation we use D'Alembert's principle, a mechanical equivalent of Kirchhoff's node equation. It states that the sum of all the forces acting on a body, including the inertia force, is zero. This is shown in the free-body diagram of the figure, and expressed by

$$-f_m(t) - f_c(t) + f(t) = 0$$

The elementary relationships for mass and damper, given in Figure 4.3 as

$$f_m(t) = m \frac{dv(t)}{dt} \quad \text{and} \quad f_c(t) = cv(t)$$

permit us to express the above relationship in terms of $f(t)$ and $v(t)$ as

$$m \frac{dv(t)}{dt} + cv(t) = f(t) \tag{4.12}$$

which is a first-order differential equation of the form found in relationships 4.11 and can be solved by the same scheme. With $a_1 = m$, $a_0 = c$ and $b_0 = 1$, this yields the responses

$$h(t) = \frac{1}{m} e^{-\frac{c}{m}t} u(t) \quad \left\{ \begin{array}{l} \xleftrightarrow{\mathscr{L}} \quad H(s) = \dfrac{1}{ms + c} \\[2mm] \xleftrightarrow{\mathscr{F}} \quad H(j\omega) = \dfrac{1}{j\omega m + c} \end{array} \right. \tag{4.13}$$

Although physically very different, these responses are formally identical to those of the electrical RC system, and the two systems are said to be **analogous**.

4.2.4 General first-order system

The time and frequency representations of the preceding systems had some formal similarities, which we now set out to identify and generalize.

Time domain

The most general first-order system is represented by the general first-order differential equation

$$a_1 y'(t) + a_0 y(t) = b_1 x'(t) + b_0 x(t) \tag{4.14}$$

whose coefficients a_n and b_n are real constants, some of which may be zero. The right-hand side is often called the **forcing function** $x_f(t)$ **of the differential equation**, that is,

$$x_f(t) = b_1 x'(t) + b_0 x(t) \tag{4.15}$$

In general terms, this function is not the same as the input signal $x(t)$, and loose usage of terminology can lead to confusion.

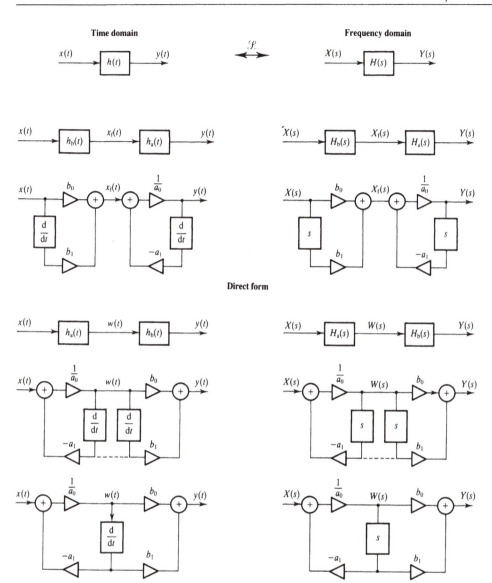

Figure 4.8   Equivalent block diagrams of general first-order system.

The differential equation 4.14 lends itself to two equivalent block diagram representations, shown in the time domain half of Figure 4.8. The upper half of the figure visualizes the system $h(t)$ as a series connection of two subsystems $h_b(t)$ and $h_a(t)$. Of these, $h_b(t)$ modifies the input $x(t)$ according to Equation 4.15, producing the forcing function $x_f(t)$ as its output. This becomes the input to $h_a(t)$, which converts it into $y(t)$, thus interpreting Equation 4.14 as

$$a_1 y'(t) + a_0 y(t) = x_f(t)$$

The complete diagram therefore represents the full differential equation (4.14). But an implementation of this form would require the use of two differentiators.

Exchanging the order in which the two subsystems are connected produces the same results, but leads to the more compact block diagram shown in the bottom row of Figure 4.8. The time domain validation of this exchange would involve a slightly awkward manipulation of differential equations. In contrast, the validity of the exchange is obvious in the frequency domain, as we show next. We anticipate the result of Equation 4.20 and state that the overall impulse response $h(t)$ is the convolution of the partial responses $h_b(t)$ and $h_a(t)$, and use the commutative property of convolution to justify the exchange,

$$h(t) = h_b(t) * h_a(t) = h_a(t) * h_b(t)$$

Frequency domain

All the preceding time domain considerations have their frequency domain counterparts. Taking the Laplace transform of both sides of Equation 4.14 gives

$$a_1 s Y(s) + a_0 Y(s) = b_1 s X(s) + b_0 X(s) \qquad (4.16)$$

Similarly, the forcing function yields

$$X_f(s) = b_1 s X(s) + b_0 X(s)$$

Gathering terms of Equation 4.16, the transfer function definition 4.7 gives the general form of the first-order system function

$$H(s) = \frac{Y(s)}{X(s)} = \frac{b_1 s + b_0}{a_1 s + a_0} \qquad (4.17)$$

This is a rational polynomial, whose coefficients a_n and b_n are those of the differential equation (4.14). The single root $z_1 = -b_0/b_1$ of the numerator is a zero of the system function, and the root $p_1 = -a_0/a_1$ of the denominator is a pole of the function. We can rewrite Equation 4.17 in terms of these roots as

$$H(s) = \frac{c_1 s + c_0}{s - p_1} = c_1 \frac{s - z_1}{s - p_1} \qquad (4.18)$$

where $c_1 = b_1/a_1$ and $c_0 = b_0/a_1$ are real-valued coefficients.

Since a first-order system can have only one pole and one zero, to give a real-valued impulse response, these must lie on the real axis, which in turn requires that the coefficients a_n and b_n be real. In addition, for system stability, the pole p_1 must lie in the open half plane $\sigma < 0$. This restriction does not extend to the zero z_1, which can be located anywhere on the real axis. Typical s-plane locations can be found in the later Figures 4.9 and 4.12.

We return to Figure 4.8 for the frequency domain block diagrams of the system function. Multiplying and dividing Equation 4.17 by $X_f(s)$, we factorize the system function as

$$H(s) = \frac{Y(s)}{X(s)} = \frac{X_f(s)}{X(s)} \frac{Y(s)}{X_f(s)} \qquad (4.19)$$

and associate the factors with the subsystems $H_b(s)$ and $H_a(s)$,

$$H_b(s) = \frac{X_f(s)}{X(s)} = b_1 s + b_0$$

$$H_a(s) = \frac{Y(s)}{X_f(s)} = \frac{1}{a_1 s + a_0}$$

which take the numerator and denominator terms of Equation 4.17, respectively. Each of these system functions contributes one loop to the block diagram of the upper half of Figure 4.8, and their product represents the overall system function $H(s)$.

The other block diagram form is a consequence of the commutative law of multiplication

$$H(s) = H_b(s) \cdot H_a(s) = H_a(s) \cdot H_b(s) \qquad (4.20)$$

which states that the sequence of the two loops can be exchanged, as shown in the lower half of Figure 4.8. But the two differentiators s are now adjacent and have common inputs $W(s)$ and the same outputs $sW(s)$, making one redundant. This leads to the compact block diagram of the bottom row of Figure 4.8, the form used most frequently in this text.

In the literature these block diagrams are called **direct form I** and **direct form II**, respectively. We will call the first **direct form** and the second **monolithic form**, as indicated in Figure 4.8, giving motives in Section 4.4.1.

Note that corresponding block digrams of the two domains of Figure 4.8 are formally identical, differing only in the symbols identifying the operators and the signals. Strictly speaking, these diagrams should be kept separate and used as shown, but, to avoid duplication, it is expedient and customary to draw one diagram for both domains and freely interchange symbols.

Frequency response

Since all the poles of a stable system lie in the left half of the s-plane, the region of convergence of $H(s)$ includes the imaginary axis. All the results obtained for the generalized frequency variable s are therefore valid on the imaginary axis. In particular, substituting $j\omega$ for s in the system function 4.17 yields the frequency response $H(j\omega)$ as

$$H(j\omega) = H(s)|_{s=j\omega} = \frac{b_1 j\omega + b_0}{a_1 j\omega + a_0} \qquad (4.21)$$

Impulse response

The system's impulse response $h(t)$ is the inverse Laplace transform of the system function $H(s)$. Writing Equation 4.18 as the sum

$$H(s) = \frac{c_1 s}{s - p_1} + \frac{c_0}{s - p_1}$$

shows that the first term is of the same form as the second multiplied by s. Their time domains are therefore related by differentiation. Interpreting the Laplace pair 3.29, for $a = p_1$, we have

$$p_1 e^{p_1 t} u(t) + \delta(t) \quad \overset{\mathscr{L}}{\longleftrightarrow} \quad \frac{s}{s - p_1}$$

which leads to the general first-order impulse response expression

$$h(t) = (c_1 p_1 + c_0) e^{p_1 t} u(t) + c_1 \delta(t) \qquad (4.22)$$

The general first-order system representations are thus related by the scheme

$$a_1 y'(t) + a_0 y(t) = b_1 x'(t) + b_0 x(t) \quad \overset{\mathscr{L}}{\longleftrightarrow} \quad a_1 s Y(s) + a_0 Y(s)$$
$$= b_1 s X(s) + b_0 X(s)$$

$$\downarrow$$

$$\begin{array}{c} \text{integrate for} \\ x(t) = \delta(t) \end{array} \downarrow$$

$$h(t) = c_1(p_1 - z_1) e^{p_1 t} u(t) + c_1 \delta(t) \left\{ \begin{array}{l} \overset{\mathscr{L}}{\longleftrightarrow} \quad H(s) = \dfrac{Y(s)}{X(s)} = \dfrac{c_1 s + c_0}{s - p_1} \\[2mm] \qquad\qquad s = j\omega \uparrow \\[2mm] \overset{\mathscr{F}}{\longleftrightarrow} \quad H(j\omega) = \dfrac{c_1 j\omega + c_0}{j\omega - p_1} \end{array} \right.$$

$$(4.23)$$

where $p_1 = -a_0/a_1$, $z_1 = -b_0/b_1 = -c_0/c_1$, $c_1 = b_1/a_1$ and $c_0 = b_0/a_1$.
    The general character of the system's behaviour is fully deter-

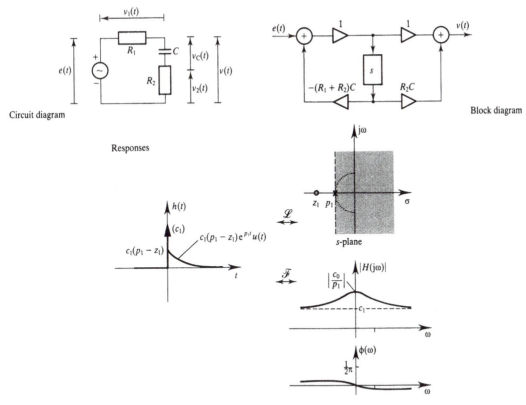

Circuit diagram

Block diagram

Responses

Figure 4.9 Characterization of RCR system.

mined by the location of its pole p_1 and zero z_1, both on the imaginary axis. These locations are themselves determined by the values of the coefficients a_n and b_n. We show typical shapes of first-order responses in Section 4.2.6.

Example 4.2

We find the responses of the series RCR system of Figure 4.9. Our first objective is the system's differential equation, expressed in terms of the input and output variables $e(t)$ and $v(t)$. Proceeding in the time domain, we write Kirchhoff's mesh equation as

$$v_1(t) + v_C(t) + v_2(t) = e(t)$$

The current $i(t)$ is common to all element characterizations,

$$v_1(t) = R_1 i(t) \qquad v_2(t) = R_2 i(t) \qquad i(t) = C \frac{dv_C(t)}{dt}$$

which, writing $v_1(t) = e(t) - v(t)$, give the two relationships

$$v_2(t) = \frac{R_2}{R_1} v_1(t) = \frac{R_2}{R_1} [e(t) - v(t)]$$

$$e(t) - v(t) = v_1(t) = R_1 C \frac{dv_C(t)}{dt}$$

But the derivative of the capacitor voltage results from

$$v_C(t) = v(t) - v_2(t) = \frac{R_1 + R_2}{R_1} v(t) - \frac{R_2}{R_1} e(t)$$

$$\frac{dv_C(t)}{dt} = \frac{R_1 + R_2}{R_1} \frac{dv(t)}{dt} - \frac{R_2}{R_1} \frac{de(t)}{dt}$$

thus leading to the desired differential equation

$$(R_1 + R_2)C \frac{dv(t)}{dt} + v(t) = R_2 C \frac{de(t)}{dt} + e(t)$$

This is of the general form of Equation 4.14, with coefficient values

$$a_1 = (R_1 + R_2)C \qquad a_0 = 1 \qquad b_1 = R_2 C \qquad b_0 = 1$$

and can be expressed in one of the block diagram forms of Figure 4.8, for example that repeated in Figure 4.9. The expressions summarized in scheme 4.23 provide all the desired responses.

4.2.5 Frequency domain formulation

Despite the circuit's simplicity, the laborious time domain derivation of the preceding example obscures the process. We now outline a more effective frequency domain method to make the distinction between time domain and frequency domain formulations.

The frequency domain approach in the scheme 4.23 relies on converting the system's differential equation into an algebraic equation that is easier to solve. But the transition to the frequency domain can be made earlier, at circuit diagram level.

This is effectively the approach found in phasor analysis of a.c. networks, which uses the Fourier transform concepts of system impedance $Z(\omega)$ and admittance $Y(\omega)$ to determine the system's response to harmonic excitation. This approach is readily extended to the generalized frequency s, leading to the concept of Laplacian impedance $Z(s)$ and admittance $Y(s)$, and formally identical methods and rules.

In Section 4.1.2 we defined the Laplacian impedance $Z(s)$ of an element as the ratio between the Laplace transforms of the element's

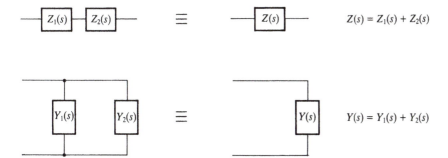

Figure 4.10   Addition rules for Laplacian impedance and admittance.

voltage and current

$$Z(s) = \frac{V(s)}{I(s)}$$

We also found the impedances of some typical system elements, such as

$$Z_R(s) = R \qquad Z_L(s) = Ls \qquad Z_C(s) = \frac{1}{Cs}$$

Frequency domain counterparts of applicable laws and methods are similarly defined. For instance, Kirchhoff's voltage and current laws become the transform pairs

$$\sum_{i=1}^{n} v_i(t) = 0 \quad \overset{\mathscr{L}}{\longleftrightarrow} \quad \sum_{i=1}^{n} V_i(s) = 0$$

$$\sum_{i=1}^{n} i_i(t) = 0 \quad \overset{\mathscr{L}}{\longleftrightarrow} \quad \sum_{i=1}^{n} I_i(s) = 0$$

The rules for manipulating Laplacian impedances and admittances are similar to those applicable to resistor networks. For instance, series or parallel connection of systems leads to familiar sums of impedances or admittances, as indicated in Figure 4.10. Even the circuit diagram can be Laplace transformed, as illustrated in the following example.

Example 4.3

We reformulate Example 4.2 in the frequency domain. The circuit diagram elements of Figure 4.9 are replaced by their Laplacian impedances, and the variables by their transforms, as shown in Figure 4.11. Interpreting the system as a voltage divider, with $Z_1(s) = R_1$ and $Z_2(s) = R_2 + 1/(Cs)$, the output $V(s)$ becomes

$$V(s) = \frac{Z_2(s)}{Z_1(s) + Z_2(s)} E(s)$$

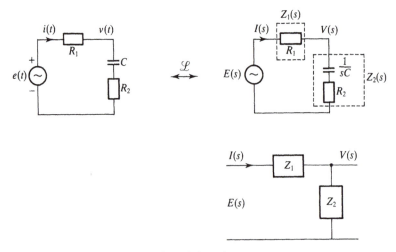

Figure 4.11 Frequency domain formulation of system.

and the system function

$$H(s) = \frac{V(s)}{E(s)} = \frac{R_2 + 1/(Cs)}{(R_1 + R_2) + 1/(Cs)}$$

which is easily rearranged to any of the forms of Equations 4.17 or 4.18. It leads to the same responses as Example 4.2.

This frequency domain derivation is much simpler and more intuitive than the equivalent time domain approach, because it involves algebraic rather than differential expressions.

4.2.6 Typical characteristics

The relationships between time domain and frequency domain responses of a general first-order system were given in the scheme 4.23, which also relates them to the coefficients a_n and b_n of the system's differential equation. Except for a scaling factor, the system function $H(s)$ is also determined by the positions of its pole p_1 and zero z_1, as expressed in Equation 4.18, and these positions fully describe the shape of the response.

To gain insight into the effects of the relative pole–zero positions, in the upper half of Figure 4.12 we show some typical responses for fixed pole position. For a natural progression responses are positioned on a backdrop of an enlarged and rotated s-plane. The position on that backdrop is given by a small s-plane insert on the real axis, which reflects the s-plane location of the zero.

Figure 4.12 (*Opposite*) Typical first-order characteristics.

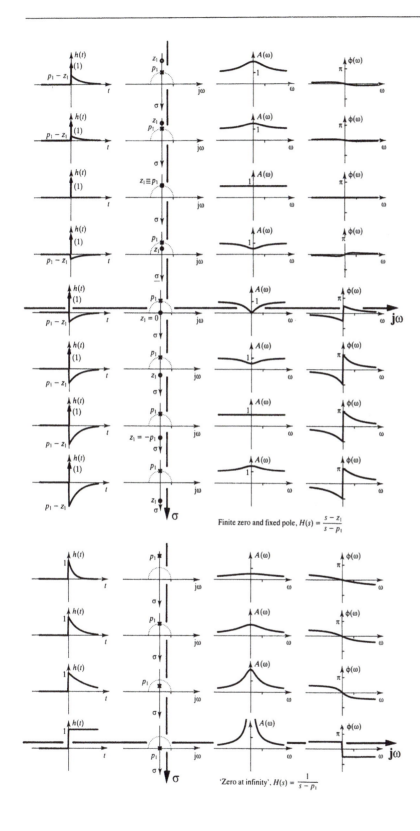

Finite zero and fixed pole, $H(s) = \dfrac{s - z_1}{s - p_1}$

'Zero at infinity', $H(s) = \dfrac{1}{s - p_1}$

It is customary and expedient to associate the frequency response shape with that of the appropriate ideal filter, such as lowpass, highpass, allpass, etc., to be introduced in Chapter 10. The case $z_1 = 0$ serves as reference and represents a basic highpass characteristic.

As the zero moves towards the pole, $z_1 < 0$, the amplitude value $A(0)$ increases, until the trivial case of the strictly allpass characteristic is reached, where $A(\omega) = 1$ and $\phi(\omega) = 0$. Progressing further, as z_1 becomes more negative, the response becomes predominantly lowpass.

Moving the zero in the positive direction, $z_1 > 0$, replicates the earlier amplitude responses, but the corresponding phase characteristics are different. The important case $z_1 = -p_1$, for which $A(\omega) = 1$, represents the first-order allpass filter used extensively for phase equalization. Also shown are cases with 'zero at infinity' for various pole positions.

Zero at infinity

The absence of the coefficient b_1 would indicate the absence of a system zero. But for $b_1 = 0$ and large values of s, the magnitude of $H(s)$ decays towards zero, in inverse proportion to frequency. For intuitive interpretations there are advantages in assuming that the missing zero exists, but is notionally 'located at infinity', where it pulls the magnitude of $H(s)$ to zero.

Complex variable literature handles this situation by defining the **point at infinity**, denoted by ∞, which has the property that any point s of the **finite s-plane** has a magnitude $|s| < \infty$. Adding the **ideal** point at infinity to the finite s-plane gives the **extended complex s-plane**, to which is associated the convention that **the behaviour of a function $F(s)$ at $s = \infty$ is identified with the behaviour of $F(1/s)$ at $s = 0$.**

For example, the behaviour at $s = \infty$ of the system function $H(s) = s/(s - 1)$ is given by the behaviour of $H(1/s)$ at the origin,

$$H(s)\big|_{s=\infty} \equiv H\left(\frac{1}{s}\right)\bigg|_{s=0} = \frac{1}{1 - s}\bigg|_{s=0} = 1$$

Similarly, the function $H(s) = 1/(s - 1)$ behaves at $s = \infty$ as

$$H(s)\big|_{s=\infty} \equiv H\left(\frac{1}{s}\right)\bigg|_{s=0} = \frac{s}{1 - s}\bigg|_{s=0} = 0$$

so that the **zero at the origin** of $H(1/s)$ behaves like a **zero at infinity** of $H(s)$.

The point at infinity is not an ordinary point in the sense of other s-plane points. It surrounds the entire finite s-plane, in all directions, as visualized in terms of a **stereographic projection**. This involves the **Riemann sphere**, a sphere of **unit radius** centred on the s-plane origin $s = 0$, as shown in Figure 4.13.

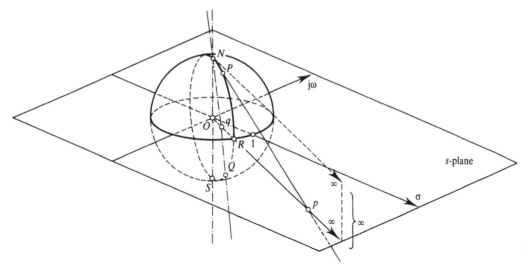

Figure 4.13   Riemann sphere and point at infinity.

A point P of the northern hemisphere defines a line with the north pole N that intersects the s-plane at a point p. Points R of the equator project onto themselves, that is onto the unit circle of the s-plane. Points Q of the southern hemisphere project onto points q inside the unit circle, with the south pole S projecting onto the s-plane origin O.

Points of the northern hemisphere project outside the unit circle. The closer P is to the north pole, the further p falls. In the limit, when P approaches N the intersection p moves to infinity $|s| = \infty$. And this applies to any direction from which P approaches N, hence the point at infinity surrounds the finite s-plane.

An alternative version of the Riemann sphere, one of **unit diameter** placed **on top** of the s-plane, gives identical results.

4.3   Second-order systems

Each of the preceding systems contained one energy-storing component whose first-order differential or integral characterization determined the degree of the associated polynomial and therefore the order of the system. In this section we examine systems with two energy-storing elements which lead to second-order descriptions. We also formalize the equivalence between differential and integral equation representations.

Second-order systems are at the core of all oscillating systems and represent the main building modules of more elaborate systems. They are often incorporated as subsystems of practical realizations. Even when a system is not realized explicitly in terms of such subsystems, its overall behaviour can be analysed by representing it in terms of a series or parallel connection of such modules. We will elaborate in Section 4.4.

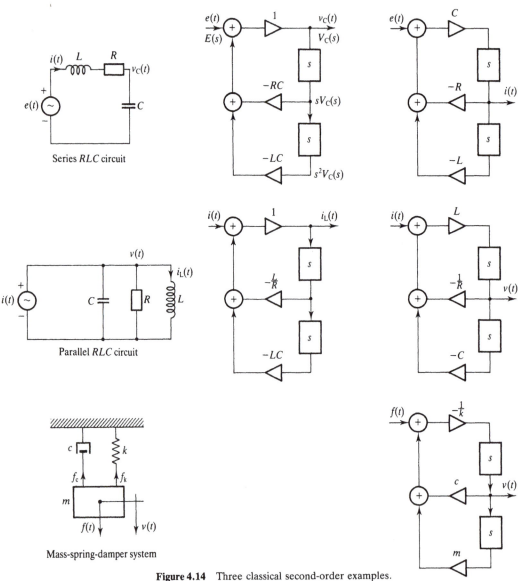

Figure 4.14 Three classical second-order examples.

4.3.1   Classical examples

The classical examples of second-order systems are the series RLC and the parallel RLC circuits of electrical engineering, and the mass-spring-damper system of mechanical engineering.

Series RLC circuit

The system most commonly associated with this circuit is that shown in the upper row of Figure 4.14, where the voltage source $e(t)$ is the input variable and the capacitor voltage $v_C(t)$ is taken as the output variable. Kirchhoff's voltage law relates the element voltages to the input $e(t)$ as

$$v_L(t) + v_R(t) + v_C(t) = e(t) \tag{4.24}$$

With the elemental integro-differential relationships

$$v_L(t) = L \frac{di(t)}{dt} \qquad v_R(t) = Ri(t) \qquad i(t) = C \frac{dv_C(t)}{dt}$$

it leads directly to the second-order differential equation

$$LC \frac{d^2 v_C(t)}{dt^2} + RC \frac{dv_C(t)}{dt} + v_C(t) = e(t) \tag{4.25}$$

Taking the Laplace transform of both sides gives

$$(LCs^2 + RCs + 1)V_C(s) = E(s) \tag{4.26}$$

and the system function

$$H_C(s) = \frac{V_C(s)}{E(s)} = \frac{1}{LCs^2 + RCs + 1} \tag{4.27}$$

Expressions 4.25 and 4.26 can also be represented in block diagram form as indicated next to the circuit in the upper row of Figure 4.14.
    Had we chosen the current $i(t)$ as the system's output, Kirchhoff's equation (4.24) would have led to the integro-differential equation

$$L \frac{di(t)}{dt} + Ri(t) + \frac{1}{C} \int i(t)\, dt = e(t) \tag{4.28}$$

which, differentiated on both sides, gives the second-order differential equation

$$L \frac{d^2 i(t)}{dt^2} + R \frac{di(t)}{dt} + \frac{1}{C} i(t) = \frac{de(t)}{dt} \tag{4.29}$$

whose Laplace transform

$$(Ls^2 + Rs + 1/C)I(s) = sE(s)$$

yields the system function

$$H_I(s) = \frac{I(s)}{E(s)} = \frac{s}{Ls^2 + Rs + 1/C} \tag{4.30}$$

The corresponding block diagram is also shown in the figure.

Parallel RLC circuit

The same elements are now connected in parallel and driven by a current source $i(t)$, as shown in the middle row of Figure 4.14. Choosing the inductor current $i_L(t)$ as the output variable produces a formally similar set of equations. Kirchhoff's current law applies, giving

$$i_C(t) + i_R(t) + i_L(t) = i(t) \tag{4.31}$$

This leads to the differential equation

$$LC\frac{d^2 i_L(t)}{dt^2} + \frac{L}{R}\frac{di_L(t)}{dt} + i_L(t) = i(t) \tag{4.32}$$

and the system function

$$H_L(s) = \frac{I_L(s)}{I(s)} = \frac{1}{LCs^2 + (L/R)s + 1} \tag{4.33}$$

which are formally similar to Equations 4.25 and 4.27 of the series RLC circuit, and this similarity extends to the block diagram, as seen in Figure 4.14.

Had we chosen the system voltage $v(t)$ as the output variable, the results would have been formally similar to Equations 4.29 and 4.30, namely

$$C\frac{d^2 v(t)}{dt^2} + \frac{1}{R}\frac{dv(t)}{dt} + \frac{1}{L}v(t) = \frac{di(t)}{dt} \tag{4.34}$$

and

$$H_V(s) = \frac{V(s)}{I(s)} = \frac{s}{Cs^2 + (1/R)s + 1/L} \tag{4.35}$$

and represented by a similar block diagram, also indicated in Figure 4.14.

Mechanical mass-spring-damper system

The classical mechanical vibrations example is the mass-spring-damper system of the lower row of Figure 4.14. D'Alambert's principle is stated as

$$f_m(t) + f_c(t) + f_k(t) + f(t) = 0$$

With the elemental relationships from Section 4.1.1,

$$f_m(t) = m \frac{dv(t)}{dt} \qquad f_c(t) = cv(t) \qquad f_k(t) = k \int v(t) \, dt$$

it yields

$$-m \frac{dv(t)}{dt} - cv(t) - k \int v(t) \, dt = f(t) \tag{4.36}$$

Differentiating both sides yields the second-order differential equation

$$-m \frac{d^2v(t)}{dt^2} - c \frac{dv(t)}{dt} - kv(t) = \frac{df(t)}{dt} \tag{4.37}$$

which is similar to Equations 4.29 and 4.34. Its Laplace transform yields

$$-(ms^2 + cs + k)V(s) = sF(s)$$

and the system function

$$H(s) = \frac{V(s)}{F(s)} = \frac{s}{-ms^2 - cs - k} \tag{4.38}$$

which is formally similar to Equation 4.30. The corresponding block diagram, not surprisingly, is of the second form of the preceding cases, as seen in Figure 4.14. For a formulation in terms of the associated displacement $x(t)$ see Exercise 4.3.

4.3.2 General second-order system

The classical examples examined above had two independent energy-storing elements, which identified them as second-order systems. Their time domain behaviour was described by second-order differential equations belonging to the general form

$$a_2 y''(t) + a_1 y'(t) + a_0 y(t) = b_2 x''(t) + b_1 x'(t) + b_0 x(t) \tag{4.39}$$

whose right side is also the system's forcing function $x_f(t)$,

$$x_f(t) = b_2 x''(t) + b_1 x'(t) + b_0 x(t) \tag{4.40}$$

The Laplace transform of Equation 4.39 i.

$$a_2 s^2 Y(s) + a_1 s Y(s) + a_0 Y(s) = b_2 s^2 X(s) + b_1 s X(s) + b_0 X(s) \tag{4.41}$$

and leads to the general expression of the second-order system function

$$H(s) = \frac{Y(s)}{X(s)} = \frac{b_2 s^2 + b_1 s + b_0}{a_2 s^2 + a_1 s + a_0} \tag{4.42}$$

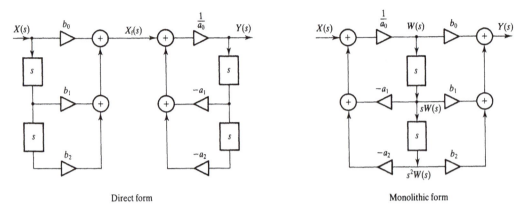

Direct form   Monolithic form

Figure 4.15   Two block diagrams of general second-order systems.

For stable systems the usual substitution $s = j\omega$ yields the general second-order frequency response

$$H(j\omega) = \frac{Y(j\omega)}{X(j\omega)} = \frac{-b_2\omega^2 + b_1 j\omega + b_0}{-a_2\omega^2 + a_1 j\omega + a_0} \tag{4.43}$$

Block diagram

These analytical relationships are usually expressed in direct form or monolithic form by one of the block diagrams of Figure 4.15, which are downward extensions of those presented in Figure 4.8 for first-order systems.

As before, it is expedient to represent both domains by the same diagram, and to examine it in the frequency domain. The two loops of the left side of Figure 4.15 represent the two sides of Equation 4.41, and these are linked by the forcing function $X_f(s)$.

The right side of Figure 4.15 results from exchanging the order of the two loops and eliminating redundant differentiators. Interpreting the remaining differentiators as belonging to the left half of the diagram, the intermediate output $W(s)$ becomes

$$W(s) = \frac{1}{a_0} X(s) - \frac{a_1}{a_0} sW(s) - \frac{a_2}{a_0} s^2 W(s)$$

and defines the subsystem function

$$H_a(s) = \frac{W(s)}{X(s)} = \frac{1}{a_2 s^2 + a_1 s + a_0} \tag{4.44}$$

Interpreting the differentiators as belonging to the right half, with $W(s)$ as the input, the output $Y(s)$ becomes

$$Y(s) = (b_2 s^2 + b_1 s + b_0)W(s)$$

which defines the second subsystem function

$$H_b(s) = \frac{Y(s)}{W(s)} = b_2 s^2 + b_1 s + b_0 \tag{4.45}$$

As in the first-order case, the series connection of the two subsystems yields the overall system function

$$H(s) = \frac{Y(s)}{X(s)} = \frac{Y(s)}{W(s)} \frac{W(s)}{X(s)} = H_b(s) H_a(s)$$

which verifies that these block diagrams also represent Equation 4.42.

Characteristic properties

The general oscillatory character of the system is determined by the roots of the system function's denominator. In Chapter 5 we will associate this with the natural response of the system's output in the absence of a forcing function $x_f(t)$, usually a decaying oscillation.

It is convenient to express the system function 4.42 in terms of its poles $p_1 = \sigma_1 + j\omega_1$ and $p_2 = \sigma_2 + j\omega_2$, by first dividing all coefficients by a_2

$$H(s) = \frac{c_2 s^2 + c_1 s + c_0}{s^2 + (a_1/a_2)s + a_0/a_2} = \frac{c_2 s^2 + c_1 s + c_0}{(s - p_1)(s - p_2)} \tag{4.46}$$

where $b_0/a_2 = c_0$, etc., and

$$p_1, p_2 = -\frac{a_1}{2a_2} \pm \sqrt{\frac{a_1^2}{4a_2^2} - \frac{a_0}{a_2}}$$

Since all the coefficients are real-valued, the radicand is real, and this offers three possibilities. When $4a_0 a_2 = a_1^2$ the radicand is zero and the two poles are real and coincide at $p_1 = p_2 = -a_1/2a_2$. When $4a_0 a_2 < a_1^2$ the radicand is positive, and leads to two distinct real poles.

The third case $4a_0 a_2 > a_1^2$ is the most common in engineering practice. The radicand is real and negative, the square root gives pure imaginary values and gives rise to a pair of complex conjugate poles,

$$p_1 = \sigma_1 + j\omega_1 \qquad p_2 = p_1^* = \sigma_1 - j\omega_1$$

where $\sigma_1 = -a_1/2a_2$ and $\omega_1 = \sqrt{4a_0 a_2 - a_1^2}/2a_2$. Such poles are characteristic of oscillating systems, whose system function $H(s)$ can be expressed more specifically in one of the forms

$$H(s) = \frac{c_2 s^2 + c_1 s + c_0}{(s - p_1)(s - p_1^*)} = \frac{c_2 s^2 + c_1 + c_0}{s^2 - 2\sigma_1 s + \omega_r^2} = \frac{c_2 s^2 + c_1 s + c_0}{(s - \sigma_1)^2 + \omega_1^2}$$

$$\tag{4.47}$$

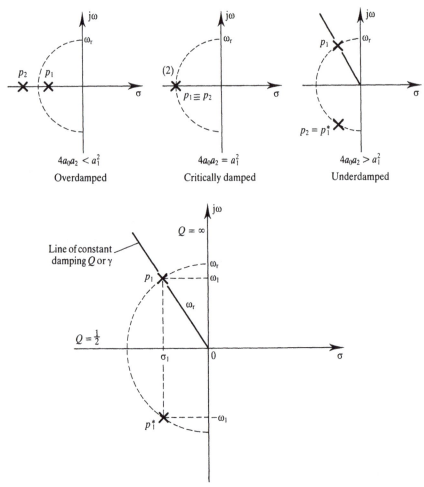

Figure 4.16   Pole parameters of oscillating system.

where $\omega_r^2 = \sigma_1^2 + \omega_1^2$, so that ω_r represents the distance of the poles from the origin and is called the **resonant frequency**, whereas ω_1 is the **natural frequency** associated with the pole pair.

Typical s-plane locations of poles encountered in the three cases are shown in the upper half of Figure 4.16. In electrical networks it is customary to substitute ω_r/Q for the factor $-2\sigma_1$ of Equation 4.47, yielding the form

$$H(s) = \frac{c_2 s^2 + c_1 s + c_0}{s^2 + (\omega_r/Q)s + \omega_r^2} \tag{4.48}$$

where ω_r and Q are termed the **pole-frequency** and **pole-Q** respectively. Keeping the pole-frequency constant defines a circle of radius ω_r. A constant pole-Q, in the range $0.5 < Q < \infty$, defines a radial line through the s-plane origin, see Figure 4.16. For the lower limit the line coincides with the negative real axis while for the upper limit it coincides with the positive imaginary axis.

In mechanical vibrations it is customary to define a damping ratio $\gamma = c/c_k$ where c is the actual damping coefficient and c_k represents the damping coefficient that would give critical damping, that is, for the case $4a_0a_2 = a_1^2$. The concept is equivalent to the electrical definition, the two parameters being related as $\gamma = 1/(2Q)$.

4.3.3   Typical responses

Whereas the denominator roots control the natural response of the system, the numerator coefficients c_i control the response type. We now

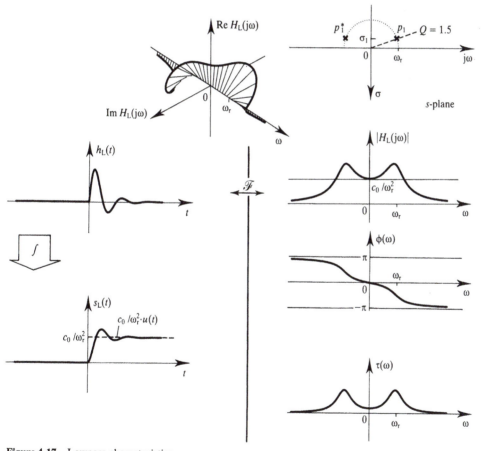

Figure 4.17   Lowpass characteristics.

examine the three basic types obtained by setting all but one coefficient to zero and the case when c_1 is missing. It is customary to use filter terminology to identify these as lowpass, bandpass, highpass and notch filter sections, as will be seen in Chapter 10.

Lowpass section

The case when c_0 is the non-zero coefficient provides the most familiar response, that of the lowpass filter

$$H_L(s) = \frac{c_0}{s^2 - 2\sigma_1 s + \omega_r^2} \tag{4.49}$$

In the literature this is often frequency normalized to $\omega_r = 1$, or amplitude normalized, by setting $c_0 = \omega_r^2$, so that $H(0) = 1$, or normalized in both.

Because the numerator of $H_L(s)$ is constant, this case is ideal for characterizing the natural response of all systems described by the same pole locations.

The pole locations of a typical case are shown in the s-plane of Figure 4.17. This was rotated by 90° to line up its imaginary axis with the ω-axes of the magnitude and phase representations $|H_L(j\omega)|$ and $\phi(\omega)$ of the frequency response $H(j\omega)$. The system's **group delay** $\tau(\omega)$ is also shown. It is a measure of the rate of change of phase with frequency, defined as

$$\tau(\omega) = -\frac{d\phi(\omega)}{d\omega}$$

The three-dimensional representation of $H_L(j\omega)$ shown in the top left of Figure 4.17 provides a visual link between the two alternative representations by magnitude-and-phase and real-and-imaginary parts. It shows the amplitude $|H_L(j\omega)|$ of each frequency component rotated by the corresponding phase value $\phi(\omega)$. For instance, the phase decreases from $\phi(-\infty) = \pi$, passing through $\phi(0) = 0$ and reaching $\phi(\infty) = -\pi$. Projected on the real and imaginary planes, these components give the corresponding parts,

$$H_L(j\omega) = \text{Re } H_L(j\omega) + j\,\text{Im } H_L(j\omega)$$

The s-plane representation is identical to that found in Figure 3.11 for the second of the Laplace pairs 3.13, repeated here

$$e^{\sigma_1 t} \sin \omega_1 t\, u(t) \quad \xleftrightarrow{\;\mathscr{L}\;} \quad \frac{\omega_1}{(s - \sigma_1)^2 + \omega_1^2} \qquad \text{Re } s > \sigma_1$$

The denominator equivalence of Equation 4.47 shows that, apart from an amplitude scaling factor, Figure 3.11 also represents the system function 4.49.

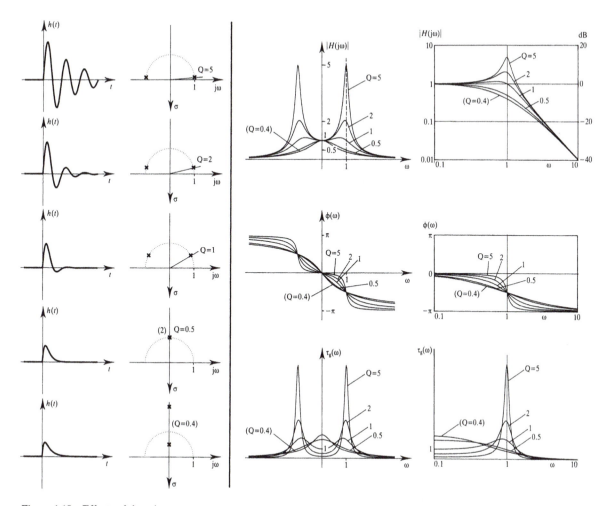

Figure 4.18   Effects of damping.

The impulse response of the lowpass filter results from scaling the above pair,

$$h_L(t) = \frac{c_0}{\omega_1} e^{\sigma_1 t} \sin \omega_1 t \, u(t) \quad \overset{\mathscr{L}}{\longleftrightarrow} \quad H_L(s) = \frac{c_0}{(s - p_1)(s - p_1^*)}$$

$$(4.50)$$

In Figure 4.17, $h_L(t)$ is linked by the Fourier transform to $H_L(j\omega)$, which in turn represents a slice of the Laplace transform $H_L(s)$. The system's step response $s_L(t)$ is related to $h_L(t)$ by integration, and this is indicated in the time domain of the figure.

Effects of damping

Damping affects the shape of the responses, as shown in Figure 4.18 for various pole positions identified by their pole-Q values. Note that for an amplitude-normalized frequency response, the value Q equals $|H(j\omega_r)|$. Also shown are phase and group delay characteristics, as well as the log–log representation of $|H(j\omega)|$. The latter exhibits a 12 dB/octave attenuation at high frequencies, which is characteristic of two zeros at infinity.

Bandpass section

Making c_1 the non-zero coefficient leads to the bandpass characteristic

$$H_B(s) = \frac{c_1 s}{s^2 - 2\sigma_1 s + \omega_r^2} \tag{4.51}$$

This case introduces a zero at the s-plane origin, as shown in the appropriate sector of Figure 4.19. This figure builds up a comparative chart of responses, each sector being laid out as the lowpass case of Figure 4.17, repeated in the top left sector of the chart.

Since $H_B(s)$ is of the same form as $H_L(s)$ multiplied by s, we conclude from the Laplace property 3.27 that the impulse response $h_B(t)$ is related to $h_L(t)$ by differentiation,

$$h_B(t) = \frac{c_1}{c_0} \frac{dh_L(t)}{dt} \quad \overset{\mathscr{L}}{\longleftrightarrow} \quad H_B(s) = \frac{c_1}{c_0} s H_L(s)$$

Differentiating $h_L(t)$ of Equation 4.50, we have

$$\frac{dh_L(t)}{dt} = \frac{c_0}{\omega_1} (\sigma_1 \sin \omega_1 t + \omega_1 \cos \omega_1 t) e^{\sigma_1 t} u(t)$$

$$+ \frac{c_0}{\omega_1} e^{\sigma_1 t} \sin \omega_1 t \delta(t)$$

But from the impulse property 1.12, $f(t)\delta(t) = f(0)\delta(t)$, the last term vanishes, because $e^0 = 1$ and $\sin 0 = 0$, so that

$$h_B(t) = \frac{c_1}{c_0} \frac{dh_L(t)}{dt} = \frac{c_1}{\omega_1} (\sigma_1 \sin \omega_1 t + \omega_1 \cos \omega_1 t) e^{\sigma_1 t} u(t) \tag{4.52}$$

The term in brackets is the sum of sinusoids of the same frequency, hence

$$\sigma_1 \sin \omega_1 t + \omega_1 \cos \omega_1 t = \omega_r \cos (\omega_1 t - \theta_1) \qquad \theta_1 = \arctan \frac{\sigma_1}{\omega_1}$$

Figure 4.19 (*Opposite*) Six common second-order sections with zeros on imaginary axis.

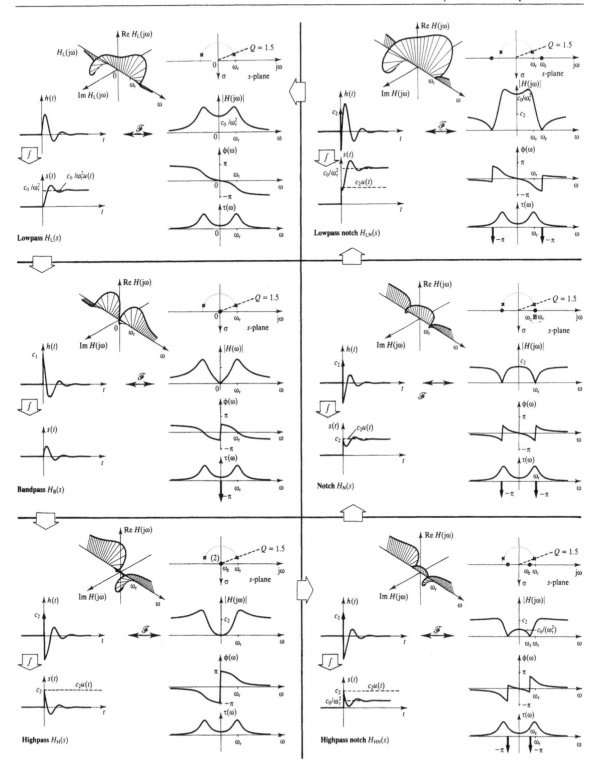

Lowpass $H_\mathrm{L}(s)$

Lowpass notch $H_\mathrm{LN}(s)$

Bandpass $H_\mathrm{B}(s)$

Notch $H_\mathrm{N}(s)$

Highpass $H_\mathrm{H}(s)$

Highpass notch $H_\mathrm{HN}(s)$

giving the alternative expression

$$h_B(t) = \frac{c_1}{\omega_1} \omega_r \cos(\omega_1 t - \theta_1) e^{\sigma_1 t} u(t)$$

Highpass section

The case where only c_2 is non-zero represents the highpass response

$$H_H(s) = \frac{c_2 s^2}{s^2 - 2\sigma_1 s + \omega_r^2} \tag{4.53}$$

It introduces two zeros at the s-plane origin, so that both $H_H(s)$ and its first derivative are zero at that point, as shown in the frequency response magnitude of Figure 4.19.

The impulse response $h_H(t)$ is now related to $h_B(t)$ by differentiation,

$$h_H(t) = \frac{c_2}{c_1} \frac{dh_B(t)}{dt} \xleftrightarrow{\mathscr{L}} H_H(s) = \frac{c_2}{c_1} s H_B(s)$$

and differentiating $h_B(s)$ of Equation 4.52 yields

$$\frac{dh_B(t)}{dt} = \frac{c_1}{\omega_1} [(\sigma_1^2 - \omega_1^2) \sin \omega_1 t + 2\sigma_1 \omega_1 \cos \omega_1 t] e^{\sigma_1 t} u(t)$$

$$+ \frac{c_1}{\omega_1} (\sigma_1 \sin \omega_1 t + \omega_1 \cos \omega_1 t) e^{\sigma_1 t} \delta(t)$$

where, by the impulse property 1.12, the last term reduces to $c_1 \delta(t)$, because $e^0 = 1$, $\sin 0 = 0$ and $\cos 0 = 1$. The highpass impulse response becomes

$$h_H(t) = \frac{c_2}{c_1} \frac{dh_B(t)}{dt}$$

$$= \frac{c_2}{\omega_1} [(\sigma_1^2 - \omega_1^2) \sin \omega_1 t + 2\sigma_1 \omega_1 \cos \omega_1 t] e^{\sigma_1 t} u(t) + c_2 \delta(t)$$

$$\tag{4.54}$$

The impulse at the time origin represents the transform of the constant frequency domain value c_2 to which the system function tends for high values of s, as seen in the magnitude of the frequency response $H_H(j\omega)$ in Figure 4.19.

The brackets of Equation 4.54 can again be replaced by a single sinusoid, yielding

$$h_H(t) = \frac{c_2}{\omega_1} \omega_r^2 \cos(\omega_1 t - \theta_2) \qquad \theta_2 = \arctan \frac{\sigma_1^2 - \omega_1^2}{2\sigma_1 \omega_1}$$

Notch responses

Another important category arises when only the coefficient c_1 is missing, so that

$$H_N(s) = \frac{c_2 s^2 + c_0}{s^2 - 2\sigma_1 s + \omega_r^2} \tag{4.55}$$

When c_2 and c_0 have the same sign, the two system zeros fall on the imaginary axis, making $H_N(s)$ zero at the points

$$z_1, z_2 = \pm j \sqrt{\frac{c_0}{c_2}}$$

It is customary to distinguish three cases, according to where the notch falls. The reference case occurs for $c_0 = c_2 \omega_r^2$, when the zeros are located at a distance ω_r from the origin, at the points $z_1, z_2 = \pm j\omega_r$. This represents the proper **notch section**, which has the property

$$H_N(0) = H_N(\infty) = c_2$$

as seen in the corresponding sector of Figure 4.19.

The situation $c_0 < c_2 \omega_r^2$ leads to the **highpass notch section** $H_{HN}(s)$, while $c_0 > c_2 \omega_r^2$ produces the **lowpass notch section** $H_{LN}(s)$. The nomenclature is obvious from Figure 4.19.

Relationship between types

The layout of Figure 4.19 leads to a unifying interpretation based on the concept that the poles of a second-order system are always matched by two zeros, which are allowed to be located at infinity, as interpreted at the end of Section 4.2.6.

The highpass section $H_H(s)$ has both zeros located at the origin. Adding progressively increasing values of the second coefficient c_0 gradually separates the zeros, as shown in the anticlockwise sequence of Figure 4.19, thus generating the highpass notch $H_{HP}(s)$, the notch $H_N(s)$ and eventually the lowpass notch $H_{LN}(s)$.

If we continue separating the zeros, taking them notionally to infinity, at the same time scaling the amplitudes to keep the d.c. value $H(0)$ constant, takes us back to the lowpass section $H_L(s)$ as the limiting case. This demonstrates the convenience of the concept of 'zeros at infinity' when interpreting the behaviour of lowpass and bandpass sections.

We can now re-examine the sequence of the left half of Figure 4.19 in this light. At high frequency values the amplitude of the lowpass section $H_L(s)$ decays as $1/s^2$, which is consistent with two zeros at ∞. On log–log scales this represents a rate of 12 dB/octave, as shown in Figure 4.18. The bandpass section $H_B(s)$, with one zero at the origin and the other at infinity, gives a **first-order rise** (6 dB/octave) at the

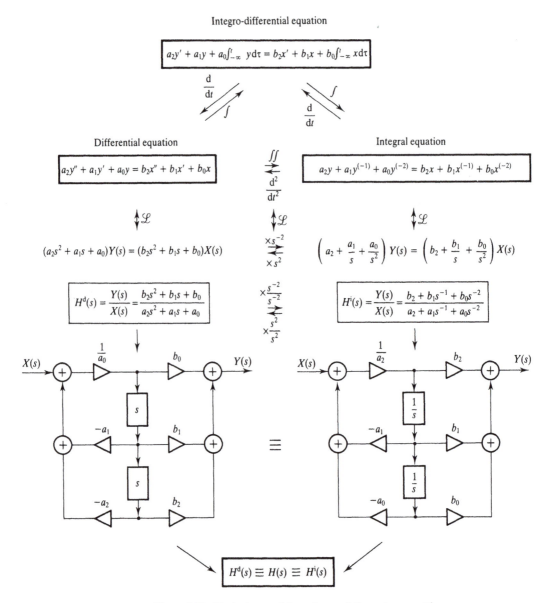

Figure 4.20 Equivalence of three forms of the system equation.

origin and a **first-order decay**, proportional to $1/s$, at high frequencies. Finally, the two zeros at the origin of the highpass section $H_H(s)$ give a **second-order rise** (12 dB/octave), leaving no zeros to pull the response down at infinity.

Note that the phase response also reflects this notion of zeros at infinity. Consider the case of the bandpass section $H_B(s)$. As frequency increases, phase tends to the value $-\frac{1}{2}\pi$ (negative imaginary axis). The zero at infinity adds the value π, so that, as the frequency 'wraps round' to minus infinity, phase emerges with a value $+\frac{1}{2}\pi$. Similar conclusions apply to the other cases.

4.3.4 Equivalence of differential and integral equations

We have previously indicated the equivalence between differential and integral representations of a given system. We now formalize the relationship in both time and frequency. The results derived for this second-order system are also valid for other orders.

Assume we possess an integro-differential representation of the system, as indicated at the top of the chart of Figure 4.20. A sufficient number of successive differentiations or integrations, suggested by arrows, yields a purely differential or a purely integral system equation. The latter are equivalent, being related by N differentiations of both sides ($N = 2$ in our example), or by N integrations, as indicated.

Taking the Laplace transforms of these representations leads to a pair of equivalent algebraic equations, mutually related through multiplication by s^{-2} or s^2, as appropriate. These multiplication factors are themselves the Laplace transforms of double integration and double differentiation.

These equations lead to two system functions $H^d(s)$ and $H^i(s)$ which are rational polynomials in s and s^{-1}, respectively, where the indices d and i indicate the differential or integral path followed for their derivation. The two system functions are in fact the same and can be derived from each other by multiplying both numerator and denominator by either s^{-2} or s^2, as appropriate. We can therefore drop the index.

The block diagrams of the resulting system functions are shown at the bottom of Figure 4.20 and use either the differentiator s or the integrator $1/s$ as elements. These frequency domain diagrams are also valid for the time domain; as we have shown earlier, both block diagrams are formally identical.

In conclusion, the two parallel derivation paths, having originated in the same equation, converge to equivalent system functions and equivalent block diagrams, as expected.

The differential representation tends to be favoured in regard to analytical manipulations. In hardware implementations, for instance when employing operational amplifiers, the differentiator tends to be more difficult to realize than the integrator. In such cases the integral representation is more practical.

4.4   Nth-order systems

The order N of a system reflects the number of independent energy-storing elements. In the preceding sections this number was either one or two. Each element contributes an elementary first-order differential or integral relationship between its local variables. Combined by the applicable laws, these lead to an Nth-order integro-differential equation which determines the order of the system. Thus, the mere presence of N independent energy-storing elements makes the system of order N.

In this section we first extend earlier findings to the general Nth-order system, to examine ways of partitioning an arbitrary system into smaller, interconnected units. Two structures are of particular interest, namely the cascaded or series connection of second-order subsystems and the parallel connection of similar units.

Our main objective is to present the analytical tools associated with each type of partitioning, the block diagrams of the corresponding structures and the physical realizations these structures suggest.

4.4.1   Monolithic structure

The differential equations of earlier systems had terms in the input and output variables and in their derivatives whose highest order equalled that of the system. Extended to the general system of order N, this representation takes the form of the general Nth-order differential equation

$$a_N y^{(N)}(t) + a_{N-1} y^{(N-1)}(t) + \ldots + a_1 y'(t) + a_0 y(t) = x_f(t)$$
$$= b_N x^{(N)}(t) + b_{N-1} x^{(N-1)}(t) + \ldots + b_1 x'(t) + b_0 x(t)$$

$$(4.56)$$

which can be written more concisely as

$$\sum_{n=0}^{N} a_n y^{(n)}(t) = \sum_{n=0}^{N} b_n x^{(n)}(t) \tag{4.57}$$

where (n) represents nth-order differentiation. Taking the Laplace transform of both sides of Equation 4.56 gives a polynomial in s of degree N,

$$a_N s^N Y(s) + a_{N-1} s^{N-1} Y(s) + \ldots + a_1 s Y(s) + a_0 Y(s) = X_f(s)$$
$$= b_N s^N X(s) + b_{N-1} s^{N-1} X(s) + \ldots + b_1 s X(s) + b_0 X(s)$$

$$(4.58)$$

or, in concise form

$$\sum_{n=0}^{N} a_n s^n Y(s) = \sum_{n=0}^{N} b_n s^n X(s) \tag{4.59}$$

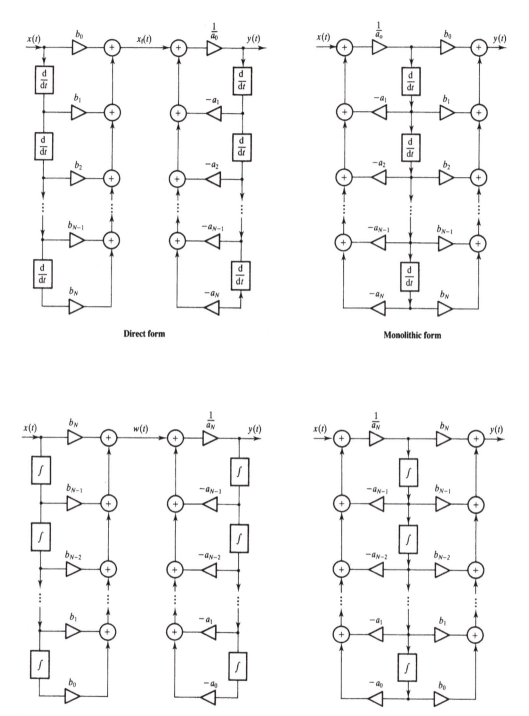

Figure 4.21 Block diagrams, equivalence using differentiators and integrators.

These lead directly to the general form of the Nth-order system function

$$H(s) = \frac{Y(s)}{X(s)} = \frac{b_N s^N + b_{N-1} s^{N-1} + \ldots + b_1 s + b_0}{a_N s^N + a_{N-1} s^{N-1} + \ldots + a_1 s + a_0} \quad (4.60)$$

or

$$H(s) = \frac{\sum_{n=0}^{N} b_n s^n}{\sum_{n=0}^{N} a_n s^n} \quad (4.61)$$

We concluded earlier that the time domain and frequency domain block diagrams were formally identical, and that they can be used interchangeably. Thus, extending the progression started in Figures 4.8 and 4.15, we obtain the general structures shown in the upper half of Figure 4.21.

The conclusions of Section 4.3.4, regarding equivalence of differential and integral equations, are also valid for a system of order N, whose general Nth-order integral equation can be written compactly as

$$\sum_{n=0}^{N} a_n y^{-(N-n)} = \sum_{n=0}^{N} b_n x^{-(N-n)} \quad (4.62)$$

where the negative indices of y and x denote orders of integration. Similarly, the appropriate form of the system function becomes

$$H(s) = \frac{\sum_{n=0}^{N} b_n s^{-(N-n)}}{\sum_{n=0}^{N} a_n s^{-(N-n)}} \quad (4.63)$$

where the negative powers of s represent integration too. The block diagrams associated with Equations 4.62 and 4.63 are again formally identical, both being represented in the lower half of Figure 4.21.

As already mentioned, in the literature the structure forms of the left of Figure 4.21 are often called **direct form I**, while those on the right are called **direct form II**, and sometimes **canonical form**. But because of the compact monolithic structure of the latter, we call this the **monolithic form**. This compactness is also reflected in the single time domain differential equation and in the single frequency domain polynomial, but is best conveyed by the block diagrams of Figure 4.21. The structures of the left half of the figure we call simply **direct form**, as they give the most direct representation of the differential equation 4.56, and give a direct meaning to the forcing function $x_f(t)$.

The monolithic form is not always the best structure that can be associated with the system. Operationally simpler structures result when the single differential equation 4.56, or some equivalent monolithic representation, is partitioned into an assemblage of smaller units.

4.4.2   Cascade partitioning

Although it affects all the other representations, partitioning is most clearly visualized in the frequency domain. The system function 4.60 is the ratio of two polynomials in s. The roots of the denominator polynomial $P(s)$ provide N system poles p_n, while the numerator provides up to N finite system zeros z_n, depending on the degree of the polynomial $Q(s)$. We can therefore factorize Equation 4.60 in terms of these roots as

$$H(s) = \frac{Q(s)}{P(s)}$$

$$= k \frac{(s - z_1)(s - z_2) \ldots (s - z_{N-1})(s - z_N)}{(s - p_1)(s - p_2) \ldots (s - p_{N-1})(s - p_N)} \tag{4.64}$$

where k is the overall scaling factor $k = b_N/a_N$.

If N is even, pairs of poles are combined with pairs of zeros into $\frac{1}{2}N$ second-order sections $H_n(s)$, such that

$$H(s) = k H_1(s) H_2(s) \ldots H_{N/2}(s) = k \prod_{n=1}^{N/2} H_n(s) \tag{4.65}$$

This represents a series or cascade connection of sections of the form

$$H_1(s) = k \frac{(s - z_1)(s - z_2)}{(s - p_1)(s - p_2)} \tag{4.66}$$

examined at length in Section 4.3.

These sections could be realized as self-contained physical units. For real systems, all complex roots must appear in conjugate pairs, which provides a natural grouping. The remaining roots, which must be real, can be artificially paired into overdamped second-order sections.

If N is odd, the system function 4.65 has an additional first-order factor representing a first-order section of the bilinear form

$$H_0(s) = \frac{s - z_0}{s - p_0} \tag{4.67}$$

where z_0 and p_0 are real roots.

This process is illustrated in Figure 4.22, where a fifth-order system (an elliptic filter taken from Section 10.5) is partitioned into one first-order and two second-order sections. The constituent block diagrams and responses are shown reconstructing those of the original system. According to Equation 4.65 the system response $H(s)$ is the product

$$H(s) = H_0(s) H_1(s) H_2(s)$$

represented symbolically by the s-plane shorthand notation of the bottom row of Figure 4.22. Evaluating each subsystem function $H_n(s)$ for $s = j\omega$ yields the frequency response

$$H(j\omega) = H_0(j\omega) H_1(j\omega) H_2(j\omega)$$

represented in Figure 4.22 by the product of their amplitudes $A_n(\omega)$,

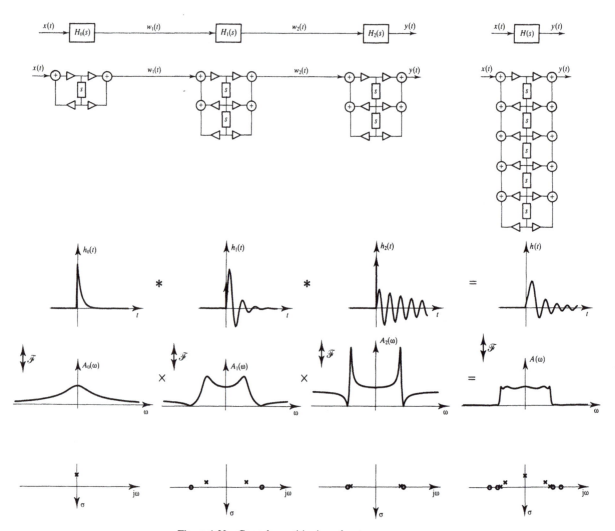

Figure 4.22   Cascade partitioning of system.

which, incidentally, on logarithmic scales would be represented by the sum

$$\log A(\omega) = \log A_0(\omega) + \log A_1(\omega) + \log A_2(\omega)$$

The corresponding impulse response is the result of the convolution of the component responses

$$h(t) = h_0(t) * h_1(t) * h_2(t)$$

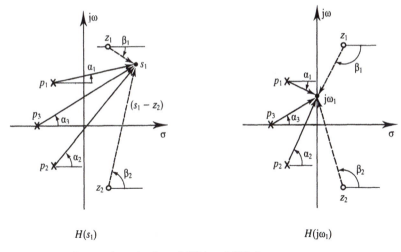

$H(s_1)$ $H(j\omega_1)$

Figure 4.23   Geometric evaluation of $H(s)$ and $H(j\omega)$.

Geometric interpretation of $H(s)$ and $H(j\omega)$

The factorized form in Equation 4.64 gives a geometric meaning to the system function $H(s)$, and therefore to $H(j\omega)$, and suggests a means for evaluating those functions from the s-plane locations of poles and zeros.

Interpreting a typical numerator factor $(s - z_n)$ as a vector, represented in the s-plane by its magnitude $|s - z_n|$ and angle β_n, see Figure 4.23, the contribution of the zero z_n to the value of $H(s)$ is the factor

$$(s - z_n) = |s - z_n|e^{j\beta_n}$$

Writing Equation 4.64 in the compact form

$$H(s) = k \frac{\prod_{n=1}^{N}(s - z_n)}{\prod_{n=1}^{N}(s - p_n)} = k \frac{\prod_{n=1}^{N}|s - z_n|e^{j\beta_n}}{\prod_{n=1}^{N}|s - p_n|e^{j\alpha_n}} = |H(s)|e^{j\theta(s)}$$

$$(4.68)$$

by the rules of multiplication and division of complex numbers (Equations 1.27 and 1.28) the magnitude of a value of $H(s)$ for an arbitrary point $s = s_1$ is obtained by multiplying the absolute value of the constant k by the product of the N numerator magnitudes $|s_1 - z_n|$ and dividing by the product of the N denominator magnitudes $|s_1 - p_n|$, that is,

$$|H(s_1)| = |k| \frac{\prod_{n=1}^{N}|s_1 - z_n|}{\prod_{n=1}^{N}|s_1 - p_n|}$$

while the argument $\theta(s)$ of $H(s)$ is the result of adding the angles β_n from the N zeros and subtracting the angles α_n from the poles,

$$\theta(s_1) = \theta_k + \sum_{n=1}^{N} \beta_n - \sum_{n=1}^{N} \alpha_n$$

where $\theta_k = 0$ for $k > 0$ and $\theta_k = \pi$ for $k < 0$.

For stable systems the frequency response $H(j\omega)$ represents a special case of Equation 4.68 evaluated on the imaginary axis.

4.4.3 Parallel partitioning

Series partitioning of a system involves factorizing $H(s)$ into a **product** of first- and second-order responses, seen in time as the equivalent convolution of the partial impulse responses.

An alternative way of partitioning the system involves a parallel connection of first- and second-order sections. This implies expanding $H(s)$ into partial fractions, thus replacing it by a **sum** of simpler rational fractions $T_i(s)$, seen in time as a similar replacement of the impulse response $h(t)$ by a sum of partial responses $t_i(t)$.

Partial fraction expansion

This familiar analytical method applies to rational polynomials, and more specifically to proper fractions, whose numerator polynomial is of lower degree than the denominator polynomial. This does not limit the scope of the method, as long division permits reducing an improper fraction, such as the general second-order form of Equation 4.60, to a proper fraction plus a constant.

Consider the proper fraction

$$H(s) = \frac{Q(s)}{P(s)} = \frac{b_{N-1}s^{N-1} + b_{N-2}s^{N-2} + \ldots + b_1 s + b_0}{s^N + a_{N-1}s^{N-1} + \ldots + a_k s^k + \ldots + a_0}$$

whose denominator $P(s)$ can be factorized in terms of the N poles as

$$H(s) = \frac{Q(s)}{P(s)} = \frac{b_{N-1}s^{N-1} + b_{N-2}s^{N-2} + \ldots + b_1 s + b_0}{(s - p_1)(s - p_2) \ldots (s - p_k) \ldots (s - p_N)}$$

(4.69)

If all the poles p_i are distinct, that is, no two pole locations coincide, this function can also be written as the sum of N first-order terms

$$H(s) = \frac{c_1}{(s - p_1)} + \frac{c_2}{(s - p_2)} + \ldots + \frac{c_k}{(s - p_k)}$$

$$+ \ldots + \frac{c_N}{(s - p_N)}$$

(4.70)

which represents a parallel connection of first-order systems,

$$H(s) = T_1(s) + T_2(s) + \ldots + T_k(s) + \ldots + T_N(s) = \sum_{i=1}^{N} T_i(s)$$

$$(4.71)$$

Symbols were changed to indicate that the modules $T_i(s)$ are different from the cascaded modules $H_i(s)$.

Linear equations approach

The coefficients c_i can be obtained by eliminating the fractions, gathering terms in powers of s and equating corresponding coefficients with $Q(s)$. This leads to a system of N linear equations in the coefficients c_i, and hence to their values. We illustrate with a simple example.

Example 4.4

The proper fraction

$$H(s) = \frac{2s + 7}{s^2 + 5s + 6}$$

has poles at $p_1 = -2$ and $p_2 = -3$, so that the denominator can be factorized as $(s + 2)(s + 3)$ and the function expanded in partial fractions as

$$H(s) = \frac{c_1}{s + 2} + \frac{c_2}{s + 3}$$

Clearing fractions yields

$$H(s) = \frac{(c_1 + c_2)s + (3c_1 + 2c_2)}{(s + 2)(s + 3)}$$

Equating coefficients with the original function leads to the linear system

$$\begin{cases} c_1 + c_2 = 2 \\ 3c_1 + 2c_2 = 7 \end{cases}$$

which yields the coefficient values $c_1 = 3$, $c_2 = -1$ and the desired result

$$H(s) = \frac{3}{s + 2} - \frac{1}{s + 3}$$

Alternative approach

For higher-order systems the above process becomes cumbersome. A simpler approach recognizes that each denominator factor $(s - p_i)$ is common to both the function $H(s)$ and one term $H_i(s)$ of the expansion. Thus, multiplying both sides of Equation 4.70 by this factor removes it from $H(s)$ and from $T_i(s)$, but introduces it into the numerators of all other terms,

$$\{H(s)(s - p_k)\} = \frac{c_1(s - p_k)}{(s - p_1)} + \frac{c_2(s - p_k)}{(s - p_2)} + \ldots + c_k$$
$$+ \ldots + \frac{c_N(s - p_k)}{(s - p_N)}$$

If the value of s is now set to $s = p_k$, all the terms on the right vanish, except c_k, which takes the value of the product on the left, i.e.

$$c_k = \{H(s)(s - p_k)\}_{s=p_k} \tag{4.72}$$

The product is shown enclosed in brackets to signify that $(s - p_k)$ is not an additional factor that would also vanish, but that it neutralizes an identical factor of the denominator of $H(s)$. This process, carried out separately for each root p_i, yields the N coefficients c_i.

Example 4.5

We wish to expand the function

$$H(s) = \frac{2s^2 + 3s - 3}{s^3 + 6s^2 + 11s + 6} = \frac{2s^2 + 3s - 3}{(s + 1)(s + 2)(s + 3)}$$

with roots $p_1 = -1$, $p_2 = -2$ and $p_3 = -3$, into the sum of partial fractions

$$H(s) = \frac{c_1}{s + 1} + \frac{c_2}{s + 2} + \frac{c_3}{s + 3}$$

The first method would have led to a third-order linear system in the coefficients c_i. But using the second method we have

$$c_1 = \{H(s)(s + 1)\}_{s=-1} = \left\{ \frac{2s^2 + 3s - 3}{(s + 2)(s + 3)} \right\}_{s=-1} = -2$$

$$c_2 = \{H(s)(s + 2)\}_{s=-2} = \left\{ \frac{2s^2 + 3s - 3}{(s + 1)(s + 3)} \right\}_{s=-2} = 1$$

$$c_3 = \{H(s)(s + 3)\}_{s=-3} = \left\{ \frac{2s^2 + 3s - 3}{(s + 1)(s + 2)} \right\}_{s=-3} = 3$$

and the desired result

$$H(s) = -\frac{2}{s+1} + \frac{1}{s+2} + \frac{3}{s+3}$$

It can be appreciated that the second method involves considerably less algebraic manipulation than the first, and also that each term of the expansion can be obtained in isolation, an important consideration if only a small number of the terms are of interest.

Complex conjugate roots

The above method is valid for both real and complex roots. However, in the latter case the resulting coefficients c_i too can take complex values.

Example 4.6

Consider the system function

$$H(s) = \frac{s+2}{s^2 + 2s + 2}$$

with complex conjugate poles $p_1 = -1+j$ and $p_2 = -1-j$, and a real zero $z_1 = -2$. The partial fraction expansion of Equation 4.70 has two terms, the method of Equation 4.72 yields the coefficients $c_1 = \frac{1}{2}(1-j)$ and $c_2 = \frac{1}{2}(1+j)$, and we can write

$$H(s) = T_1(s) + T_2(s) = \frac{\frac{1}{2}(1-j)}{s - (-1+j)} + \frac{\frac{1}{2}(1+j)}{s - (-1-j)}$$

Complex conjugate poles always lead to complex conjugate coefficients $c_2 = c_1^*$.

The individual functions $T_i(s)$ of the above example are rational polynomials in s with complex coefficients. They are valid for analytical purposes, see Section 11.3.2, but do not represent realizable systems. The polynomials of real systems have real coefficients, and any complex roots always appear in conjugate pairs.

At the system realization stage, such roots are kept together and the corresponding pair of nonrealizable first-order sections are combined as

$$\bar{T}_1(s) = \frac{c_1}{s - p_1} + \frac{c_2}{s - p_2} = \frac{(c_1 + c_2)s - (c_1 p_2 + c_2 p_1)}{(s - p_1)(s - p_2)}$$
$$= \frac{d_1 s + d_0}{(s - p_1)(s - p_2)}$$

whose coefficients d_1 and d_0 can easily be shown to be real. In such case the relationship 4.71 becomes

$$H(s) = \sum_{i=1}^{N/2} \bar{T}_i(s)$$

when N is even, to which a first-order term $T_0(s)$ is added if N is odd. This is a parallel connection of $\frac{1}{2}N$ second-order sections, plus a possible first-order section, as represented in Figure 4.24.

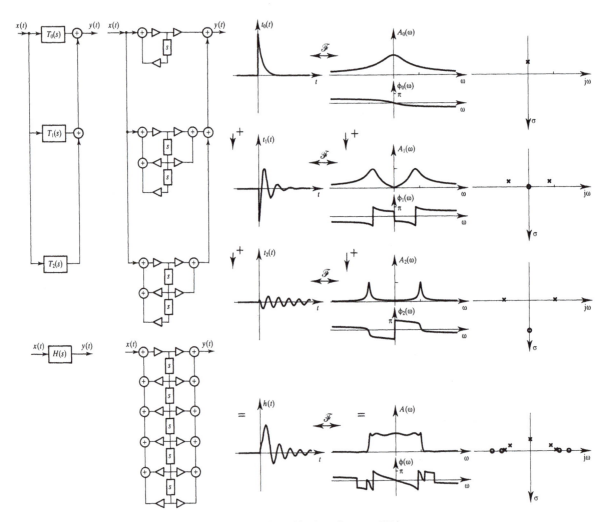

Figure 4.24   Parallel partitioning of system $H(s)$.

Multiple roots

To simplify presentation we assumed that all poles, real or complex, are distinct, and this assumption is kept for the remaining examples of this book. We only mention, without proof, that when an s-plane location has more than one pole the method requires a slight modification. A pole of multiplicity m gives rise to m terms, with denominators of increasing powers of $(s - p_i)$,

$$H(s) = \frac{Q(s)}{(s - p_i)^m} = \frac{c_1}{(s - p_i)} + \frac{c_2}{(s - p_i)^2} + \dots + \frac{c_m}{(s - p_i)^m}$$

whose coefficients are evaluated by the expression

$$c_{m-r} = \frac{1}{r!} \frac{\mathrm{d}^r}{\mathrm{d}s^r} [H(s)(s - p_i)^m]_{s=p_i}$$

Exercises

4.1 Fully characterize the system of Figure 4.25, taking $e(t)$ as the system input and $v(t)$ as its output. Arrange the results according to the scheme of Equations 4.23. Draw the block diagram and sketch the responses. Formulate the description in the frequency domain, using the concept of Laplacian impedance and treating the system as a voltage divider.

4.2 Compare the results of Exercise 4.1 and Example 4.2. Bearing in mind that resistor and capacitor values are positive quantities, for each system determine the location of the pole and zero relative to the s-plane origin and to each other. Locate both systems on the chart of Figure 4.12.

4.3 Find the system function $H(s) = X(s)/F(s)$ of the mass-spring-damper system of Figure 4.14, when the excitation force $f(t)$ is the system input, but the displacement $x(t) = \int v(t)\,\mathrm{d}t$ is taken as the system output. Draw the missing block diagram of Figure 4.14 and on it locate the points representing the displacement $x(t)$ and its first and second derivatives, that is, the velocity $v(t)$ and acceleration $a(t)$.

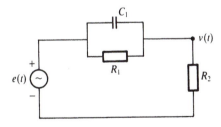

Figure 4.25   Circuit for Exercise 4.1.

4.4 Derive the two system functions $H_C(s) = V_C(s)/E(s)$ and $H_I(s) = I(s)/E(s)$ associated with the series RLC circuit of Figure 4.14. In each case associate the resulting polynomial coefficients with those of the general second-order expression 4.42 and write the corresponding differential equation and frequency response as subsets of Equations 4.39 and 4.43. Also, draw the applicable block diagrams as subsets of Figure 4.15.

4.5 For the values $L = 1\,\text{mH}$ and $C = 10\,\mu\text{F}$ and three different resistor values, $R = 100\,\Omega$, $R = 20\,\Omega$ and $R = 5\,\Omega$ establish whether the series RLC systems of Exercise 4.4 oscillate. Plot the corresponding poles and zeros for $H_C(s)$ and $H_I(s)$.

4.6 Take the oscillating case of Exercise 4.5, determine what types of response the system functions $H_C(s)$ and $H_I(s)$ represent and find the corresponding impulse responses $h_C(t)$ and $h_I(t)$. Sketch all frequency and impulse responses.

4.7 With the aid of Figure 4.19 discuss the effects of moving first one and then both of the system zeros of a second-order lowpass system from infinity to the s-plane origin, and then separating the zeros along the imaginary axis.

4.8 Extend both domains of the differential and integral representations of the second-order system of Figure 4.20 to the case of a general third-order system. Discuss the equivalences they reveal.

4.9 Express the following system functions as sums of partial fractions

$$H_1(s) = \frac{5s + 7}{s^2 + 3s + 2} \qquad H_2(s) = \frac{s + 1}{s^2 + 2s + 5}$$

$$H_3(s) = \frac{2s^2 - 9s + 6}{s(s - 2)(s - 3)}$$

Use the results to write the corresponding impulse responses $h(t)$, assuming these to be causal.

4.10 Use the geometric interpretation of the system response $H(s)$ to evaluate and sketch the magnitude $A(\omega)$ and phase $\phi(\omega)$ of the frequency response $H(j\omega)$ corresponding to the function $H_2(s)$ of Exercise 4.9.

4.11 When a system is implemented as a parallel connection of simple sections complex poles are kept in conjugate pairs to form realizable second-order sections of the form

$$T_i(s) = \frac{d_1 s + d_0}{(s - p_i)(s - p_i^*)} = \frac{c_1}{s - p_i} + \frac{c_2}{s - p_i^*}$$

whose coefficients d_1 and d_0 are real. Show that the partial fraction coefficients c_1 and c_2 must also form a complex conjugate pair.

4.12 Discuss the procedures for partitioning a system into first- and second-order sections connected either in series or in parallel, and the analytical processes applicable in each case.

CHAPTER 5

System Solution

In the preceding chapter we examined suitable means for **describing** a given physical system. In particular, we characterized the system by its impulse response $h(t)$, frequency response $H(j\omega)$ and system function $H(s)$, and found these to be different forms of the same characterization, related by the Fourier and Laplace transforms.

The characterizations were expressed at an abstract systems level, in terms of real system coefficients a_n and b_n. We thus distanced ourselves from the physical components and their values, in the knowledge that their contributions made their way into those coefficients.

In this chapter we derive the **solution methods** associated with the above responses, which permit finding the system's response $y(t)$ to an arbitrary input signal $x(t)$. One of our main objectives is to show that equivalent characterizations also lead to equivalent solution methods.

For this we first derive the convolution method associated with the impulse response $h(t)$, and use this time domain result to derive the frequency domain methods. We show that the latter are merely Fourier or Laplace interpretations of the impulse response method involving transform properties of the time domain convolution of two functions.

In the derivations of this chapter we assume that systems are initially at rest. The one exception is Section 5.4, where we examine the complete solution of systems with initial conditions and the role of the unilateral Laplace transform.

Excepting the occasional hint, no attempt is made in this chapter to associate particular systems with functional properties, such as filtering, integration, etc. Nor do we attempt to synthesize, design or simulate systems. Such topics are covered systematically in Part 3.

5.1 Elementary inputs and responses

The methods available for finding the system output $y(t)$ consist of subdividing the given input signal $x(t)$ into elementary components of a type consistent with the particular method, finding the corresponding elementary outputs and combining these to build up the complete output signal.

For this the system has to be linear, so that the various elementary components do not interact as they traverse the system, and time-invariant, so that individual shifted inputs give rise to similarly shifted responses, and the superposition principle can be applied to the output.

Each method is based on a particular type of elementary input signal which should have a simple representation, produce a simple system characterization and be capable of representing a wide class of input signals. These properties should also be easily transformed from one domain to the other.

Elementary input types

The most familiar types of elementary signals meet the above requirements. We briefly introduce those used in this chapter, which cover the great majority of practical applications.

When a system, described generically by H in Figure 5.1, is excited with an elementary input, it produces an elementary output, or response, that characterizes the system for the particular type of input.

Thus, when the input is the impulse function $\delta(t)$, as indicated in the figure, the output is, by definition, the system's impulse response $h(t)$, which provides one possible way of describing the system.

Another useful time domain function is the unit step $u(t)$, which gives rise to the step response $r(t)$, much favoured in control systems. This is related by integration to $h(t)$, as suggested in Figure 5.1 and substantiated in Example 5.1.

The exponential function $e^{j\omega_1 t}$, examined in detail in Section 1.3, is another favourite. We show in Section 5.3.1 that it produces a response that is an amplitude scaled and phase shifted version of the input exponential, as shown in Figure 5.1. The scaling factor $H(j\omega_1)$ represents a value of the system's frequency response.

The cosine function $\cos \omega_1 t$ being the sum of two counter-rotating exponentials, as seen in Section 1.3, also leads to a characterization in terms of the frequency response $H(j\omega_1)$.

Finally, the generalized complex exponential $e^{s_1 t}$ produces a scaled and shifted version of the input, with scaling factor $H(s_1)$.

In Figure 5.1 all these functions are shown represented in the

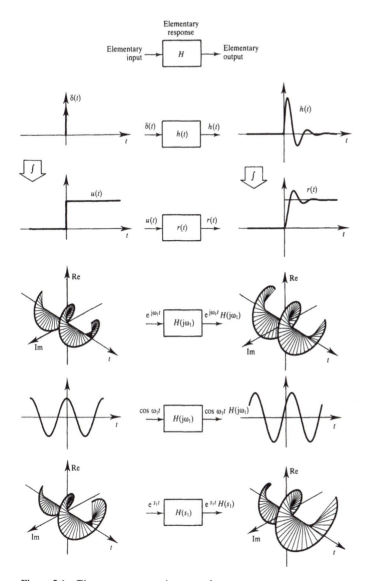

Figure 5.1   Elementary system inputs and responses.

time domain. Of these, the impulse function $\delta(t)$ and the unit step $u(t)$ can be said to belong to the time domain, while the responses of the other three are identified with the frequency domain. But each possesses an equivalent representation in the alternative domain. To emphasize this, in this book we constantly keep both domains together, to give a fuller understanding of all concepts and to benefit from the many dualities and other analogies that arise frequently in signals and systems.

We will also show that the most fundamental elemental signals are the time domain impulse $\delta(t)$ and the frequency domain impulse $\delta(\omega)$, which are related by Fourier transform duality. The other types are derived from these, be it by integration or differentiation, or by transforms.

5.2 Time domain methods

Two time domain methods are in common use, namely the **impulse response method** and the **step response method**. These result from expanding the system's input function $x(t)$ into a sum of weighted and shifted impulses, or of unit steps, so that the output $y(t)$ is expressed as a superposition of the system's response to such impulses or steps. Each case leads to a convolution integral, for which reason they are known as **convolution methods**, although this term is most commonly used for the impulse response method. The two variants are related by integration.

5.2.1 Superposition of impulses and their responses

We first examine how a linear time-invariant system responds to scaled and shifted impulses. Consider the system shown in the upper part of Figure 5.2, which we assume is initially de-energized. Probing this system with the trial function $x(t) = \delta(t)$ produces the impulse response $h(t)$ as output, as seen in Chapter 4.

Because the system is linear, a weighted impulse $a\delta(t)$ elicits a similarly weighted response $ah(t)$. Since the system is time invariant, a shifted impulse $\delta(t - t_0)$ gives rise to an equally shifted response $h(t - t_0)$. An input consisting of an arbitrary superposition of such impulses leads to a corresponding addition of responses, as indicated in the lower part of Figure 5.2.

This simple result forms the basis of the impulse response method. But we need to be able to interpret an arbitrary time function $f(t)$ as an infinite sum of infinitesimal impulses. Applied to both the input signal $x(t)$ and to the resulting sum of impulse responses, this interpretation leads to the output signal $y(t)$, and thereby to the convolution method.

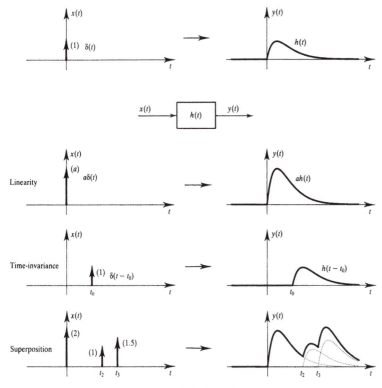

Figure 5.2 Response to scaled and shifted impulses.

5.2.2 Arbitrary function as convolution with impulse

An arbitrary function $f(t)$ can be interpreted as a superposition of finite segments $f_n(t)$,

$$f(t) = \sum_{n=-\infty}^{\infty} f_n(t)$$

as indicated in Figure 5.3. Each segment is the result of a product

$$f_n(t) = f(t)p_T(t - nT)$$

where $p_T(t - nT)$ is a pulse of unit height and width T centred on $t = nT$. Approximating each segment by a rectangle of height $f(nT)$,

$$\bar{f}_n(t) = f(nT)p_T(t - nT)$$

leads to an approximation of the whole function $f(t)$ as

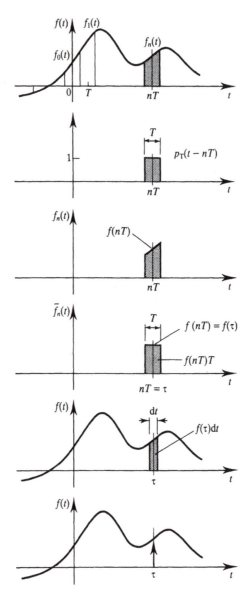

Figure 5.3   Impulse interpretation of arbitrary function $f(t) = f(t) * \delta(t)$.

$$f(t) \simeq \bar{f}(t) = \sum_{n=-\infty}^{\infty} \bar{f}_n(t) = \sum_{n=-\infty}^{\infty} f(nT) p_T(t - nT)$$

This approximation improves as T decreases. In the limit, as $T \to 0$, both representations coincide. We write this limit as

$$f(t) = \lim_{T \to 0} \sum_{n=-\infty}^{\infty} f(nT) \left[\frac{1}{T} p_T(t - nT) \right] T$$

where the factor $[1/T p_T(t - nT)]$, of unit area, represents one possible generating function for the shifted impulse, see Section 1.1.5,

$$\lim_{T \to 0} \frac{1}{T} p_T(t - nT) = \lim_{T \to 0} \delta(t - nT)$$

and permits writing $f(t)$ as

$$f(t) = \lim_{T \to 0} \sum_{n=-\infty}^{\infty} f(nT) \delta(t - nT) T \qquad (5.1)$$

Interpreting T as a time increment, in the limit it becomes a differential $d\tau$. As T decreases, the number n of segments required to reach a time value τ increases in inverse proportion, so that the product $nT = \tau$ remains constant. The limit (Equation 5.1) thus becomes the integral

$$f(t) = \int_{-\infty}^{\infty} f(\tau) \delta(t - \tau) \, d\tau \qquad (5.2)$$

This is the integral of Equation 1.10 used in Section 1.1.5 to define the impulse function $\delta(t)$, where we also described it as the sifting property of the impulse, because it sifts out the value of the function $f(\tau)$ at the particular point $\tau = t$.

In the present context we are more interested in the way it collectively describes the time function $f(t)$ as a continuum of displaced impulses $\delta(t - \tau)$, each scaled by the infinitesimal area $f(\tau) \, d\tau$.

Since Equation 5.2 has the form of a convolution, it also interprets a function as the convolution of itself with the impulse function, that is,

$$f(t) = f(t) * \delta(t) \qquad (5.3)$$

Note that this property of the impulse $\delta(t)$ is sometimes used to describe it as the **identity element of convolution.**

5.2.3   Impulse response method

Consider a system characterized by its impulse response $h(t)$. An arbitrary input $x(t)$ can be described in the form of Equation 5.1 as the limit

$$x(t) = \lim_{T \to 0} \sum_{n=-\infty}^{\infty} x(nT) T \delta(t - nT) \qquad (5.4)$$

which interprets the function $x(t)$ as an infinite sum of displaced impulses $\delta(t - nT)$, each weighted by an area $x(nT)T$.

Since an impulse $\delta(t)$ at the input gives rise to the elementary response $h(t)$, indicated as the input/output relationship

$$x(t) = \delta(t) \xrightarrow{h(t)} y(t) = h(t)$$

then, according to the observations of Section 5.2.1, each weighted and shifted impulse $x(nT)T\delta(t - nT)$ produces an equally weighted and shifted elementary output, that is,

$$x_n(t) = x(nT)T\delta(t - nT) \xrightarrow{h(t)} y_n(t) = x(nT)Th(t - nT)$$

(5.5)

as suggested in the upper half of Figure 5.4. Consequently, the limit in Equation 5.4, taken as an input sum, produces a corresponding output sum

$$y(t) = \lim_{T \to 0} \sum_{n=-\infty}^{\infty} x(nT)Th(t - nT)$$

(5.6)

This limit is formally identical to Equation 5.1, and similarly converges to the integral

$$y(t) = \int_{-\infty}^{\infty} x(\tau)h(t - \tau)\,d\tau$$

(5.7)

This yields the fundamental relationship

$$x(t) = \int_{-\infty}^{\infty} x(\tau)\delta(t - \tau)\,d\tau \xrightarrow{h(t)} y(t) = \int_{-\infty}^{\infty} x(\tau)\,h(t - \tau)\,d\tau$$

(5.8)

which now interprets the input $x(t)$ as an infinite sum of impulses $\delta(t - \tau)$ weighted by the infinitesimal values $x(\tau)\,d\tau$.

By the commutative property of convolution, we can also write Equation 5.8 as

$$x(t) = \int_{-\infty}^{\infty} x(t - \tau)\delta(\tau)\,d\tau \xrightarrow{h(t)} y(t) = \int_{-\infty}^{\infty} x(t - \tau)\,h(\tau)\,d\tau$$

(5.9)

The two expressions 5.8 and 5.9 are written more compactly as

$$\boxed{x(t) = x(t) * \delta(t) \xrightarrow{h(t)} y(t) = x(t) * h(t)}$$

(5.10)

and represent the **impulse response method**, or convolution method. This

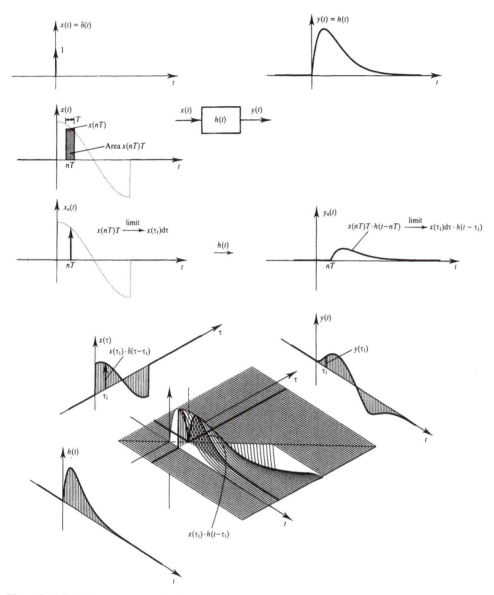

Figure 5.4   Impulse response method.

is expressed succinctly as: **if the input $x(t)$ is interpreted as the convolution of itself with the impulse function $\delta(t)$, then the output $y(t)$ is similarly interpreted as the convolution of the input $x(t)$ with the impulse response $h(t)$ of the system.**

The form given in expression 5.8 is interpreted graphically in the lower half of Figure 5.4. Each elementary input component $x(\tau_1)\delta(\tau - \tau_1)$ gives rise to an elementary response $x(\tau_1)h(t - \tau_1)$ indicated in the figure. Collectively, these build up a surface on the $t\tau$-plane. To find a value of $y(t)$ for a particular point $t = \tau_1$ we slice that surface with a plane of constant t and integrate for all τ, as specified in expression 5.8. Repeating the process for every t builds up the complete output function $y(t)$.

This construction gives a **systems interpretation of convolution**, an alternative to the more mathematically oriented interpretation given earlier in Figure 1.4. Both methods generate the same surface on the $t\tau$-plane, which is also sliced and integrated in the same way, giving identical results.

The difference lies in the process by which this surface is built up. In the case of Figure 1.4 it was assembled from slices of constant t, that is, from functions of the variable τ. In the case of Figure 5.4 it is built up from slices of constant τ, which are functions of t.

Example 5.1

We find the output of a system characterized by its impulse response $h(t)$, when the input is the step function, i.e. $x(t) = u(t)$. Applying the result 5.9 yields

$$y(t) = u(t) * h(t) = \int_{-\infty}^{\infty} u(t - \tau)h(\tau)\,d\tau = \int_{-\infty}^{t} h(\tau)\,d\tau$$

because $u(t - \tau)$ is zero for $\tau > t$ and unity elsewhere. This shows that the step response $r(t)$ of a system is the integral of the impulse response

$$r(t) = \int_{-\infty}^{t} h(\tau)\,d\tau \tag{5.11}$$

If $h(t)$ is causal, then $h(\tau) = 0$ for $\tau < 0$ and the lower integration limit can also be restricted, leaving

$$r(t) = \int_{0}^{t} h(\tau)\,d\tau$$

The cases of Section 4.3.3 provide specific examples of the relationship between step responses and impulse responses, see Figure 4.19. Furthermore, given the similarities of the system functions $H_L(s)$, $H_B(s)$ and $H_H(s)$, the lowpass impulse response is proportional to the bandpass step response, while the bandpass impulse response is proportional to the highpass step response.

5.3   Frequency domain methods

We now examine the system's behaviour when the input is either a complex exponential $x(t) = e^{j\omega t}$ or a generalized exponential $x(t) = e^{st}$. To stress the close relationship between methods, we will apply the impulse response approach to each case, allowing the frequency domain methods to emerge as consequences.

5.3.1   Response to an isolated exponential

Given a system of known impulse response $h(t)$, we probe it with a complex exponential of fixed frequency ω_1,

$$x(t) = e^{j\omega_1 t}$$

Using the form of expression 5.9 of the convolution method, the system output is expressed by the integral

$$y(t) = e^{j\omega_1 t} * h(t) = \int_{-\infty}^{\infty} e^{j\omega_1(t-\tau)} h(\tau) \, d\tau$$

The factor $e^{j\omega_1 t}$ is not a function of the integration variable τ, hence

$$y(t) = e^{j\omega_1 t} \int_{-\infty}^{\infty} h(\tau) e^{-j\omega_1 \tau} \, d\tau$$

The resulting integral is the formal definition of the Fourier transform of the impulse response $h(t)$, that is, the frequency response $H(j\omega)$. More precisely, it represents the value for the frequency $\omega = \omega_1$ of the exponential. One isolated exponential thus gives rise to the input/output correspondence

$$x(t) = e^{j\omega_1 t} \quad \xrightarrow{h(t)} \quad y(t) = e^{j\omega_1 t} H(j\omega_1) \tag{5.12}$$

The output is therefore a scaled version of the input signal. The scaling factor $H(j\omega_1)$ is a complex number, which, written in polar form

$$H(j\omega_1) = A(\omega_1) e^{j\phi(\omega_1)}$$

is seen to scale the exponential's magnitude by $A(\omega_1)$ and shift its phase by $\phi(\omega_1)$,

$$y(t) = A(\omega_1) e^{j[\omega_1 t + \phi(\omega_1)]} \tag{5.13}$$

without affecting the frequency ω_1.

It is worth mentioning that a function that is only scaled when passing through a system, is called an **eigenfunction** of the system, and the associated scaling factor is its **eigenvalue**.

Graphical interpretation

To get the full meaning of the input/output relationship 5.12, and anticipating the frequency response method, we interpret it in both domains. Taking the Fourier transform of both sides yields the group

$$x(t) = e^{j\omega_1 t} \qquad \xrightarrow{h(t)} \qquad y(t) = e^{j\omega_1 t} H(j\omega_1)$$
$$\updownarrow \mathcal{F} \qquad\qquad\qquad\qquad \updownarrow \mathcal{F} \qquad\qquad\qquad (5.14)$$
$$X(j\omega) = 2\pi\delta(\omega - \omega_1) \quad \xrightarrow{H(j\omega)} \quad Y(j\omega) = 2\pi\delta(\omega - \omega_1)H(j\omega_1)$$

This group is interpreted in Figure 5.5, which shows the time domain and frequency domain representations of the system and of the complex exponential signal as it enters and leaves the system.

The system, represented at the centre of each domain by the impulse response $h(t)$ and frequency response $H(j\omega)$, is a typical second-order lowpass section of the type examined earlier in Figure 4.17.

The input function is placed to the left of the system. In the time domain we show the exponential $x(t) = e^{j\omega_1 t}$ itself, as well as its real and imaginary parts. Its transform is simply the shifted impulse $X(j\omega) = 2\pi\delta(\omega - \omega_1)$ of strength 2π and zero phase.

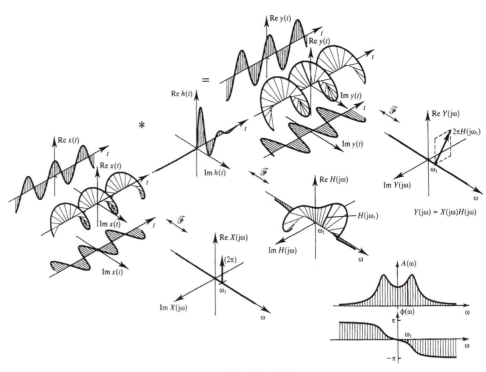

Figure 5.5 Response to exponential $x(t) = e^{j\omega_1 t}$.

The interpretation of the frequency domain of expressions 5.14 is now trivial: the impulse $2\pi\delta(\omega - \omega_1)$, located at $\omega = \omega_1$, is simply multiplied by the corresponding value of $H(j\omega)$. This involves scaling its strength by the amplitude $A(\omega_1)$, and shifting its angle by the phase $\phi(\omega_1)$.

In the time domain, this scaling manifests itself in a larger magnitude and a negative phase shift of the output exponential $y(t)$, as expressed by Equation 5.13 and clearly visible in the real and imaginary components.

We see that this frequency domain interpretation sheds additional light on the system's response to a complex exponential formulated earlier in the time domain. It also prepares the ground for the frequency response method.

Effect on real and imaginary parts

The input exponential is interpreted by Euler's formula as

$$x(t) = e^{j\omega_1 t} = \cos \omega_1 t + j \sin \omega_1 t$$

and the output as

$$y(t) = e^{j\omega_1 t} H(j\omega_1) = \cos \omega_1 t H(j\omega_1) + j \sin \omega_1 t H(j\omega_1)$$

This shows that the real and imaginary parts of the input are scaled and shifted by the same coefficient as the exponential, as seen in the appropriate time domain projections of Figure 5.5.

The two components can thus be treated independently, a consequence of the system's linearity and time-invariance. This is stated explicity as

$$\mathrm{Re}\, e^{j\omega_1 t} = \cos \omega_1 t$$
$$\xrightarrow{\ h(t)\ } \mathrm{Re}\{e^{j\omega_1 t} H(j\omega_1)\} = A(\omega_1) \cos[\omega_1 t + \phi(\omega_1)]$$

$$\mathrm{Im}\, e^{j\omega_1 t} = \sin \omega_1 t$$
$$\xrightarrow{\ h(t)\ } \mathrm{Im}\{e^{j\omega_1 t} H(j\omega_1)\} = A(\omega_1) \sin[\omega_1 t + \phi(\omega_1)]$$

The sine and cosine functions too are eigenfunctions of the system.

5.3.2   Sum of exponentials

Having established the response to one exponential, the superposition principle provides the system's response to input signals $x(t)$ that are sums of exponentials of different amplitudes and frequencies. For instance, two exponentials of frequencies ω_1 and ω_2 lead to

$$x(t) = X_1 e^{j\omega_1 t} + X_2 e^{j\omega_2 t}$$

$$\xrightarrow{h(t)} \quad y(t) = X_1 e^{j\omega_1 t} H(j\omega_1) + X_2 e^{j\omega_2 t} H(j\omega_2)$$

$$(5.15)$$

The coefficients X_i can take complex values. In such case they also define the amplitude and the phase of individual input components, both parameters being affected by the scaling factors $H(j\omega_i)$.

Example 5.2

We interpret the cosine function of the preceding section as the semi-sum of exponentials of opposite frequencies,

$$x(t) = \cos \omega_1 t = \tfrac{1}{2}[e^{j\omega_1 t} + e^{-j\omega_1 t}]$$

This leads to the output

$$y(t) = \tfrac{1}{2}[e^{j\omega_1 t} H(j\omega_1) + e^{-j\omega_1 t} H(-j\omega_1)]$$

If the system is real, the two factors $H(j\omega_1)$ and $H(-j\omega_1)$ are complex conjugates, and lead back to the earlier result

$$y(t) = A(\omega_1)\cos[\omega_1 t + \phi(\omega_1)]$$

We now interpret arbitrary inputs as sums of exponentials, starting with periodic time functions, whose frequency representations involve only discretely spaced impulses. The results will be extended in Section 5.3.3 to aperiodic time inputs with continuous frequency spectra.

Periodic input

A periodic input signal $\tilde{x}(t)$, with period T_0 and fundamental frequency $\omega_0 = 2\pi/T_0$, can be interpreted by means of the Fourier series synthesis equation (1.18) as a sum of exponentials,

$$\tilde{x}(t) = \sum_{k=-\infty}^{\infty} X_k e^{j\omega_k t}$$

where the magnitude and the phase associated with each exponential $e^{j\omega_k t}$ are provided by the coefficient X_k.

Extending the result of Equation 5.15 to this infinite sum leads to a similar interpretation of the output as a sum of time domain exponentials. Both input and output are more readily visualized in the frequency domain, where they are represented by impulses, as expressed by the Fourier relationships

$$\tilde{x}(t) = \sum_{k=-\infty}^{\infty} X_k e^{j\omega_k t} \qquad \xrightarrow{h(t)} \qquad \tilde{y}(t) = \sum_{k=-\infty}^{\infty} X_k H(j\omega_k) e^{j\omega_k t}$$

$$\downarrow \mathscr{F} \qquad\qquad\qquad\qquad \downarrow \mathscr{F}$$

$$X_s(j\omega) = 2\pi \sum_{k=-\infty}^{\infty} X_k \delta(\omega - \omega_k) \xrightarrow{H(j\omega_k)} Y_s(j\omega) = 2\pi \sum_{k=-\infty}^{\infty} X_k H(j\omega_k) \delta(\omega - \omega_k)$$

$$(5.16)$$

where the magnitude and the phase of each output exponential, or of the associated impulse, are determined by the weighting factor $Y_k = X_k H(j\omega_k)$.

The group 5.16 is interpreted in Figure 5.6, which shows clearly how each weighted input impulse $2\pi X_k \delta(\omega - \omega_k)$ is further scaled and phase shifted by the scaling factor $H(j\omega_k)$. Indicated by a heavy line, this factor represents the value taken by the system's frequency response $H(j\omega)$ at the corresponding frequency value $\omega_k = k\omega_0$. Note that the impulses and the weighting factors are functions of the continuous frequency variable ω.

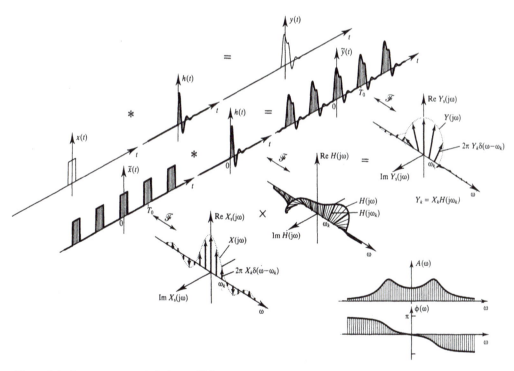

Figure 5.6   Response to periodic input $\tilde{x}(t)$.

The dotted line envelopes of the frequency domain impulses are the Fourier transforms of associated aperiodic functions $x(t)$ and $y(t)$, and will be explained later in the context of the discrete-frequency Fourier transform.

5.3.3   Frequency response method

An aperiodic input signal $x(t)$ can be similarly treated, by means of the **inverse Fourier transform** expression 2.6,

$$x(t) = \mathcal{F}^{-1}\{X(j\omega)\} = \frac{1}{2\pi}\int_{-\infty}^{\infty} X(j\omega)e^{j\omega t}\,d\omega \tag{5.17}$$

where $X(j\omega)$ represents the continuous-frequency spectrum of $x(t)$. But here we interpret this integral as an **infinite sum of exponentials**, each of which is amplitude scaled and phase shifted by an infinitesimal weighting factor $\frac{1}{2\pi}X(j\omega)\,d\omega$.

According to the result 5.12, one such infinitesimal input component, of frequency $\omega = v$, elicits an infinitesimal response

$$\left[\frac{1}{2\pi}X(jv)\,d\omega\right]e^{jvt} \xrightarrow{\;h(t)\;} \left[\frac{1}{2\pi}X(jv)\,d\omega\right]H(jv)e^{jvt} \tag{5.18}$$

so that the sum 5.17 of such exponential components, of all frequencies, produces the collective response

$$y(t) = \frac{1}{2\pi}\int_{-\infty}^{\infty}[X(j\omega)H(j\omega)]e^{j\omega t}\,d\omega \tag{5.19}$$

This too has the form of an inverse Fourier transform, that of the product $X(j\omega)H(j\omega)$, thus relating the time domain and frequency domain descriptions of the system's output. We conclude that the frequency representations of the input signal in Equation 5.17 and of the output signal in Equation 5.19 lead to the frequency domain input/output relationship

$$\boxed{\;X(j\omega) \xrightarrow{\;H(j\omega)\;} Y(j\omega) = X(j\omega)H(j\omega)\;} \tag{5.20}$$

which represents the **frequency response method** for solving a system. It is no more than the frequency domain interpretation of the impulse response method, a relationship expressed in the Fourier equivalence

$$
\begin{array}{ccc}
x(t) = x(t) * \delta(t) & \xrightarrow{\;\;h(t)\;\;} & y(t) = x(t) * h(t) \\[2pt]
\updownarrow \mathcal{F} & \Big\downarrow \mathcal{F} \quad \updownarrow \mathcal{F} & \\[2pt]
X(j\omega) = X(j\omega) \cdot 1 & \xrightarrow[\;\;H(j\omega)\;\;]{} & Y(j\omega) = X(j\omega)H(j\omega)
\end{array}
\tag{5.21}
$$

This result should not come as a surprise. Each side of expression 5.21 merely confirms the transform property 2.35 of convolution in time. The input $x(t)$, interpreted as a convolution of itself with the impulse $\delta(t)$, transforms to a multiplication with unity. Consequently, since the frequency response $H(j\omega)$ is the Fourier transform of the impulse response $h(t)$, the time domain convolution with $x(t)$ transforms to a frequency multiplication with $X(j\omega)$.

The equivalence 5.21 is visualized in Figure 5.7, which extends the construction of Figure 5.6 to aperiodic time signals. In the frequency domain this involves replacing the sum of discretely spaced impulses by

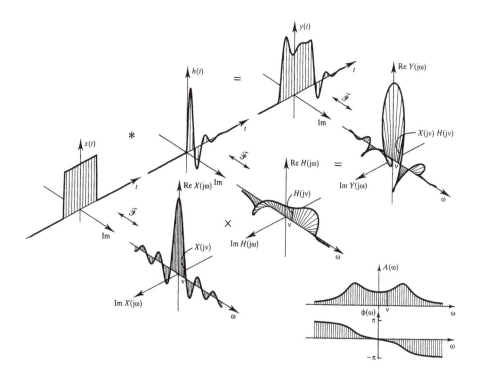

Figure 5.7   Equivalent time and frequency methods.

a continuous spectrum. The process is highlighted for a typical frequency value $\omega = v$, for which the input $X(jv)$ is magnitude-scaled and phase-shifted by $H(jv)$ to yield the output value $Y(jv)$. The collective effect is a distortion of the frequency representation of the input signal, which results in a corresponding distortion of the time domain.

5.3.4 Duality of impulse and frequency response methods

A comparison of Figures 5.4 and 5.7 reveals a remarkable duality between the impulse response method and the frequency response method: in each case one domain of the input signal is interpreted as a sum of impulses.

The Fourier transform dualities presented in Chapter 2 help to understand these similarities. Let us re-examine the response to the elementary signals associated with each method.

Duality of elementary signals

The time domain exponential e^{jvt} of the frequency response method has the shifted impulse $2\pi\delta(\omega - v)$ as its frequency domain counterpart, as shown earlier in Figure 5.5 where we identified v by ω_1. The system scaled and rotated this impulse according to the corresponding scaling factor $H(jv)$, and this scaling and rotation was also observed in the time domain of the figure, as expressed by the relationship 5.12.

The system's response to the shifted time domain impulse $\delta(t - \tau)$, which is the basis of the impulse response method, can be similarly interpreted. Taken as the input $x(t)$ of the input/output relationship 5.21 it yields

$$
\begin{array}{ccc}
x(t) = \delta(t - \tau) & \xrightarrow{\ h(t)\ } & y(t) = \delta(t - \tau) * h(t) = h(t - \tau) \\
\updownarrow \mathcal{F} & & \updownarrow \mathcal{F} \\
X(j\omega) = e^{-j\omega\tau} & \xrightarrow{\ H(j\omega)\ } & Y(j\omega) = e^{-j\omega\tau} H(j\omega)
\end{array}
\tag{5.22}
$$

as represented in Figure 5.8. The shifted time domain impulse is now convolved with the impulse response $h(t)$ to produce a correspondingly scaled and shifted response $h(t - \tau)$ as the system's output. The frequency domain expresses the same process in terms of complex exponentials and of multiplication.

We conclude that the two methods exchange the role of the impulse in one domain and the exponential in the other. This role

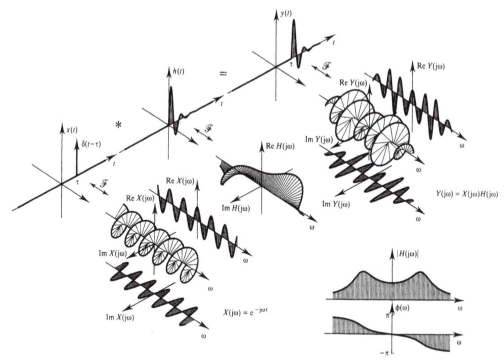

Figure 5.8   Response to displaced impulse $x(t) = \delta(t - \tau)$.

reversal also applies to the interpretation of arbitrary input signals, thus unifying the two methods.

Duality of methods

We can now re-examine Figure 5.7 in this light. As a preliminary to solution, the system is characterized by its response to either a time domain impulse, yielding $h(t)$, or to frequency domain impulses, yielding $H(j\omega)$. The input signal is then decomposed into a sum of impulses in one or the other domain.

The impulse response method implies a decomposition of $x(t)$ into a **sum of weighted time domain impulses,** conveniently provided by the convolution integral definition of an impulse,

$$x(t) = x(t) * \delta(t) = \int_{-\infty}^{\infty} x(\tau)\delta(t - \tau)d\tau \qquad (5.23)$$

Each infinitesimal impulse, represented in the frequency domain by a similarly scaled exponential, gives rise to a corresponding infinitesimal output, expressed according to the relationship 5.22 as

$$
\begin{array}{ccc}
[x(\tau)\,\mathrm{d}t]\delta(t-\tau) & \xrightarrow{\;h(t)\;} & [x(\tau)\,\mathrm{d}t]h(t-\tau) \\
\Big\updownarrow \mathcal{F} & & \Big\updownarrow \mathcal{F} \\
[x(\tau)\,\mathrm{d}t]e^{-j\omega\tau} & \xrightarrow{\;H(j\omega)\;} & [x(\tau)\,H(j\omega)\,\mathrm{d}t]e^{-j\omega\tau}
\end{array}
\tag{5.24}
$$

which is interpreted as in Figure 5.8. The infinite sum 5.23 of such components turns the terms of expression 5.24 into the integrals implied in expression 5.21.

The frequency domain method too can be viewed as implying a similar decomposition of $X(j\omega)$ into a **sum of weighted frequency domain impulses**,

$$
X(j\omega) = X(j\omega) * \delta(\omega) = \int_{-\infty}^{\infty} X(jv)\delta(\omega - v)\,\mathrm{d}v
\tag{5.25}
$$

whose weighting factors $X(jv)$ are conveniently provided by the Fourier integral. Each infinitesimal impulse, represented now in the time domain by a similarly scaled exponential, gives rise to a corresponding infinitesimal output. Their correspondence is expressed, according to the relationship 5.18 and its Fourier transform, as

$$
\begin{array}{ccc}
\left[\dfrac{1}{2\pi}X(jv)\,\mathrm{d}\omega\right]e^{jvt} & \xrightarrow{\;h(t)\;} & \left[\dfrac{1}{2\pi}X(jv)\,\mathrm{d}\omega\right]H(jv)e^{jvt} \\
\Big\updownarrow \mathcal{F} & & \Big\updownarrow \mathcal{F} \\
[X(jv)\,\mathrm{d}\omega]\delta(\omega - v) & \xrightarrow{\;H(j\omega)\;} & [X(jv)\,\mathrm{d}\omega]H(jv)\delta(\omega - v)
\end{array}
\tag{5.26}
$$

which is interpreted as in Figure 5.5.

The duality between expressions 5.24 and 5.26 shows that the two methods are not only equivalent, but that they are almost identical in their use of impulses and exponentials.

Note that the values $X(j\omega)$ and $Y(j\omega)$ are in general complex, so that the weighting factors in expression 5.26 must be interpreted as directed infinitesimal areas. When the signal is real, as in the case of $X(jv)$ of Figure 5.7, this area only carries amplitude information, as indicated in the upper left of Figure 5.9. But an arbitrary signal, such as $Y(jv)$ of Figure 5.7, also carries phase information, which determines the orientations of the differential area and of the equivalent impulse, as suggested in the lower right of Figure 5.9.

Amplitude modulation

The above duality is not absolute, in that the operation of convolution remains firmly associated with the time domain, while multiplication remains in the frequency domain. For complete symmetry these two operations would also need to be exchanged, which is precisely the case with **amplitude modulation**.

In this class of system, the time domain output is the product of two time functions, one called the carrier $f_c(t)$, the other the modulating function $x(t)$. According to the convolution property 2.36 of the Fourier transform, the frequency domain representation of the output is the result of convolving the corresponding transforms, that is,

$$
\begin{array}{ccc}
x(t) & \xrightarrow{\;f_c(t)\;} & y(t) = x(t)f_c(t) \\[4pt]
\Big\downarrow \mathscr{F} & & \Big\downarrow \mathscr{F} \\[6pt]
X(j\omega) & \xrightarrow{\;F_c(j\omega)\;} & Y(j\omega) = \dfrac{1}{2\pi}X(j\omega) * F_c(j\omega)
\end{array}
\qquad (5.27)
$$

If the carrier is a cosine function of frequency ω_c, then the frequency domain convolution replicates the baseband signal $X(j\omega)$ at the carrier

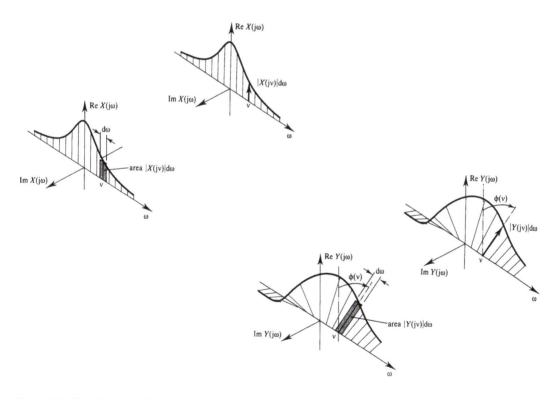

Figure 5.9   Signal as sum of impulses.

frequencies ω_c and $-\omega_c$. An example is shown in Figure 5.10, which is also an illustration of the dual relationship with Figure 5.7.

5.3.5 System response method

We now extend the arguments employed in deriving the frequency response method to the generalized frequency variable s. This involves formally similar expressions and leads to the system response method, or Laplace method, as a further alternative for solving systems.

Response to single generalized exponential

If the system characterized by $h(t)$ is probed with the elementary input signal $x(t) = e^{s_1 t}$, where s_1 is a complex value, the impulse response method yields the output

$$y(t) = e^{s_1 t} * h(t) = \int_{-\infty}^{\infty} e^{s_1(t-\tau)} h(\tau)\, d\tau = e^{s_1 t} \int_{-\infty}^{\infty} h(\tau) e^{-s_1 \tau}\, d\tau$$

The last integral is formally the bilateral Laplace transform of $h(t)$, evaluated for $s = s_1$, and leads to the time domain input/output relationship

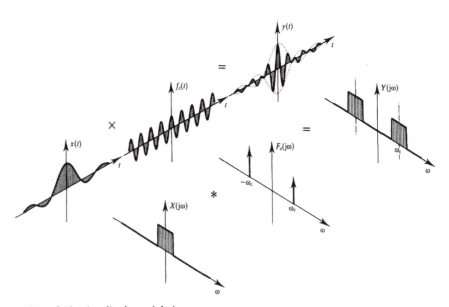

Figure 5.10 Amplitude modulation.

$$x(t) = e^{s_1 t} \quad \xrightarrow{h(t)} \quad y(t) = e^{s_1 t} H(s_1)$$ (5.28)

This result is formally similar to expression 5.12, but the complex scaling factor $H(s_1)$ now represents the value of the system function $H(s)$ at the s-plane location $s = s_1$. We see that the generalized exponential e^{st} is also an eigenfunction of the system, and $H(s)$ is its eigenvalue.

Graphical interpretation

An unconventional but expedient interpretation of the relationship 5.28 is given in Figure 5.11, which extends the frequency domain of Figure 5.5 to the s-plane. The frequency representations are presented mainly in terms of their magnitudes, but the concepts apply to phase and to real and imaginary parts as illustrated in the foreground of the frequency domain.

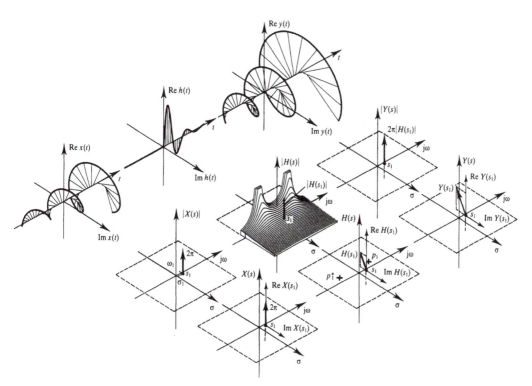

Figure 5.11   Response to generalized exponential $e^{s_1 t}$.

We showed in Figure 3.13 that the exponential $f(t) = e^{s_1 t}$ lacks a region of convergence and in the strict sense does not possess a bilateral Laplace transform. The Fourier interpretation (Equation 3.3) of the Laplace transform,

$$F(\sigma + j\omega) = \mathcal{F}\{f(t)e^{-\sigma t}\}$$

did not include modified functions $f(t)e^{-\sigma t}$ whose Fourier integral converged only in the limit. If we push this interpretation to include such functions, we can also interpret the Laplace transform on the boundary of the region of convergence.

In this broader sense, the bilateral Laplace transform of the input exponential $x(t) = e^{s_1 t}$ only exists on the line $\sigma_1 = \operatorname{Re} s_1$. Furthermore, it is zero-valued along that line, except at the point $s = s_1$, where it is an impulse of strength 2π. Indeed, calling $\operatorname{Im} s_1 = \omega_1$, on the line $\sigma_1 = \operatorname{Re} s_1$ the modified function becomes

$$f(t)e^{-\sigma_1 t} = e^{(\sigma_1 + j\omega_1)t}e^{-\sigma_1 t} = e^{j\omega_1 t}$$

whose Fourier transform exists in the limit and according to expression 2.34 is

$$e^{j\omega_1 t} \quad \overset{\mathcal{F}}{\longleftrightarrow} \quad 2\pi\delta(\omega - \omega_1)$$

so that

$$x(t) = e^{s_1 t} \quad \overset{(\mathcal{L})}{\longleftrightarrow} \quad X(s) = 2\pi\delta(s - s_1)$$

This is represented in Figure 5.11 by an increasing exponential ($\sigma_1 > 0$) in the time domain and by an impulse located on the s-plane of $X(s)$ at $s = s_1$. In other words, the function $X(s)$ is zero everywhere in the s-plane, except in the infinitesimal region surrounding $s = s_1$, where it is an impulse.

The Laplace representation of the output $Y(s) = X(s)H(s)$ is a magnitude-scaled and phase-shifted version of the impulse $X(s)$, where the scaling factor is the complex number $H(s_1)$. It is an impulse of strength $2\pi|H(s_1)|$ and phase equal to $\arg H(s_1)$.

The unilateral Laplace transform would give a more conventional interpretation, but the conciseness of the impulse notation would be lost.

Sum of generalized exponentials

Proceeding as in Section 5.3.3, an arbitrary input function $x(t)$ can now be interpreted by means of the inverse Laplace transform as an infinite sum of infinitesimal exponentials,

$$x(t) = \mathcal{L}^{-1}\{X(s)\} = \frac{1}{2\pi j}\int_{\sigma - j\infty}^{\sigma + j\infty} X(s)e^{st}\, ds \tag{5.29}$$

According to expression 5.28 an exponential of frequency $s = p$ and infinitesimal weighting factor $\frac{1}{2\pi j} X(p)\, ds$ causes a similarly weighted output

$$\left[\frac{1}{2\pi j} X(p)\, ds\right] e^{pt} \quad \xrightarrow{\;h(t)\;} \quad \left[\frac{1}{2\pi j} X(p) H(p)\, ds\right] e^{pt}$$

and the full input of Equation 5.39 gives the collective output

$$y(t) = \frac{1}{2\pi j} \int_{\sigma-j\infty}^{\sigma+j\infty} X(s) H(s) e^{st}\, ds \tag{5.30}$$

As in the Fourier case, the frequency representations lead to the **system response method** or **Laplace method**, expressed as the input/output relationship

$$\boxed{X(s) \quad \xrightarrow{\;H(s)\;} \quad Y(s) = X(s) H(s)} \tag{5.31}$$

This is linked by the Laplace transform to the impulse response method as

$$\boxed{\begin{array}{ccc}
x(t) = x(t) * \delta(t) & \xrightarrow{\;h(t)\;} & y(t) = x(t) * h(t) \\
\Big\downarrow \mathscr{L} & \Big\downarrow \mathscr{L} \quad \Big\uparrow \mathscr{L} & \\
X(s) = X(s) \cdot 1 & \xrightarrow{\;H(s)\;} & Y(s) = X(s) H(s)
\end{array}} \tag{5.32}$$

The relationships 5.32 are interpreted graphically in Figure 5.12, which is an extension to the s-plane of the Fourier interpretation of Figure 5.7.

5.4   Initial conditions and complete solution

The solutions provided by the methods developed so far are only complete when the system is initially at rest, this is, when the system's output and its derivatives are zero at the origin of time, when the input signal is applied.

In terms of physical elements, this implies that at time $t = 0$ none of the system's components are storing energy. Capacitors are discharged, no currents are flowing through inductors, masses are at rest and springs are untensioned. We will show in this section that these conditions are a prerequisite of the general purpose characterizations of Chapter 4.

In contrast, if no signal is applied to the input, but the system is

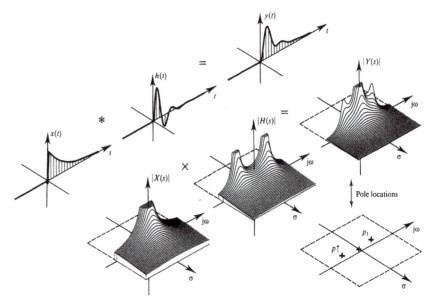

Figure 5.12   Equivalence between impulse response method and system response method.

still energized from an earlier excitation, the associated output signal decays naturally, by a law that is a characteristic of the system.

In linear time-invariant systems the two situations may coexist. If an input signal is applied while the output is still recovering from a previous excitation, the two effects carry on independently, and their effects are superimposed to give the complete response $y_c(t)$,

$$y_c(t) = y(t) + y_i(t)$$

where $y(t)$ represents the forced output due to the input function $x(t)$ and $y_i(t)$ is the contribution from the initial conditions.

Different approaches exist for determining $y_i(t)$ as a complement to the forced output $y(t)$, but the **unilateral Laplace transform method** stands out from these. It makes use of the transform properties of the derivative of a function to incorporate the system's initial conditions explicitly into the frequency domain expressions, thus providing a simple and clear formulation of the complete solution process.

5.4.1   Complete system description

The general Nth-order system was represented earlier by the Nth-order differential equation (4.56), which we repeat here with the notation of the complete output $y_c(t)$, using a second-order system for illustration,

$$a_2 y_c''(t) + a_1 y_c'(t) + a_0 y_c(t)$$
$$= x_f(t) = b_2 x''(t) + b_1 x'(t) + b_0 x(t) \quad (5.33)$$

The intermediate function $x_f(t)$ represents the **forcing function of the differential equation**, which effectively describes how the system assimilates the input signal $x(t)$ and its first N derivatives.

The full expression 5.33 is represented in Figure 5.13 by a block diagram of the **direct form**. In Chapter 4 we favoured the **monolithic form**, see Figure 4.21, because it was more compact and economical in its use of differentiators, but the direct form describes the general solution process more effectively. It not only gives a **direct description** of Equation 5.33, but also provides a place for the forcing function $x_f(t)$ and reserves specific places for initial conditions, as we presently show.

To finalize the system's description it is necessary to specify its initial conditions. Three questions arise regarding their nature, number

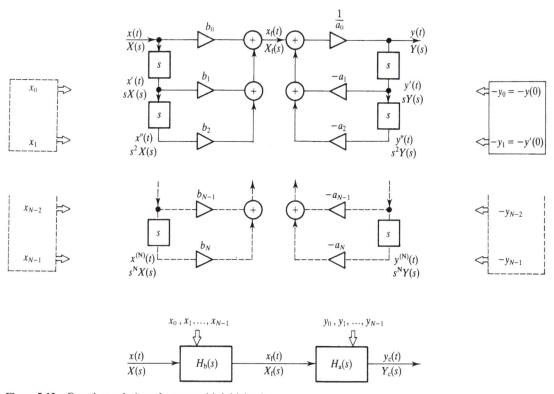

Figure 5.13 Complete solution of system with initial values.

diagram. We will see that these are automatically answered by the unilateral Laplace transform method.

Time domain descriptions of causal signals often have discontinuities at the origin, which involve the impulse function and its derivatives. To include such functions into the unilateral Laplace transform, it is necessary to use the definition 3.41, which uses $t = 0^-$ as its lower integration limit.

If the complete output signal is represented by the unilateral Laplace transform pair

$$y_c(t) \xleftrightarrow{\mathcal{L}_1} Y_c(s)$$

then, according to expression 3.38, the mth derivative of $y_c(t)$ is represented by the pair

$$y_c^{(m)}(t) \xleftrightarrow{\mathcal{L}_1} s^m Y_c(s) - s^{m-1} y_0 - s^{m-2} y_1 -$$
$$\cdots - s y_{m-2} - y_{m-1} \qquad (5.34)$$

The coefficients $y_0 = y_c(0)$, $y_1 = y_c'(0)$, $\ldots$, $y_{m-1} = y_c^{(m-1)}(0)$ are the values taken by the output signal and its first $m - 1$ derivatives at the time value $t = 0$, and represent the **initial values of the output signal**.

Taking the unilateral Laplace transform of the input signal $x(t)$ yields an expression formally similar to expression 5.34, with initial values $x_0 = x(0)$, etc. When the input signal is causal, all initial values are zero, leaving only the term $s^m X(s)$. Note that the Laplace definition 3.41 allows discontinuities at the origin, and includes their effects into $X(s)$.

5.4.2   Unilateral Laplace transform method

Assuming that $x(t)$ is causal, the unilateral Laplace transform of the individual terms of Equation 5.33 yields

$$a_2[s^2 Y_c(s) - s y_0 - y_1] + a_1[s Y_c(s) - y_0] + a_0 Y_c(s)$$
$$= X_f(s) = b_2 s^2 X(s) + b_1 s X(s) + b_0 X(s) \qquad (5.35)$$

which is now an algebraic expression. Gathering terms in $Y_c(s)$ we have

$$(a_2 s^2 + a_1 s + a_0) Y_c(s) - [(a_2 s + a_1) y_0 + a_2 y_1]$$
$$= (b_2 s^2 + b_1 s + b_0) X(s)$$

from which we obtain an expression for the output

$$Y_c(s) = \frac{b_2 s^2 + b_1 s + b_0}{a_2 s^2 + a_1 s + a_0} X(s) + \frac{(a_2 s + a_1) y_0 + a_2 y_1}{a_2 s^2 + a_1 s + a_0} \qquad (5.36)$$

This expression provides the complete solution of the system and neatly separates two responses. Indeed, since the coefficients a_n and b_n are fixed parameters of the system, the first term is fully determined by the input signal $X(s)$ and represents the **forced response** $Y(s)$, while the second term depends entirely on the system's initial conditions y_n and represents the system's **response to initial conditions** $Y_i(s)$. We can therefore write Equation 5.36 as

$$Y_c(s) = Y(s) + Y_i(s)$$

The second term answers the questions posed earlier regarding the system's initial conditions. Their nature is that of the output signal's initial values. From Equation 5.35 it follows that their number corresponds to that of the highest derivative appearing in Equation 5.33. For instance, the second-order system used for illustration has room for two initial conditions, $y_c(0^-)$ and $y_c'(0^-)$. Finally, regarding their rightful places, these are given explicitly in the numerator polynomial of Equation 5.36, and are indicated on the output side of the block diagram of Figure 5.13, as explained below.

It is this ability to explicitly and unambiguously incorporate initial **time-domain** conditions into the frequency domain expression, and hence into the frequency domain solution of the system, that gives the unilateral Laplace transform method its conceptual clarity and practical relevance.

Generalization

We generalize the result in Equation 5.36 for Nth-order systems by interpreting the system of Figure 5.13 as two cascaded subsystems $H_b(s)$ and $H_a(s)$, as seen earlier in the context of Figure 4.8. Their system functions are

$$H_a(s) = \frac{1}{\sum_{n=0}^{N} a_n s^n}$$

and

$$H_b(s) = \sum_{n=0}^{N} b_n s^n$$

whose product is the overall system function

$$H(s) = H_b(s) H_a(s) = \frac{\sum_{n=0}^{N} b_n s^n}{\sum_{n=0}^{N} a_n s^n}$$

The result 5.36 can thus be written compactly as

$$Y_c(s) = H(s) X(s) + H_a(s) Y_0(s) \tag{5.37}$$

The function $Y_0(s)$ is a generalization of the numerator of the second term of Equation 5.36. It can be written as a polynomial in s, whose coefficients are weighted sums of the initial conditions y_n, but for didactic purposes we write it explicitly as a sum of y_n as

$$Y_0(s) = \sum_{n=0}^{N-1} c_n(s) y_n \tag{5.38}$$

whose coefficients are the polynomials

$$c_n(s) = \sum_{k=n+1}^{N} a_k s^{k-(n+1)}$$

which only depend on the feedback loop coefficients a_n, see Figure 5.13, and effectively describe the paths of the initial values y_n back into the input of the subsystem $H_a(s)$.

Link to other methods

If the system is initially at rest, the initial value polynomial $Y_0(s)$ vanishes and the complete output reduces to

$$Y_c(s) = H(s)X(s)$$

which is identical to the bilateral Laplace transform result 5.31. We conclude that the system function definition 4.7 used in that result characterizes the system at rest, that is with all initial conditions set to zero,

$$H(s) = \frac{Y(s)}{X(s)} = \left.\frac{Y_c(s)}{X(s)}\right|_{y_n=0} \tag{5.39}$$

This also confirms the requirement that the system be initially at rest when deriving any of the other characterizations of Chapter 4.

Time domain representation

For a time domain representation of the complete output, the result 5.37 is inverted by taking the inverse unilateral Laplace transform, to give the general form

$$y_c(t) = y(t) + y_i(t) \tag{5.40}$$

The first term is the forced response, which would also have been obtained by any of the earlier methods of this chapter, to be summarized in Section 5.5. The second represents the time response due to initial conditions, and this emerges naturally from the unilateral Laplace transform method. Using any other method would require a separate evaluation and addition of this term. For instance, the impulse response method would yield

$$y_c(t) = x(t) * h(t) + y_i(t)$$

We illustrate the concepts involved by an example.

Example 5.3

Consider the second-order mechanical system of Figure 5.14, whose frequency response $H(j\omega)$ and impulse response $h(t)$ are characterized in terms of the displacement $x(t)$, see Exercise 4.3. The forcing function $f(t)$ is a cosinusoidal force applied at time $t = 0$,

$$f(t) = \cos \omega_c t \, u(t)$$

when the system is still recovering from an earlier excitation, represented by the initial displacement $x_0 = x(0)$ and initial velocity $x_1 = x'(0) = v(0)$.

The complete response $x_c(t)$ is the sum of two independent terms,

$$x_c(t) = x(t) + x_i(t)$$

where $x(t)$ represents the forced response due to the forcing function $f(t)$ and $x_i(t)$ represents the initial condition response which takes the shape of some right-sided segment of the impulse response $h(t)$. The forcing function and the outputs are shown in Figure 5.14.

Note that there are two transients involved, one represented by the decay of the initial conditions response, the other by the growth of the forced response. The latter builds up to a steady state response for $t \gg 0$, characterized by the magnitude amplification $A(\omega_c)$ and phase shift $\phi(\omega_c)$, by which time both transients have died down.

5.5   Relationship between methods

To develop this chapter we first derived the impulse response method, and used this to derive the two frequency domain methods, thereby revealing their interdependence and the role of transforms.

The equivalence between methods was expressed in the relationships 5.21 and 5.32 and visualized in Figures 5.7 and 5.12. We now combine these into a single graphical statement, shown at the top of Figure 5.15, which establishes a useful shorthand notation.

The three clusters represent the input signal, the system

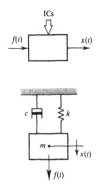

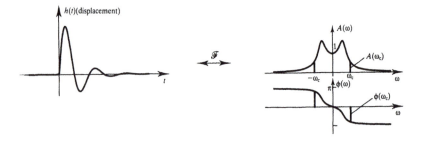

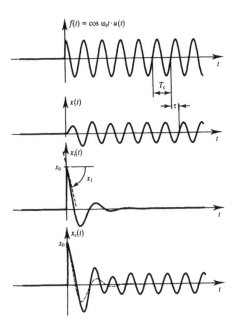

Figure 5.14 Complete response of system $x_c(t) = x(t) + x_i(t)$.

characterization and the output signal, in that order, while each cluster relates alternative representations in three domains by appropriate transforms. In addition, the central cluster, which represents the system, also indicates the operation applicable to the solution method associated with each domain.

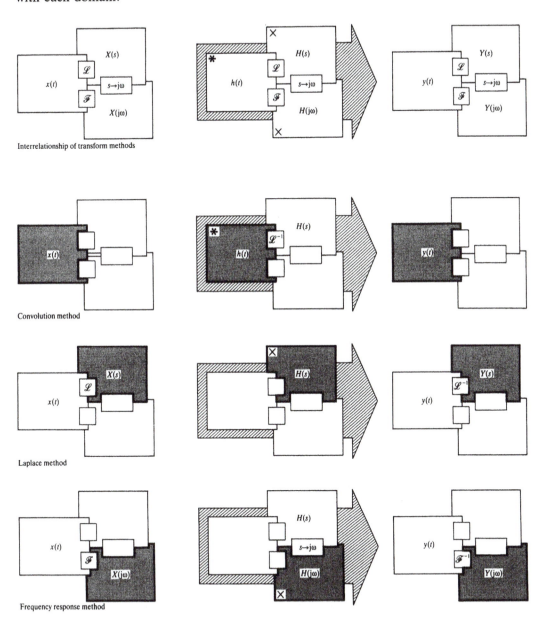

Figure 5.15   Three equivalent methods of obtaining $y(t)$, when $x(t)$ and $H(s)$ are known.

The remainder of Figure 5.15 gives an example that highlights three different paths for solving the same problem. In each case the input signal is available as the time function $x(t)$, the system is given in the form of a Laplacian system function $H(s)$ and the output $y(t)$ is also required in time.

To proceed in the time domain, we first obtain the impulse response $h(t)$ from $H(s)$, as indicated by the inverse Laplace transform, and then apply the convolution method. To use the Laplace method the input $x(t)$ must be transformed to the frequency domain, and the resulting output $Y(s)$ converted back to time. Similarly for the frequency response method, where $H(j\omega)$ is obtained from $H(s)$ by the substitution $s = j\omega$.

Transform techniques thus offer the freedom to choose the path along which the signal is to be processed. The actual choice depends, to a great extent, on the form in which the signals are specified, on the character of the system and on the available processing tools.

Exercises

5.1 Explain the purpose of interpreting an arbitrary input signal $x(t)$ as a convolution of itself with the impulse function $\delta(t)$, and the relevance of such interpretation to the impulse response method.

5.2 Contrast the two graphical interpretations of convolution presented in this text by applying the system interpretation of Figure 5.4 to the functions used in the more analytical interpretation of Figure 1.4.

5.3 The impulse response of the ideal integrator is a unit step $h(t) = u(t)$. Using the impulse response method, find the output $y(t)$ of the integrator when the input is: (a) another unit step $x(t) = u(t)$ and (b) a unit ramp $x(t) = t\,u(t)$. Verify results by the system response method.

5.4 With the aid of Figure 5.5 discuss the time domain and frequency domain implications of passing an exponential $x(t) = e^{j\omega_1 t}$ through a system $H(j\omega)$.

5.5 A system is described by the system function

$$H(s) = \frac{2}{(s^2 + 2s + 2)}.$$

What is its time response $y(t)$ to the following (acausal) inputs

$$x_1(t) = e^{j2t} \quad x_2(t) = e^{(0.2+j)t}$$
$$x_3(t) = \cos 2t \quad x_4(t) = \sin(-3t)$$

Sketch the inputs and outputs, comment on magnitude and phase changes and evaluate all inputs and outputs for the instant $t = 1$ s.

5.6 A ramp function $x(t) = 2t\, u(t)$ is applied to a system characterized by an impulse response $h(t) = e^{-3t}\, u(t)$. Using the impulse response method find the system output $y(t)$. Verify the result by the Laplace method.

5.7 A series RC network has a frequency response

$$H(j\omega) = \frac{1}{1 + j\omega RC}$$

Using the frequency response method find the output $y(t)$ when the input is an exponential $x(t) = e^{-at}\, u(t)$, with $a = 1/RC$.

5.8 Use the Laplace method to find the output $y(t)$ from the system

$$H(s) = \frac{6}{s^2 + 5s + 6}$$

when the input is the exponential $x(t) = e^{-4t}\, u(t)$.

5.9 With the aid of Figures 5.5, 5.7 and 5.8 discuss the equivalence between the impulse response method and the frequency response method. Comment on their formal differences.

5.10 Find the complete output $y_c(t)$ of a system described by the differential equation

$$y_c''(t) + 3y_c'(t) + 2y_c(t) = e^{-3t} u(t)$$

with initial conditions $y_c(0) = 1$ and $y_c'(0) = -1$. Separate the result into forced and initial condition components.

5.11 Use the unilateral Laplace method to give the complete solution of the differential equation

$$f'(t) + a\, f(t) = e^{-\alpha t} u(t)$$

for an arbitrary initial condition $f(0) = f_0$.

5.12 Consider the complete response of a system with initial conditions to show that a characterization at system level (such as the responses derived in Chapter 4) requires the system to be initially at rest.

5.13 The step response method (not developed in the text) can be expressed succinctly by the input/output relationship

$$x(t) = x'(t) * u(t) \xrightarrow{\ r(t)\ } y(t) = x'(t) * r(t)$$

which effectively interprets the input $x(t)$ as a superposition of scaled and shifted step functions and the output $y(t)$ as a superposition of similarly scaled and shifted step responses $r(t) = \int h(t)\mathrm{d}t$. Show in the Laplace domain that this is consistent with the impulse response method.

PART 2

Analysis of Discrete-time Signals and Systems

We now extend the continuous-time transforms, system descriptions and solution methods of Part 1 to discrete-time signals and systems. As far as practicable we follow the same layout and use the same or related examples, to stress similarities and equivalences.

We first derive the applicable transforms by interpreting discrete-time signals with the aid of equivalent sampled continuous-time functions and applying the continuous Fourier or Laplace transform, as appropriate. In Chapter 6 we also exploit transform dualities to derive all the discrete forms of the Fourier transform in a consistent manner.

In Chapter 7 we make use of the formal similarities between Fourier and Laplace transforms to derive the discrete-time Laplace transform and then express the result with the notation of the z-transform. The discrete-time Fourier transform is then reinterpreted as a subset of both the discrete-time Laplace transform and the z-transform. This process unifies the derivations of all the time domain to frequency domain transforms, relates them and stresses their common ancestry in the Laplace transform.

Chapters 8 and 9 are presented as discrete-time counterparts of Chapters 4 and 5, whose concepts, terminology and methods are extended with only minor modifications. The aim is to emphasize similarities to be exploited in the discrete-time simulations presented in Chapter 11.

To unify the treatment of all function classes, we develop a consistent notation reflecting the class and domain of the function and conveying the ancestry of the transform. The full notation is found in the chart of Figure 7.17 in the context of signal classes. The elaborate symbols required for presentation are later discarded in routine applications.

CHAPTER 6

Discrete Classes of the Fourier Transform

In Chapter 1 we introduced the Fourier Series as a starting point for developing the Fourier transform of Chapter 2. We will now employ the latter to derive expressions for the discrete classes of the Fourier transform, applicable to signals that are discrete either in time or in frequency, or in both domains.

Historically, the different Fourier transform classes evolved via different disciplines, leaving a confusing array of descriptors in the literature. In the remainder of this text we reserve the descriptor **Fourier transform** $\mathcal{F}$ for the most general class, where both domains are of continuous variable. Other classes are qualified according to the discreteness of the appropriate domain, as **discrete-time Fourier transform** $\mathcal{F}_{dt}$, **discrete-frequency Fourier transform** $\mathcal{F}_{df}$ (usually associated with the Fourier series) and the **discrete Fourier transform** $\mathcal{F}_d$, which is discrete in both domains. The regions of validity of the four classes of transform are shown in Figure 6.1, superimposed on the classes of functions to which they apply.

To derive the expressions of each discrete class of the Fourier transform, we first construct a continuous-variable sampled function, apply the appropriate Fourier transform and then replace the continuous samples by equivalent discrete samples. By using the same procedure for each class we can exploit the time–frequency dualities of the Fourier transform, thus leading to a unified and consistent set of expressions for the various transforms and clear interrelationships.

A systematic notation is needed for development. We use the subscripts s and d to identify functions of a sampled variable and of a discrete variable, and the overscript ˘ to indicate that a function is periodic. Parentheses ( ) enclose a continuous variable, square brackets [ ] a discrete variable. The complete Fourier set is summarized in Figure 6.22.

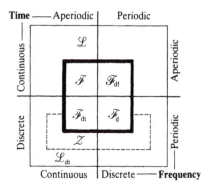

Figure 6.1 Discrete classes of the Fourier transform.

6.1 Sampling and periodicity

In this section we examine the operation of sampling a function. The analytical instruments involved are the ideal time-sampler $s_T(t)$ and the ideal frequency-sampler $S_\omega(j\omega)$, both of which are essentially impulse trains.

We first show that the frequency domain representations of a time domain impulse train $\delta_T(t)$ is another impulse train, that is,

$$\delta_T(t) = \sum_{n=-\infty}^{\infty} \delta(t - nT) \quad \overset{\mathscr{F}}{\longleftrightarrow} \quad \omega_s \sum_{k=-\infty}^{\infty} \delta(\omega - k\omega_s) \tag{6.1}$$

where each infinite sum represents a superposition of shifted impulses spaced at equal intervals T and $\omega_s = 2\pi/T$, respectively.

Each domain of this function can be interpreted from two viewpoints: either as an essentially discrete function, or as a periodic function. These concurrent attributes come into play when sampling one domain causes the other to become periodic.

We then consider amplitude sampling and area sampling, and define the sampling operators used in the remainder of this chapter, namely the time-sampler $s_T(t)$ and the frequency-sampler $S_\omega(j\omega)$.

Samplers provide the analytical tools for bridging the formal gap between sampled-variable and discrete-variable functions. The conversion equivalence is considered in Section 6.1.3, with the criterion of leaving intact the alternative domain representation of the function. The consequences of applying these operators, and the conditions necessary to preserve information integrity, will be considered in Section 6.6.1, in the context of the sampling theorem.

6.1.1   Fourier transform of an impulse train

We now show that the Fourier transform of a time domain impulse train is a similar frequency domain impulse train. According to expression 2.10 a time-shifted impulse transforms to a frequency domain exponential

$$\delta(t - nT) \quad \overset{\mathcal{F}}{\longleftrightarrow} \quad e^{-j\omega nT}$$

so that an infinite sum of such impulses gives a sum of exponentials,

$$\delta_T(t) = \sum_{n=-\infty}^{\infty} \delta(t - nT) \quad \overset{\mathcal{F}}{\longleftrightarrow} \quad \sum_{n=-\infty}^{\infty} e^{-j\omega nT} \tag{6.2}$$

This result, though useful, does not show explicitly that the frequency representation is another impulse train.

Alternative time domain description and its transform

To reveal the remarkable symmetries of this function we note that $\delta_T(t)$ is a periodic function of time, with period T. As such it can be expressed as a Fourier series (Equation 1.18)

$$\tilde{f}(t) = \sum_{k=-\infty}^{\infty} c_k e^{jk\omega_s t} \tag{6.3}$$

whose coefficients c_k are expressed by the integral of Equation 1.19 as

$$c_k = \frac{1}{T} \int_{-T/2}^{T/2} \tilde{f}(t) e^{-jk\omega_s t} \, dt$$

$$= \frac{1}{T} \int_{-T/2}^{T/2} \sum_{n=-\infty}^{\infty} \delta(t - nT) e^{-jk\omega_s t} \, dt$$

Each shifted impulse $\delta(t - nT)$ is zero-valued everywhere, except in the infinitesimal region surrounding the point $t = nT$. Of the sum on the right only the impulse at the origin possesses a non-zero region that falls within the integration interval $(-\frac{1}{2}T, \frac{1}{2}T)$. All the products $\delta(t - nT) e^{-jk\omega_s t}$, except that for $n = 0$, vanish within that interval leaving

$$c_k = \frac{1}{T} \int_{-T/2}^{T/2} \delta(t) e^{-jk\omega_s t} \, dt$$

Since $\delta(t)$ is zero-valued outside the integration interval, the limits can be extended to infinity, so that

$$c_k = \frac{1}{T} \int_{-\infty}^{\infty} \delta(t) e^{-jk\omega_s t} = \frac{1}{T} e^0 = \frac{1}{T}$$

where we used the sifting property 1.10 of the impulse function.

Inserted in Equation 6.3 this result leads to an alternative time domain description of the impulse train as a sum of exponentials. According to Equation 2.11 each of these exponentials transforms to a frequency domain impulse of strength 2π. With $2\pi/T = \omega_s$ this leads to the pair

$$\delta_T(t) = \frac{1}{T} \sum_{k=-\infty}^{\infty} e^{jk\omega_s t} \quad \xleftrightarrow{\mathcal{F}} \quad \omega_s \sum_{k=-\infty}^{\infty} \delta(\omega - k\omega_s) \qquad (6.4)$$

We combine this result with expression 6.2 to give the symmetrical expressions

$$\tilde{f}(t) = \delta_T(t) = \sum_{n=-\infty}^{\infty} \delta(t - nT) \quad \xleftrightarrow{\mathcal{F}} \quad \tilde{F}(j\omega) = \sum_{n=-\infty}^{\infty} e^{-j\omega nT} \qquad (6.5)$$

$$\Bigg\downarrow \quad \begin{array}{c} \text{time dom. FS} \\ c_k = 1/T \end{array} \qquad \qquad \Bigg\uparrow \quad \begin{array}{c} \text{freq. dom. FS} \\ c_n = 1 \end{array}$$

$$\tilde{f}(t) = \delta_T(t) = \frac{1}{T} \sum_{k=-\infty}^{\infty} e^{jk\omega_s t} \quad \xleftrightarrow{\mathcal{F}} \quad \tilde{F}(j\omega) = \omega_s \sum_{k=-\infty}^{\infty} \delta(\omega - k\omega_s)$$

Each domain of this function can thus be interpreted as either an impulse train, when considering discreteness, or as a sum of exponentials, when stressing periodicity.

Frequency domain equivalence

To complete the symmetry or duality of expressions 6.5, we also relate the two frequency domain descriptions by means of a Fourier series. Noting that this impulse train too is periodic, with period ω_s, we can express it as a sum of frequency domain exponentials

$$\tilde{F}(j\omega) = \sum_{n=-\infty}^{\infty} c_n e^{j\omega nT}$$

which represents a **frequency domain Fourier series**, with coefficients

$$c_n = \frac{1}{\omega_s} \int_{-\omega_s/2}^{\omega_s/2} \tilde{F}(j\omega) e^{-j\omega nT} \, d\omega$$

$$= \frac{1}{\omega_s} \int_{-\omega_s/2}^{\omega_s/2} \omega_s \sum_{k=-\infty}^{\infty} \delta(\omega - k\omega_s) e^{-j\omega nT} \, d\omega = 1$$

derived by the same arguments as c_k. Symmetry in n permits writing the infinite sum as

$$\widetilde{F}(j\omega) = \sum_{n=-\infty}^{\infty} e^{j\omega nT} = \sum_{n=-\infty}^{\infty} e^{-j\omega nT}$$

which completes the schematic relationship 6.5.

Graphical visualization

The duality inherent in expression 6.5 is demonstrated in Figure 6.2, which shows the process by which the individual frequency domain impulses of expression 6.4 contribute to build up periodic time domain pulses, which, in the limit, become an impulse train. To stress that both variables are continuous the impulses were simulated by narrow pulses.

Arranging the time domain exponentials in conjugate pairs and using Euler's cosine formula, we write their sum as

$$\sum_{k=-\infty}^{\infty} e^{jk\omega_s t} = 1 + 2 \sum_{k=1}^{\infty} \cos k\omega_s t \tag{6.6}$$

whose terms represent the functions of the ωt-plane of Figure 6.2, where the cosines at the fundamental frequencies $\omega = \pm\omega_s$ determine the time domain period T.

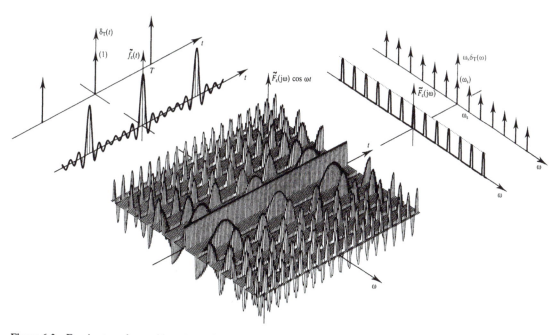

Figure 6.2   Fourier transform of impulse train.

The progressive build-up of the function 6.6 is further illustrated in Figure 6.3. It shows the process by which each additional component of

$$f_N(t) = \sum_{k=-N}^{N} e^{jk\omega_s t} = 1 + 2\sum_{k=1}^{N} \cos k\omega_s t$$

not only adds height to each pulse, but also has a narrowing effect on its base, thus concentrating the power of the signal in the vicinity of the pulses. As the number of terms N grows to infinity, in the limit the pulses grow into another impulse train.

6.1.2 Samplers

The two domains of the impulse train of expression 6.1 form the basis of a variety of samplers.

Amplitude sampling and area sampling

Multiplying a function $f(t)$ by the impulse train $\delta_T(t)$

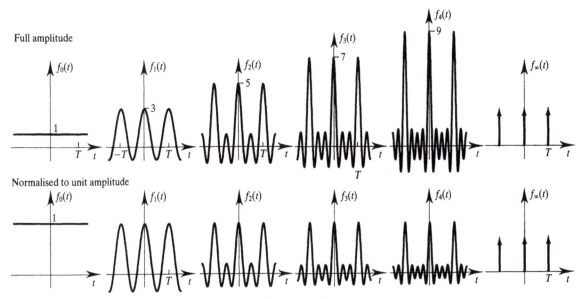

Figure 6.3   Build-up of impulse train.

$$\hat{f}(t) = f(t)\delta_T(t) = \sum_{n=-\infty}^{\infty} f(t)\delta(t - nT)$$

can be viewed as an amplitude modulation process, in which the impulse train takes the role of carrier, while $f(t)$ is the modulating function. With the impulse equivalence

$$f(t)\delta(t - nT) = f(nT)\delta(t - nT)$$

we write

$$\hat{f}(t) = \sum_{n=-\infty}^{\infty} f(nT)\delta(t - nT) \qquad (6.7)$$

The strength of each shifted impulse $\delta(t - nT)$ is determined by the value of $f(t)$ at the impulse location $t = nT$. For real signals this represents an **amplitude-sampling** operation.

If, on the other hand, we wish each sample to reflect the area under the signal within one sampling interval, then the impulse train needs to be scaled. Over a given time interval, the total area of the samples is approximately the same as that of the original function, regardless of the sampling interval used, provided the function is reasonably smooth.

We will use **area-samplers** in the developments of this chapter.

Time-samplers

The impulse train of expression 6.1 represents the time domain amplitude-sampler. Applied to a function $f(t)$ it produces the amplitude-sampled function $\hat{f}(t)$ expressed in Equation 6.7.

Scaling both domains of expression 6.1 with the value of the sampling interval T yields the area-preserving time-sampler $s_T(t)$

$$s_T(t) = T \sum_{n=-\infty}^{\infty} \delta(t - nT) \quad \overset{\mathcal{F}}{\longleftrightarrow} \quad S_T(j\omega) = 2\pi \sum_{k=-\infty}^{\infty} \delta(\omega - k\omega_s)$$

$$(6.8)$$

We illustrate its desirable properties with the aid of Figure 6.4, where we sample the function $f(t) = \cos \omega_c t$ with two different sampling intervals. In the upper half of the figure, multiplication with $s_T(t)$ yields a scaled impulse train,

$$f_s(t) = f(t)s_T(t) = T \sum_{n=-\infty}^{\infty} \cos \omega_c nT \delta(t - nT)$$

The preservation of area can be verified by integrating $f(t)$ and $f_s(t)$ over the same time interval, for instance one of the sampling intervals T indicated in the figure.

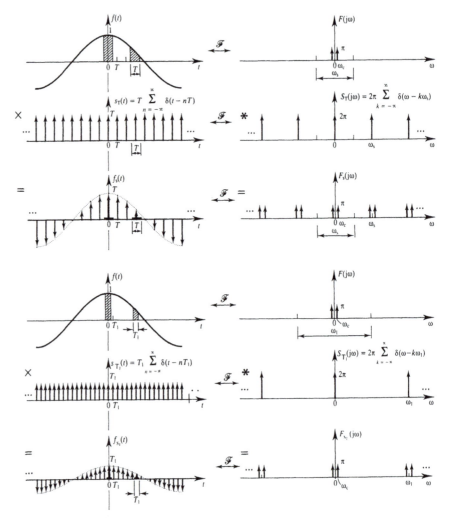

Figure 6.4 Sampling with ideal time-samplers.

According to expression 2.36 the transform $F_s(j\omega)$ is the result of the frequency domain convolution

$$F_s(j\omega) = \frac{1}{2\pi} F(j\omega) * S_T(j\omega)$$

which replicates, without scaling, the shape of the function $F(j\omega)$ at integer multiples of the sampling frequency ω_s. For a suitably band-limited function, in this case a pair of impulses, the original function can be fully recovered by filtering out all frequency components located outside the interval $\pm\frac{1}{2}\omega_s$.

Using a different sampling interval, for instance $T_1 = \frac{1}{2}T$ shown in the lower half of Figure 6.4, produces smaller but more closely spaced samples with the same area-preserving and frequency-replicating properties.

Frequency-samplers

Frequency-sampling is the dual counterpart of time-sampling. It involves a frequency domain multiplication with a function of the same form as expression 6.1, but with a **frequency-sampling interval** ω_0 and a corresponding **time domain period** $T_0 = 2\pi/\omega_0$. We describe this train as the frequency-sampler $S_\omega(j\omega)$

$$s_\omega(t) = \sum_{n=-\infty}^{\infty} \delta(t - nT_0) \quad \overset{\mathcal{F}}{\longleftrightarrow} \quad S_\omega(j\omega) = \omega_0 \sum_{k=-\infty}^{\infty} \delta(\omega - k\omega_0)$$

$$(6.9)$$

whose frequency domain is already suitably scaled by ω_0 for area-sampling. Its properties are the dual counterparts of those of the time-sampler.

A corresponding amplitude-sampler would result from dividing both sides of expression 6.9 by the frequency-sampling interval ω_0.

6.1.3   Equivalence of sampled and discrete functions

An ideal sampler has the effect of concentrating the distributed area, or the energy of a continuous function into packets located at discrete points of the sampled variable. Sampled functions are therefore essentially discrete.

To make these 'discrete' functions amenable to analytical operations, to integration in particular, it becomes necessary to represent them by functions of a continuous variable. The **continuous-variable samplers** $s_T(t)$ and $S_\omega(j\omega)$ provide the necessary links. Although essentially discrete, their component impulses are associated with special rules that make them behave as if they were of continuous variable. Thus endowed with continuity, such 'discrete' functions can be integrated, and this opens the doors to Fourier transformation and to convolution.

Once analytical manipulations in the sampled domain are completed, the resulting function can be converted to an equivalent function of discrete variable. The equivalence criterion is that the alternative domain representations of both the continuous and the discrete functions are identical.

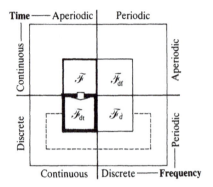

Figure 6.5   Discrete-time Fourier transform from continuous $\mathscr{F}$.

6.2   Discrete time

The ideal samplers of the preceding section provide the necessary analytical tools for developing the discrete classes of the Fourier transform. We start with the class associated with discrete-time signals, see Figure 6.5, which is most naturally associated with sampling and discreteness.

Our strategy for deriving the expressions of the discrete-time Fourier transform is to relate values $f_d[mT]$ of the discrete-time function to equivalent continuous-time samples $f(mT)$ of an auxiliary function $f(t)$, thus legitimizing the use of the continuous Fourier transform on the resulting sampled-time function $f_s(t)$.

The inverse transform is subsequently derived with a similar conversion of variables. For the initial derivation we use a suitably band-limited auxiliary function, showing later that the results are generally valid.

The section concludes by finding expressions of the discrete-time frequency sampler, in readiness for the derivation of the discrete Fourier transform in Section 6.4.

Most properties and examples of the discrete-time Fourier transform are relegated to Chapter 7, they are more revealing in the context of the z-transform.

6.2.1   Forward transform

Consider a continuous and aperiodic auxiliary function $f(t)$, whose frequency domain $F(j\omega)$ is limited to a band $(-L, L)$, as indicated at

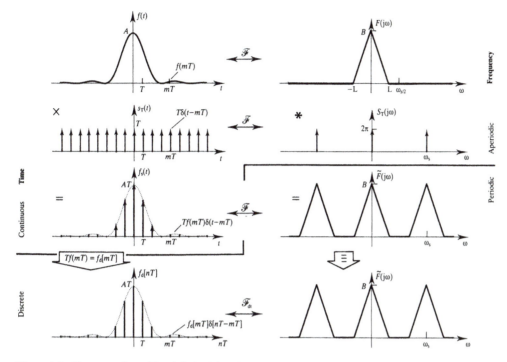

Figure 6.6 Time-sampling of band-limited function.

the top of Figure 6.6. The latter restriction, introduced to avoid aliasing problems during initial derivations, will be removed in Section 6.2.3.

We now sample $f(t)$ with the time-sampler $s_T(t)$ described by expression 6.8, and generate the sampled-time function

$$f_s(t) = f(t)s_T(t) = T \sum_{m=-\infty}^{\infty} f(t)\delta(t - mT) \tag{6.10}$$

as indicated in Figure 6.6. In the infinitesimal neighbourhood of a point $t = mT$, that is, the region in which the associated impulse $\delta(t - mT)$ takes non-zero values, the value of $f(t)$ can be considered constant, so that we can write $f(t)\delta(t - mT) = f(mT)\delta(t - mT)$, and

$$f_s(t) = T \sum_{m=-\infty}^{\infty} f(mT)\delta(t - mT) \tag{6.11}$$

This describes the sampled function $f_s(t)$ as a sum of shifted impulses, as indicated in Figure 6.6

The frequency representation of the sampled-time function $f_s(t)$ is the periodic function $\tilde{F}(j\omega)$, a result of the frequency domain convolution of the original shape $F(j\omega)$ with the impulse train $S_T(j\omega)$. This too is indicated in Figure 6.6 and will be examined later in this section.

But for our immediate purposes we take the Fourier transform of the sum 6.11, treating the factor $Tf(mT)$ associated with each impulse $\delta(t - mT)$ as a constant scaling coefficient, so that

$$Tf(mT)\delta(t - mT) \quad \overset{\mathcal{F}}{\longleftrightarrow} \quad Tf(mT)e^{-j\omega mT}$$

and

$$f_s(t) = \sum_{m=-\infty}^{\infty} Tf(mT)\delta(t - mT)$$

$$\overset{\mathcal{F}}{\longleftrightarrow} \quad \tilde{F}(j\omega) = \sum_{m=-\infty}^{\infty} Tf(mT)e^{-j\omega mT} \tag{6.12}$$

All the functions encountered so far were of continuous time and continuous frequency. It was therefore legitimate to employ the continuous Fourier transform $\mathcal{F}$.

We illustrate the implications of sampling with the graphics developed in Section 2.2 for the forward Fourier transform, using a narrow pulse function $\hat{f}_s(t)$ to simulate the impulses of $f_s(t)$. These pulses modulate the cosine surface of Figure 2.8 as shown in the upper half of Figure 6.7.

The sample at the time origin $nT = 0$ picks out the ridge of the cosine surface while the two samples at $nT = T$ and $nT = -T$ pick out the cosine of period ω_s representing the sampling frequency. Other samples pick out harmonically related cosines, which add constructively at integer multiples of ω_s and in a selectively destructive fashion elsewhere, thus building up the triangular shapes of the frequency domain. The background of both domains shows the ideal case of infinitely narrow pulses.

Discrete-time equivalence

The relationship 6.12 only uses values of $f(t)$ located at integer multiples of the sampling period T and, for the purposes of Fourier transformation, the function $f(t)$ may as well not exist elsewhere. It can be replaced by an equivalent discrete-time function $f_d[nT]$.

The equivalence between those functions is achieved by replacing each continuous-time impulse $Tf(mT)\delta(t - mT)$ with an equivalent discrete-time impulse $f_d[mT]\delta[nT - mT]$, as indicated in Figure 6.6. The equivalence criterion is that both impulses shall have the same frequency domain representation, that is,

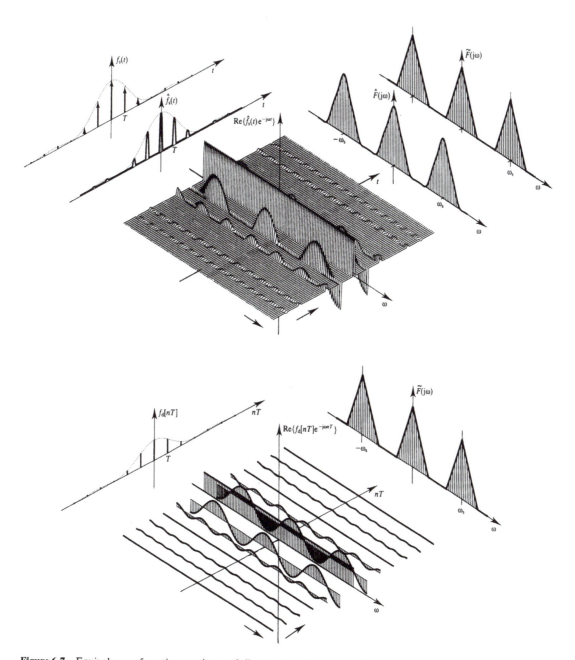

Figure 6.7 Equivalence of continuous-time and discrete-time Fourier transforms.

$$Tf(mT)\delta(t - mT) \quad \overset{\mathcal{F}}{\longleftrightarrow} \quad Tf(mT)e^{-j\omega mT}$$

$$\downarrow \equiv \qquad\qquad\qquad\qquad \downarrow \equiv \qquad\qquad (6.13)$$

$$f_d[mT]\delta[nT - mT] \quad \overset{\mathcal{F}_{dt}}{\longleftrightarrow} \quad f_d[mT]e^{-j\omega mT}$$

Analytically, this identity requires that the amplitudes of the frequency domain exponentials are the same, as expressed by the conversion law

$$\boxed{Tf(mT) = f_d[mT]} \qquad\qquad (6.14)$$

This expression must not be interpreted as an equality in the analytical sense. It provides a **numerical conversion** between two essentially different functions. Apart from the obvious scaling factor T, one side is defined for all values of time, whereas the other does not exist between adjacent samples. We thus convert all the terms of the sum 6.12, to yield

$$f_s(t) = \sum_{m=-\infty}^{\infty} Tf(mT)\delta(t - mT) \quad \overset{\mathcal{F}}{\longleftrightarrow} \quad \widetilde{F}(j\omega) = \sum_{m=-\infty}^{\infty} Tf(mT)e^{-j\omega mT}$$

$$\updownarrow \qquad\qquad Tf(mT) = f_d[mT] \qquad\qquad \updownarrow \qquad (6.15)$$

$$f_d[nT] = \sum_{m=-\infty}^{\infty} f_d[mT]\delta[nT - mT] \quad \overset{\mathcal{F}_{dt}}{\longleftrightarrow} \quad \widetilde{F}(j\omega) = \sum_{m=-\infty}^{\infty} f_d[mT]e^{-j\omega mT}$$

This interprets the frequency domain function $\widetilde{F}(j\omega)$ as a sum of either continuous-time impulses or of equivalent discrete-time impulses, and the strengths of corresponding impulses are also related by the law 6.14.

The equivalence 6.15 is fully illustrated in Figure 6.7. The upper half shows the sampled-time version, which makes use of the continuous exponential surface, as explained earlier. The lower half of the figure gives the equivalent discrete-time version, which only requires discrete slices of that exponential surface.

Formally, the sum representing $f_d[nT]$ is a discrete-time convolution (Equation 1.6),

$$f_d[nT] = f_d[nT] * \delta[nT] \qquad\qquad (6.16)$$

that interprets $f_d[nT]$ as a convolution of itself with the impulse function $\delta[nT]$.

The associated frequency representation can be taken as the defining expression for the **forward discrete-time Fourier transform**,

$$\boxed{\widetilde{F}(j\omega) = \sum_{n=-\infty}^{\infty} f_d[nT]e^{-j\omega nT}} \qquad\qquad (6.17)$$

where we changed the dummy summation variable to n.

This expression gives the frequency representation $\widetilde{F}(j\omega)$ of a discrete-time sequence $f_d[nT]$. We denote this transform by the symbol $\mathcal{F}_{dt}$ and by the pair

$$f_d[nT] \xleftrightarrow{\mathcal{F}_{dt}} \widetilde{F}(j\omega) \tag{6.18}$$

Note that the period $\omega_s = 2\pi/T$ of $\widetilde{F}(j\omega)$ is determined entirely by the sampler and does not depend on the function $f(t)$ being sampled. In contrast, the shape of $\widetilde{F}(j\omega)$ depends on both the function and the sampler.

Example 6.1

We find the transform of a causal discrete-time exponential $f[nT] = a^n u[nT]$, $|a| < 1$, for the cases $a > 0$ and $a < 0$. Applying Equation 6.17 gives

$$\widetilde{F}(j\omega) = \sum_{n=-\infty}^{\infty} a^n u[nT] e^{-j\omega nT} = \sum_{n=0}^{\infty} (ae^{-j\omega T})^n$$

This represents a geometric series of the form

$$\sum_{n=0}^{\infty} x^n = \frac{1}{1-x} \qquad |x| < 1$$

With $x = ae^{-j\omega T}$ this gives the discrete-time Fourier pair

$$a^n u[nT] \xleftrightarrow{\mathcal{F}_{dt}} \frac{1}{1 - ae^{-j\omega T}} \qquad |a| < 1 \tag{6.19}$$

whose validity condition takes into account the unit magnitude of $e^{-j\omega T}$. The two domains of the pair are illustrated in the upper half of Figure 6.8 for $a > 0$. The exponential character of $f[nT]$ is clearly recognized when the constant a is written as $a = e^{\alpha T}$, a real exponential of rate α, and expression 6.19 is written equivalently as

$$e^{\alpha nT} u[nT] \xleftrightarrow{\mathcal{F}_{dt}} \frac{1}{1 - e^{\alpha T} e^{-j\omega T}} \qquad \alpha < 0$$

Since $e^{\alpha T}$ is a positive constant, the case $a < 0$ implies $a = -e^{\alpha T}$ and the pair

$$(-e^{\alpha T})^n u[nT] \xleftrightarrow{\mathcal{F}_{dt}} \frac{1}{1 + e^{\alpha T} e^{-j\omega T}} \qquad \alpha < 0$$

illustrated in the lower half of Figure 6.8. Note that the frequency representation is a shifted version of the case $a > 0$.

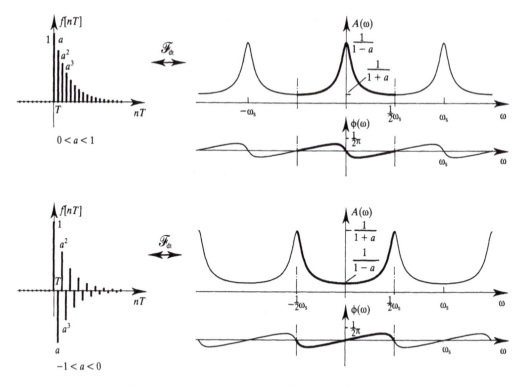

Figure 6.8   Transform of the function $f[nT] = a^n u[nT]$.

Alternative frequency domain description

It is often helpful to relate the frequency representation $\widetilde{F}(j\omega)$ to the transform $F(j\omega)$ of the original function $f(t)$. According to the convolution property 2.36 of the Fourier transform, the time domain multiplication 6.10 finds its frequency domain equivalent in a convolution of the transforms,

$$f_s(t) = f(t)s_T(t) \quad \overset{\mathscr{F}}{\longleftrightarrow} \quad \widetilde{F}(j\omega) = \frac{1}{2\pi} F(j\omega) * S_T(j\omega)$$

so that

$$\widetilde{F}(j\omega) = \frac{1}{2\pi} F(j\omega) * 2\pi \sum_{l=-\infty}^{\infty} \delta(\omega - l\omega_s)$$

$$= \sum_{l=-\infty}^{\infty} F(j\omega) * \delta(\omega - l\omega_s)$$

Each term of this sum replicates the function $F(j\omega)$ in correspondence with the shifted impulse $\delta(\omega - l\omega_s)$, that is,

$$F(j\omega) * \delta(\omega - l\omega_s) = F(j\omega - jl\omega_s)$$

and the sum becomes

$$\widetilde{F}(j\omega) = \sum_{l=-\infty}^{\infty} F(j\omega - jl\omega_s) \qquad (6.20)$$

If the original function is band-limited, so that $F(j\omega) = 0$ for $|\omega| > \frac{1}{2}\omega_s$, the terms of the sum 6.20 do not overlap, and the shifted replicas do not interfere with each other. Such were the cases illustrated in Figures 6.6 and 6.4. Otherwise a phenomenon known as aliasing occurs, which will be considered in Section 6.2.3, but was already experienced in the examples of Figure 6.8.

6.2.2   Inverse transform

We now derive the inversion formula for Equation 6.17, which permits finding the function $f_d[nT]$ when $\widetilde{F}(j\omega)$ is known. To consolidate concepts, we refer to the earlier Figure 6.6, and make use of the band-limiting restriction to simplify presentation. This restriction is lifted in the next section.

The two representations of the auxiliary function are related by the inverse Fourier transform definition 2.6 as

$$f(t) = \mathcal{F}^{-1}\{F(j\omega)\} = \frac{1}{2\pi} \int_{-\infty}^{\infty} F(j\omega)e^{j\omega t}\, d\omega$$

Having assumed that $F(j\omega)$ is frequency limited to $\pm\frac{1}{2}\omega_s$, within that period it is identical to the function $\widetilde{F}(j\omega)$, as shown in Figure 6.6. The infinite integral of $F(j\omega)$ is therefore identical to the integral of one period of $\widetilde{F}(j\omega)$, so that

$$f(t) = \frac{1}{2\pi} \int_{-\omega_s/2}^{\omega_s/2} \widetilde{F}(j\omega)e^{j\omega t}\, d\omega \qquad (6.21)$$

Since $\widetilde{F}(j\omega)$ is periodic and the integration interval spans one complete period ω_s, the alternative interval $(0, \omega_s)$ gives the same result, with the benefit of simpler notation. Evaluating this integral at sampling instants $t = nT$, yields values $f(nT)$

$$f(nT) = \frac{1}{2\pi} \int_{0}^{\omega_s} \widetilde{F}(j\omega)e^{j\omega nT}\, d\omega$$

which, combined with the conversion law 6.14 and with $2\pi/T = \omega_s$ gives the desired expression for the **inverse discrete-time Fourier transform** $\mathcal{F}_{dt}^{-1}$

$$f_d[nT] = \frac{1}{\omega_s} \int_0^{\omega_s} \widetilde{F}(j\omega)e^{j\omega nT}\, d\omega \qquad\qquad (6.22)$$

Expressions 6.17 and 6.22 are the discrete-time counterparts of the forward and inverse Fourier transform expressions 2.5 and 2.6. They provide the reversible link between the discrete-time sequence $f_d[nT]$ and the periodic continuous-frequency function $\widetilde{F}(j\omega)$ symbolized in expression 6.18.

6.2.3 Unlimited auxiliary function and aliasing

To avoid digressions we assumed that the auxiliary function $F(j\omega)$ was band-limited. As this function does not appear in the final expressions 6.17 and 6.22, we now remove that restriction and introduce the concept of aliasing.

Consider an arbitrary continuous function $g(t)$ whose Fourier transform $G(j\omega)$ is not band-limited. This is illustrated in Figure 6.9 with a function of the form $g(t) = e^{-|at|}$ whose transform was encountered in the context of Example 2.7. Proceeding as in Section 6.2.1 we multiply with the same time-sampler $s_T(t)$ to obtain the sampled-time pair

$$g_s(t) \xleftrightarrow{\ \mathcal{F}\ } \widetilde{G}(j\omega)$$

The conversion law 6.14 becomes $Tg(mT) = g_d[mT]$ and leads to the discrete-time transform pair

$$g_d[nT] = \sum_{m=-\infty}^{\infty} g_d[mT]\delta[nT - mT]$$

$$\xleftrightarrow{\ \mathcal{F}_{dt}\ } \quad \widetilde{G}(j\omega) = \sum_{m=-\infty}^{\infty} g_d[mT]e^{-j\omega mT}$$

This is illustrated in the graphical sequence of the upper half of Figure 6.9, which almost repeats Figure 6.6. None of the above steps requires that the function be band-limited.

Only the construction of $\widetilde{G}(j\omega)$ reflects the absence of such limiting. According to Equation 6.20 it is a superposition of shifted replicas of $G(j\omega)$,

$$\widetilde{G}(j\omega) = \sum_{k=-\infty}^{\infty} G(j\omega - jk\omega_s)$$

The individual cycles of the periodic $\widetilde{G}(j\omega)$ are no longer identical replicas of $G(j\omega)$. In the summation process the tails of the replicas merged into adjacent cycles, irretrievably corrupting the original signal. This is a manifestation of aliasing.

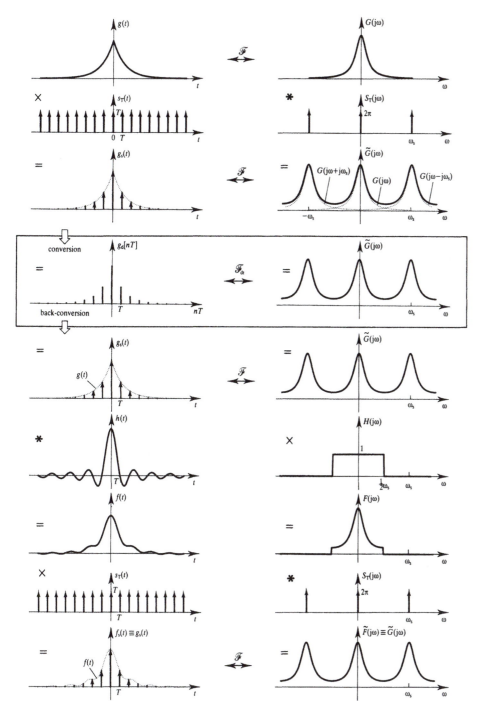

Figure 6.9 Arbitrary function and aliasing.

Aliasing

We now come to the main point of this sub-section. Using the sampled-time function $g_s(t)$ we construct another sampled-time function $f_s(t)$ with identical transform $\widetilde{F}(j\omega) = \widetilde{G}(j\omega)$ to show that samples of different functions $f(t)$ and $g(t)$ can lead to the same frequency representation.

The process is visualized in the lower half of Figure 6.9, where $g_s(t)$ was retrieved from the boxed result $g_d[nT]$ by back-converting impulses according to Equation 6.14. We now perform two complementary frequency domain operations on $\widetilde{G}(j\omega)$: a multiplication with an ideal lowpass filter $H(j\omega)$ of bandwidth $\pm\frac{1}{2}\omega_s$, followed by a convolution with the sampler $S_T(j\omega)$, used earlier on $G(j\omega)$, whose impulses are spaced by the width ω_s of the filter. This results in a mutual cancellation of the two operations, and leads back to the original function $\widetilde{G}(j\omega)$.

But the equivalent time domain convolution of $g_s(t)$ with $h(t)$ creates an intermediate function $f(t)$ whose transform $F(j\omega)$ is clearly different from $G(j\omega)$, so that it must be different from the function $g(t)$ sampled in the upper half of the figure. Some of the difference is reflected in the ripple of $f(t)$, which represents the sharp transitions of $F(j\omega)$ at the filter edges.

But Fourier transform pairs are unique, hence $f_s(t) = g_s(t)$ and the corresponding discrete-time functions must also be identical, that is to say, $f_d[nT] = g_d[nT]$.

This means that $f(t)$ and $g(t)$ are **identical at the sampling points** $t = nT$ **only**, but are not so constrained between those points. In this sense the two functions are aliases of each other. Any other function fitted to the same sampled values would provide a further alias.

We conclude that the discrete-time Fourier transform $\mathcal{F}_{dt}$ provides an exact frequency representation of the **discrete-time samples** of a continuous function $f(t)$, and that this function need not be frequency limited. We will return to aliasing in Section 6.6, in the context of the sampling theorem, where we examine the necessary relationship between $f(t)$ and the sampler $s_T(t)$ that permits recovery of the original function from the samples. We illustrate the interaction between high frequency components with a typical example.

Example 6.2

Consider the rectangular pulse $f(t) = p_\tau(t)/T$ and its Fourier transform $F(j\omega)$ shown in the top row of Figure 6.10. Sampling with $s_T(t)$, such that the pulse width τ is just greater than $2NT$, leads to the discrete-time function $f_d[nT]$ with $2N + 1$ unit-valued time samples, as shown.

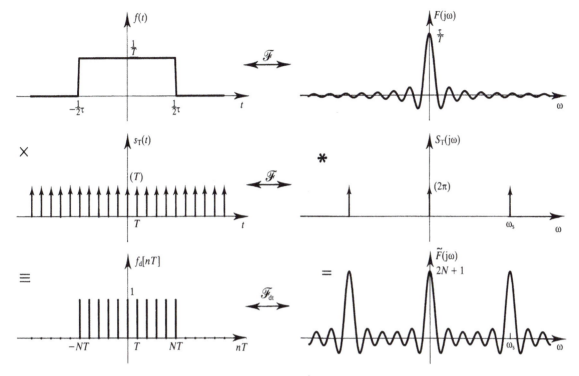

Figure 6.10   Transform of discrete-time rectangular pulse.

The corresponding frequency domain convolution yields infinite replicas of shifted $\sin x/x$ functions. These combine according to Equation 6.20 into the periodic function $\widetilde{F}(j\omega)$, whose analytical expression follows from Equation 6.17,

$$\widetilde{F}(j\omega) = \sum_{n=-N}^{N} e^{-j\omega nT}$$

Writing $m = n + N$, hence $n = m - N$, gives

$$\widetilde{F}(j\omega) = \sum_{m=0}^{2N} e^{-j\omega(m-N)T} = e^{j\omega NT} \sum_{m=0}^{2N} (e^{-j\omega T})^{m}$$

$$= e^{j\omega NT} \frac{1 - e^{-j\omega(2N+1)T}}{1 - e^{-j\omega T}}$$

where we used the geometric series result,

$$\sum_{n=0}^{M-1} x^n = \frac{1 - x^M}{1 - x}$$

Multiplying and dividing the numerator by $e^{j\omega(2N+1)T/2}$ and the denominator by $e^{j\omega T/2}$ yields

$$\widetilde{F}(j\omega) = \frac{e^{-j\omega T/2}}{e^{-j\omega T/2}} \frac{e^{j\omega(2N+1)T/2} - e^{-j\omega(2N+1)T/2}}{e^{j\omega T/2} - e^{-j\omega T/2}} = \frac{\sin\left[(2N+1)\omega T/2\right]}{\sin\left(\omega T/2\right)}$$

6.2.4   Discrete-time impulse trains and samplers

The continuous-time impulse train $f(t) = \delta_T(t)$ and its Fourier transform $\widetilde{F}(j\omega)$ were expressed in expression 6.5 as sums of impulses or, alternatively, as sums of exponentials. We now find similar expressions for an equivalent discrete-time Fourier pair $g_d[nT] \longleftrightarrow \widetilde{G}(j\omega)$, with the criterion that they have identical frequency representations.

We use the alternative of expression 6.5 that expresses $\widetilde{F}(j\omega)$ by impulses, to write

$$\widetilde{G}(j\omega) = \widetilde{F}(j\omega) = \omega_s \sum_{k=-\infty}^{\infty} \delta(\omega - k\omega_s)$$

Interpreted as a continuous-frequency periodic function, with period ω_s, the discrete-time Fourier transform 6.22 is applicable. Choosing the fundamental interval $(-\frac{1}{2}\omega_s, \frac{1}{2}\omega_s)$ for integration yields

$$g_d[nT] = \int_{-\omega_s/2}^{\omega_s/2} \sum_{k=-\infty}^{\infty} \delta(\omega - k\omega_s)e^{j\omega nT}\,d\omega$$

But the chosen interval only includes the impulse at the origin, as seen in the upper half of Figure 6.11, and the infinite sum reduces to a single term,

$$g_d[nT] = \int_{-\omega_s/2}^{\omega_s/2} \delta(\omega)e^{j\omega nT}\,d\omega = e^0 = 1$$

This is the unit function of discrete time, which can also be expressed as a sum of shifted impulses and leads to the desired transform pair

$$g_d[nT] = 1 = \sum_{m=-\infty}^{\infty} \delta[nT - mT]$$

$$\xrightarrow{\;\mathscr{F}_{dt}\;} \widetilde{G}(j\omega) = \omega_s \sum_{k=-\infty}^{\infty} \delta(\omega - k\omega_s) \qquad (6.23)$$

This is the discrete-time counterpart to the continuous-time pair 2.9, where the right side can be interpreted as a periodic version of the frequency domain impulse $\delta(\omega)$ of continuous time.

Example 6.3

We use the result 6.23 to confirm the frequency domain equivalence in expression 6.5. Writing expression 6.23 as

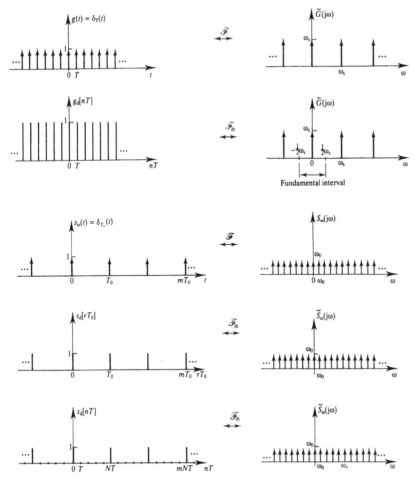

Figure 6.11 Discrete-time impulse train and frequency-sampler.

$$g[nT] = 1 \quad \xleftrightarrow{\mathscr{F}_{dt}} \quad \tilde{G}(j\omega) = \omega_s \sum_{k=-\infty}^{\infty} \delta(\omega - k\omega_s)$$

and comparing the frequency domain with that obtained by applying the Fourier expression 6.17 to the function $g[nT] = 1$,

$$\tilde{G}(j\omega) = \sum_{n=-\infty}^{\infty} g[nT]\, e^{-j\omega nT} = \sum_{n=-\infty}^{\infty} e^{-j\omega nT}$$

gives the desired equivalence

$$\sum_{n=-\infty}^{\infty} e^{-j\omega nT} = \omega_s \sum_{k=-\infty}^{\infty} \delta(\omega - k\omega_s)$$

Discrete-time frequency-sampler

Consider a different continuous-time impulse train, the frequency-sampler (6.9), $s_\omega(t) = \delta_{T_0}(t)$, whose impulses are spaced at intervals $T_0 = NT$, where N is a positive integer. The corresponding frequency impulses are proportionally smaller, and spaced at shorter intervals $\omega_0 = 2\pi/T_0$, as expressed in the Fourier pair

$$s_\omega(t) = \delta_{T_0}(t) = \sum_{m=-\infty}^{\infty} \delta(t - mT_0)$$

$$\xleftrightarrow{\;\mathcal{F}\;} \quad \tilde{S}_\omega(j\omega) = \omega_0 \sum_{k=-\infty}^{\infty} \delta(\omega - k\omega_0)$$

and shown in the lower half of Figure 6.11 for comparison. Treating $\tilde{S}_\omega(j\omega)$ as a periodic function, conversion to discrete time leads to a result similar to that of expression 6.23,

$$s_d[rT_0] = \sum_{m=-\infty}^{\infty} \delta[rT_0 - mT_0]$$

$$\xleftrightarrow{\;\mathcal{F}_{dt}\;} \quad \tilde{S}_\omega(j\omega) = \omega_0 \sum_{k=-\infty}^{\infty} \delta(\omega - k\omega_0)$$

which represents the unit function in terms of a time variable rT_0 of sampling interval T_0.

To express this function in terms of the usual time variable nT, we note that the location $t = mT_0$ of an arbitrary impulse is equivalently described by the location $t = mNT$, as seen from the two lowest rows of Figure 6.11, so that

$$s_d[nT] = \sum_{m=-\infty}^{\infty} \delta[nT - mNT]$$

$$\xleftrightarrow{\;\mathcal{F}_{dt}\;} \quad \tilde{S}_\omega(j\omega) = \omega_0 \sum_{k=-\infty}^{\infty} \delta(\omega - k\omega_0) \qquad (6.24)$$

where $\omega_0 = \omega_s/N$.

The time domain of this function is a train of discrete impulses located at integer multiples of NT. The frequency domain is a continuous-frequency impulse train, which will be used in Section 6.4 as a **discrete-time frequency-sampler** in the derivation of the discrete Fourier transform.

Frequency-shifted function

Shifting the frequency representation $\tilde{G}(j\omega)$ of a function to a frequency value $\omega = \omega_1$ generates the shifted function $\tilde{F}(j\omega) = \tilde{G}(j\omega - j\omega_1)$. Its inverse transform

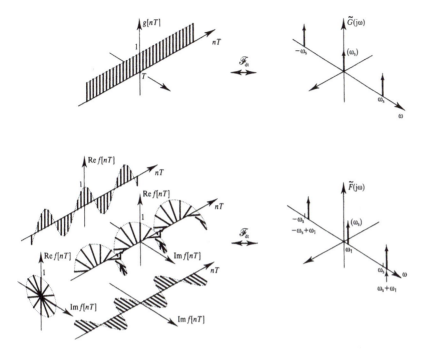

Figure 6.12   Discrete-time exponential $f[nT] = e^{j\omega_1 nT}$ as frequency-shifted impulse train.

$$f_d[nT] = \mathscr{F}_{dt}^{-1}\{\widetilde{F}(j\omega)\} = \frac{1}{\omega_s}\int_0^{\omega_s}\widetilde{G}(j\omega - j\omega_1)e^{j\omega nT}\,d\omega$$

$$= e^{j\omega_1 nT}\frac{1}{\omega_s}\int_0^{\omega_s}\widetilde{G}(j\nu)e^{j\nu nT}\,d\nu$$

where $\nu = \omega - \omega_1$, yields the frequency-shifting property

$$f_d[nT] = e^{j\omega_1 nT}g_d[nT] \overset{\mathscr{F}_{dt}}{\longleftrightarrow} \widetilde{F}(j\omega) = \widetilde{G}(j\omega - j\omega_1) \qquad (6.25)$$

This is the discrete-time counterpart of expression 2.33. Note formal similarities.

The particular case of shifting the impulse train of expression 6.23, as indicated in Figure 6.12, yields the pair

$$f_d[nT] = e^{j\omega_1 nT} \overset{\mathscr{F}_{dt}}{\longleftrightarrow} \widetilde{F}(j\omega) = \widetilde{G}(j\omega - j\omega_1)$$

$$= \omega_s \sum_{k=-\infty}^{\infty}\delta((\omega - \omega_1) - k\omega_s)$$

$$(6.26)$$

It has the time domain effect of winding up the unit envelope of the discrete impulse train into a helix, whose spokes at sampling points nT

represent the discrete-time complex exponential. This exponential can be expressed in the equivalent forms

$$f_d[nT] = e^{j\omega_1 nT} = \sum_{m=-\infty}^{\infty} e^{j\omega_1 mT} \, \delta[nT - mT]$$

also shown in the figure, together with its real and imaginary parts.

6.3   Discrete frequency

We return to the Fourier series of Chapter 1, interpreting it now as the discrete-frequency form of the Fourier transform, as suggested in Figure 6.13. We take advantage of transform dualities to derive a set of discrete-frequency expressions that are the dual counterparts of the discrete-time expressions of Section 6.2.

To stress similarities we take the dual counterpart of the auxiliary continuous function $f(t)$ of the preceding section and sample its frequency domain with the sampler $S_\omega(j\omega)$. The symmetry in scheme 6.5 of the impulse train simply exchanges the roles of the two domains, transferring all the derivation processes to the alternative domain, and leads to fully symmetric results.

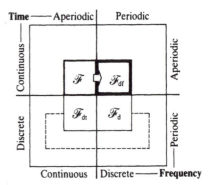

Figure 6.13   Discrete-frequency Fourier transform (Fourier series) from continuous $\mathcal{F}$.

Making allowances for notation, all the Fourier series properties, examples and graphics of Chapter 1 are applicable. The section concludes by deriving the discrete-frequency time-sampler required in Section 6.4.

6.3.1   Inverse transform

Repeating the process of Section 6.2 in the alternative domain leads first to the inverse transform. We start with a continuous and aperiodic auxiliary function $f(t)$, time-limited as shown at the top of Figure 6.14, and multiply its frequency representation $F(j\omega)$ by the frequency sampler of expression 6.9, repeated here,

$$s_\omega(t) = \sum_{n=-\infty}^{\infty} \delta(t - nT_0) \quad \overset{\mathcal{F}}{\longleftrightarrow} \quad S_\omega(j\omega) = \omega_0 \sum_{l=-\infty}^{\infty} \delta(\omega - l\omega_0)$$

$$(6.27)$$

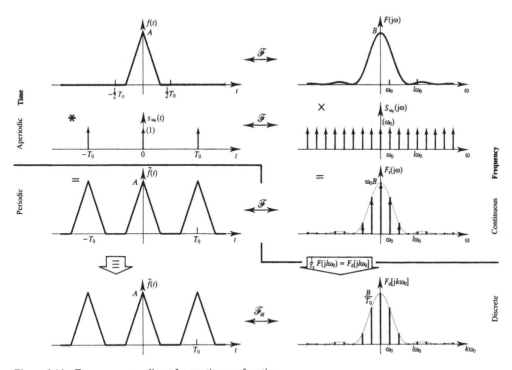

Figure 6.14   Frequency-sampling of a continuous function.

where ω_0 assumes the role of sampling interval, and $T_0 = 2\pi/\omega_0$ is the associated time domain period. This yields the sampled-frequency function

$$F_s(j\omega) = F(j\omega)\omega_0 \sum_{l=-\infty}^{\infty} \delta(\omega - l\omega_0) = \sum_{l=-\infty}^{\infty} \frac{2\pi}{T_0} F(j\omega)\,\delta(\omega - l\omega_0)$$

$$(6.28)$$

Considering $F(j\omega)$ to be constant in the infinitesimal region surrounding a point $\omega = l\omega_0$, each term of the sum can be written as

$$\frac{2\pi}{T_0} F(j\omega)\delta(\omega - l\omega_0) = \frac{2\pi}{T_0} F(jl\omega_0)\delta(\omega - l\omega_0)$$

which represents a scaled and shifted impulse, whose transform, according to expression 2.11 is

$$\frac{1}{T_0} F(jl\omega_0)e^{jl\omega_0 t} \overset{\mathcal{F}}{\longleftrightarrow} \frac{2\pi}{T_0} F(jl\omega_0)\delta(\omega - l\omega_0) \qquad (6.29)$$

so that the sum 6.28 of such impulses transforms as

$$\tilde{f}(t) = \sum_{l=-\infty}^{\infty} \frac{1}{T_0} F(jl\omega_0)e^{jl\omega_0 t}$$

$$\overset{\mathcal{F}}{\longleftrightarrow} F_s(j\omega) = 2\pi \sum_{l=-\infty}^{\infty} \frac{1}{T_0} F(jl\omega_0)\delta(\omega - l\omega_0) \quad (6.30)$$

which is the dual counterpart of expression 6.12. The process, which so far is of continuous frequency, is illustrated in the upper three rows of Figure 6.14. The relationship 6.30 is also visualized in Figure 6.15 where the impulses of $F_s(j\omega)$ are simulated by narrow pulses.

We now replace the sampled-frequency function $F_s(j\omega)$ by an equivalent discrete-time function $F_d[jk\omega_0]$, with the criterion that the periodic time function $\tilde{f}(t)$ remains intact.

Discrete-frequency equivalent

As in section 6.2, we formulate the equivalence at the elementary component level. Each continuous-frequency impulse is replaced by an equivalent discrete-frequency impulse, such that their associated time domain exponentials have the same amplitude. The frequency conversion law is

$$F_d[jl\omega_0] = \frac{1}{T_0} F(jl\omega_0) \qquad (6.31)$$

which applied to both domains of a typical frequency domain impulse expression (6.29) yields the elemental equivalence

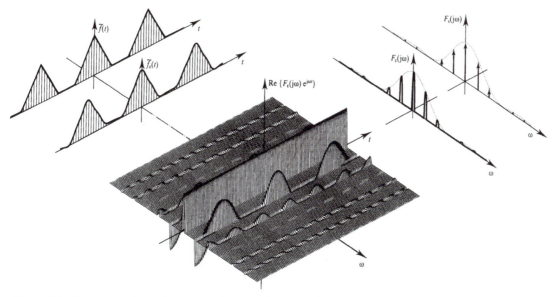

Figure 6.15 Inverse transform of a sampled-frequency function $F_s(j\omega)$. Simulation by narrow pulses $F_\varepsilon(j\omega)$.

$$\frac{1}{T_0} F(jl\omega_0)e^{jl\omega_0 t} \quad \overset{\mathcal{F}}{\longleftrightarrow} \quad \frac{2\pi}{T_0} F(jl\omega_0)\delta(\omega - l\omega_0)$$

$$\updownarrow \qquad F_d[jl\omega_0] = \frac{1}{T_0} F(jl\omega_0) \updownarrow \tag{6.32}$$

$$F_d[jl\omega_0]e^{jl\omega_0 t} \quad \overset{\mathcal{F}_{df}}{\longleftrightarrow} \quad 2\pi F_d[jl\omega_0]\delta[k\omega_0 - l\omega_0]$$

and, extended to the sum of frequency domain impulses expression 6.30, yields

$$\tilde{f}(t) = \sum_{l=-\infty}^{\infty} \frac{1}{T_0} F(jl\omega_0)e^{jl\omega_0 t} \quad \overset{\mathcal{F}}{\longleftrightarrow} \quad F_s(j\omega)=2\pi \sum_{l=-\infty}^{\infty} \frac{1}{T_0} F(jl\omega_0)\delta(\omega-l\omega_0)$$

$$\updownarrow \qquad F_d[jl\omega_0] = \frac{1}{T_0} F(jl\omega_0) \qquad \updownarrow \tag{6.33}$$

$$\tilde{f}(t) = \sum_{l=-\infty}^{\infty} F_d[jl\omega_0]e^{jl\omega_0 t} \quad \overset{\mathcal{F}_{df}}{\longleftrightarrow} \quad F_s[jk\omega_0]=2\pi \sum_{l=-\infty}^{\infty} F_d[jl\omega_0]\delta[k\omega_0-l\omega_0]$$

The resulting time domain expression, written with the usual dummy summation variable k, defines the **inverse discrete-frequency Fourier transform** of a function $F_d[jk\omega_0]$,

$$\tilde{f}(t) = \sum_{k=-\infty}^{\infty} F_d[jk\omega_0]e^{jk\omega_0 t} \tag{6.34}$$

We express this relationship with the symbol $\mathcal{F}_{df}$ as

$$\tilde{f}(t) = \mathcal{F}_{df}^{-1}\{F_d[jk\omega_0]\}$$

or in the form of a transform pair

$$\tilde{f}(t) \overset{\mathcal{F}_{df}}{\longleftrightarrow} F_d[jk\omega_0] \tag{6.35}$$

With the simplified notation $F_d[jk\omega_0] = c_k$, the expression 6.34 is no other than the synthesis equation 1.18 of the Fourier series.

Alternative time domain description

The periodic function $\tilde{f}(t)$ can also be interpreted as a sum of shifted functions. Referring back to Figure 6.14, the frequency domain sampling operation has a convolution as its time domain counterpart, expressed by the convolution property 2.35 as

$$\tilde{f}(t) = f(t) * s_\omega(t) \overset{\mathcal{F}}{\longleftrightarrow} F_s(j\omega) = F(j\omega)S_\omega(j\omega)$$

so that

$$\tilde{f}(t) = f(t) * \sum_{n=-\infty}^{\infty} \delta(t - nT_0) = \sum_{n=-\infty}^{\infty} f(t) * \delta(t - nT_0)$$

Each of the terms of this sum is a displaced replica of $f(t)$,

$$f(t) * \delta(t - nT_0) = f(t - nT_0)$$

and the infinite sum of all the replicas builds up the periodic signal

$$\tilde{f}(t) = \sum_{n=-\infty}^{\infty} f(t - nT_0) \tag{6.36}$$

If $f(t)$ is not time limited, the tails of the replicas extend into adjacent periods, causing aliasing.

6.3.2   Forward transform

We now derive the expression for the forward transform $\mathcal{F}_{df}$. The Fourier transform of the auxiliary function $f(t)$ is defined as

$$F(j\omega) = \mathcal{F}\{f(t)\} = \int_{-\infty}^{\infty} f(t)e^{-j\omega t}\, dt$$

But assuming $f(t)$ to be zero-valued outside the interval $|t| < \frac{1}{2}T_0$, see Figure 6.14, within that period it is identical to $\tilde{f}(t)$. The infinite integral of $f(t)$ is therefore identical to the integral of one period of $\tilde{f}(t)$,

$$F(j\overset{\text{l}}{\omega}) = \int_{-T_0/2}^{T_0/2} \tilde{f}(t)e^{-j\omega t}\,dt$$

We evaluate this integral for the sampling values $\omega = k\omega_0$, using the integration interval $(0, T_0)$ to simplify notation,

$$F(jk\omega_0) = \int_0^{T_0} \tilde{f}(t)e^{-jk\omega_0 t}\,dt$$

and this, combined with the frequency conversion law (6.31), yields

$$\boxed{F_d[jk\omega_0] = \frac{1}{T_0}\int_0^{T_0} \tilde{f}(t)e^{-jk\omega_0 t}\,dt} \qquad (6.37)$$

This is the desired expression for the **forward discrete-frequency Fourier transform**.

The pair of expressions 6.37 and 6.34 provide the forward and inverse relationships, symbolized in expression 6.35, which link the discrete-frequency values $F_d[jk\omega_0]$ with the periodic continuous-time function $\tilde{f}(t)$.

A similar strategy to that employed in Section 6.2.3 would show that these relationships are reversible and unique to each function $\tilde{f}(t) \longleftrightarrow F_d[jk\omega_0]$. Sampling continuous functions $F(j\omega)$ that are not suitably time-limited would similarly lead to aliasing. Different functions $\tilde{f}_i(t)$ could be fitted to the samples represented by $F_d[jk\omega_0]$, and these would represent aliases of each other.

6.3.3   Discrete-frequency impulse trains and samplers

To derive the transform of the discrete-frequency impulse train we treat the time domain of the continuous train 6.27 as a periodic function

$$\tilde{g}(t) = \sum_{m=-\infty}^{\infty} \delta(t - mT_0) \qquad (6.38)$$

and apply to it the discrete-frequency formula 6.37. Using the integration interval $(-\tfrac{1}{2}T_0, \tfrac{1}{2}T_0)$, we write

$$G_d[jk\omega_0] = \frac{1}{T_0}\int_{-T_0/2}^{T_0/2} \sum_{m=-\infty}^{\infty} \delta(t - mT_0)e^{-jk\omega_0 t}\,dt$$

But, with the exception of the impulse of index $m = 0$, all the impulses of the summation fall outside the integration interval, see the upper half of Figure 6.16, so that only the product $\delta(t)e^{-jk\omega_0 t}$ is not identically zero, leaving

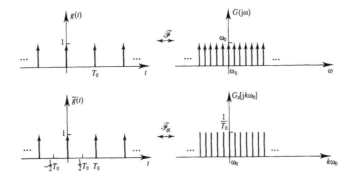

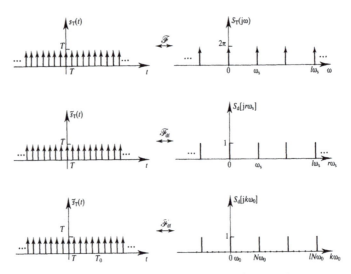

Figure 6.16 Discrete-frequency impulse train and time-sampler.

$$G_\mathrm{d}[jk\omega_0] = \frac{1}{T_0} \int_{-T_0/2}^{T_0/2} \delta(t) e^{-jk\omega_0 t}\, \mathrm{d}t = \frac{1}{T_0} e^0 = \frac{1}{T_0}$$

This is a discrete-frequency constant, which can be expressed as a sum of shifted impulses of constant value, so that

$$\tilde{g}(t) = \sum_{m=-\infty}^{\infty} \delta(t - mT_0)$$

$$\xleftarrow{\mathscr{F}_\mathrm{df}} \quad G_\mathrm{d}[jk\omega_0] = \frac{1}{T_0} \sum_{l=-\infty}^{\infty} \delta[k\omega_0 - l\omega_0] \qquad (6.39)$$

which is formally the same as its dual counterpart expression 6.23.

Discrete-frequency time-sampler

The continuous time sampler 6.8 repeated here

$$s_T(t) = T \sum_{n=-\infty}^{\infty} \delta(t - nT) \overset{\mathcal{F}}{\longleftrightarrow} S_T(j\omega) = 2\pi \sum_{l=-\infty}^{\infty} \delta(\omega - l\omega_s)$$

has its impulses spaced at intervals $T = T_0/N$ and $\omega_s = N\omega_0$ respectively, where N is a positive integer. It transforms according to expression 6.39, as

$$\tilde{s}_T(t) = T \sum_{n=-\infty}^{\infty} \delta(t - nT) \overset{\mathcal{F}_{df}}{\longleftrightarrow} S_d[jr\omega_s] = \sum_{l=-\infty}^{\infty} \delta[r\omega_s - l\omega_s]$$

whose frequency domain is a function of the variable $r\omega_s$, as shown in the lower half of Figure 6.16. Changing to the usual variable $k\omega_0$ yields the discrete-frequency time sampler

$$\tilde{s}_T(t) = T \sum_{n=-\infty}^{\infty} \delta(t - nT) \overset{\mathcal{F}_{df}}{\longleftrightarrow} S_d[jk\omega_0] = \sum_{l=-\infty}^{\infty} \delta[k\omega_0 - lN\omega_0] \quad (6.40)$$

6.4   Discrete time and discrete frequency

One would be hard pressed to find any phenomenon of the physical world that is truly discrete in both domains. But the advent of microelectronics and of fast Fourier transform (FFT) algorithms makes it very attractive to convert continuous physical signals to an equivalent discrete form and to process these digitally.

We have so far covered the discrete-time Fourier transform and the discrete-frequency Fourier transform. To derive their expressions we sampled a related continuous function, applied the continuous Fourier transform to the sampled function and then converted the result to a discrete variable. We now derive the discrete Fourier transform, building on the earlier results as suggested in Figure 6.17. We follow the

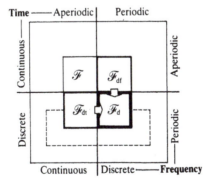

Figure 6.17   Discrete Fourier transform from discrete time and discrete frequency transforms.

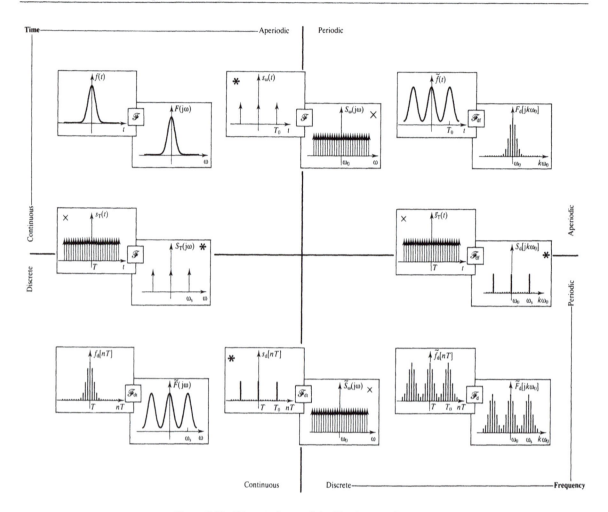

Figure 6.18 Discrete forms of the Fourier transform.

same strategy by sampling in the alternative domain, in which the function is still continuous.

6.4.1 Strategy

A graphical summary of the procedure and results of Sections 6.2 and 6.3 is given in the left and upper halves of Figure 6.18. Extending this procedure into the discrete quadrant suggests a strategy for deriving the discrete Fourier transform expressions and gives an indication of the expected results.

To highlight symmetries and transform dualities we again employ the Gaussian function used in Figure 6.9, repeated in the continuous quadrant of Figure 6.18, where both domains are shown linked by the continuous Fourier transform.

The left half of Figure 6.18 shows the function as it is modified with the aid of the time sampler $s_T(t)$ to generate a suitable discrete-time function $f_d[nT] \longleftrightarrow \tilde{F}(j\omega)$, whose domains are related by the discrete-time Fourier transform. The upper half of the figure illustrates the dual case, where the frequency sampler $S_\omega(j\omega)$ leads to the function $\tilde{f}(t) \longleftrightarrow F_d[jk\omega_0]$, whose domains are related by the discrete-frequency transform. Note the dual symmetry between the domains of the two resulting functions.

A second sampling of the function, in the alternative domain, should produce a convergence of both effects and lead to a function $\tilde{f}_d[nT] \longleftrightarrow \tilde{F}_d[jk\omega_0]$ that is discrete and periodic in both domains. But note that a discrete function can only be periodic if its period is an integer multiple of the sampling interval. Furthermore, if a function is discrete and periodic in one domain, with N samples per period, then it is also discrete and periodic in the other domain, with the same number of samples per period. For instance, if $T_0 = NT$, it follows that

$$\omega_s = 2\pi/T = 2\pi N/T_0 = N\omega_0$$

We follow alternative paths to derive the forward and inverse transforms, verifying later that the resulting expressions are consistent.

6.4.2 From discrete-frequency transform

We first derive the expression of the forward transform. We start with an auxiliary discrete-frequency function $\tilde{f}(t) \longleftrightarrow F_d[jk\omega_0]$, shown at the top of Figure 6.19, and sample its time domain with the discrete-frequency sampler expression 6.40, whose sampling interval T and sampling frequency $\omega_s = 2\pi/T$ are related by the integer N to the dual counterparts of $\tilde{f}(t)$, as $T = T_0/N$ and $\omega_s = N\omega_0$.

This produces a sampled-time periodic function,

$$\tilde{f}_s(t) = \tilde{f}(t)\tilde{s}_T(t) = T \sum_{n=-\infty}^{\infty} \tilde{f}(t)\delta(t - nT)$$

whose frequency representation is obtained by applying the discrete-frequency Fourier transform formula 6.37 to $\tilde{f}_s(t)$. On this occasion we choose the integration interval $(0^-, T_0^-)$, to avoid splitting impulses in half, and this yields

$$\tilde{F}_d[jk\omega_0] = \mathcal{F}_{df}^{-1}\{\tilde{f}_s(t)\} = \frac{T}{T_0} \int_{0^-}^{T^-} \sum_{n=-\infty}^{\infty} \tilde{f}(t)\delta(t - nT)e^{-jk\omega_0 t}\,dt$$

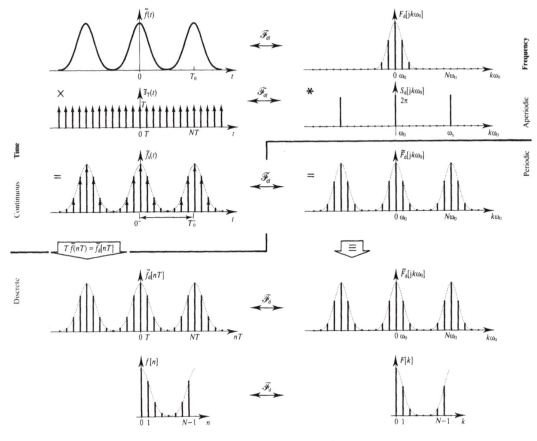

Figure 6.19 Forward transform from discrete frequency.

The chosen integration interval limits the number of affected impulses to those identified by the indices $n = 0$ to $n = N - 1$. The finite integral of the infinite sum of impulses is therefore equivalent to the infinite integral of the N affected impulses. Exchanging sum and integral, as well as their limits, and writing $T_0/T = N$, we have

$$\widetilde{F}_d[jk\omega_0] = \frac{1}{N} \sum_{n=0}^{N-1} \int_{-\infty}^{\infty} \widetilde{f}(t)\delta(t - nT)e^{-jk\omega_0 t}\,dt$$

Writing $\widetilde{f}(t)e^{-jk\omega_0 t} = g(t)$ and $nT = \tau$, the integral takes the form of the impulse definition

$$\int_{-\infty}^{\infty} g(t)\delta(t - \tau)\,dt = g(\tau)$$

and integrates to the value $g(nT) = \tilde{f}(nT)\mathrm{e}^{-jk\omega_0 nT}$, so that

$$\tilde{F}_{\mathrm{d}}[jk\omega_0] = \frac{1}{N} \sum_{n=0}^{N-1} \tilde{f}(nT)\mathrm{e}^{-jk\omega_0 nT}$$

The function $\tilde{f}(t)$, of continuous variable, is only invoked at discrete points of t, so that a conversion to the discrete variable nT is appropriate. With the substitution

$$\boxed{T\tilde{f}(nT) = \tilde{f}_{\mathrm{d}}[nT]} \qquad (6.41)$$

which is formally identical to the earlier time domain conversion of Equation 6.14, we find the desired expression for the **discrete Fourier transform**

$$\boxed{\tilde{F}_{\mathrm{d}}[jk\omega_0] = \frac{1}{NT} \sum_{n=0}^{N-1} \tilde{f}_{\mathrm{d}}[nT]\mathrm{e}^{-jk\omega_0 nT}} \qquad (6.42)$$

6.4.3 From discrete-time transform

To derive the inversion formula to Equation 6.42 we start with the discrete-time Fourier pair $f_{\mathrm{d}}[nT] \longleftrightarrow \tilde{F}(j\omega)$ shown at the top of Figure 6.20. Sampling its frequency domain with the discrete-time frequency-sampler expression 6.24, whose sampling interval ω_0 and period T_0 are again related to those of the signal as $\omega_0 = \omega_{\mathrm{s}}/N$ and $T_0 = NT$, yields the sampled-frequency function

$$\tilde{F}_{\mathrm{s}}(j\omega) = \tilde{F}(j\omega)\tilde{S}_\omega(j\omega) = \omega_0 \sum_{k=-\infty}^{\infty} F(j\omega)\delta(\omega - k\omega_0)$$

The discrete-time Fourier transform 6.22 applies, which, with the integration interval $(0^-, \omega_{\mathrm{s}}^-)$, yields

$$\begin{aligned}
\tilde{f}_{\mathrm{d}}[nT] &= \mathcal{F}_{\mathrm{dt}}^{-1}\{\tilde{F}_{\mathrm{s}}(j\omega)\} \\
&= \frac{\omega_0}{\omega_{\mathrm{s}}} \int_{0^-}^{\omega_{\mathrm{s}}^-} \sum_{k=-\infty}^{\infty} \tilde{F}(j\omega)\delta(\omega - k\omega_0)\mathrm{e}^{j\omega nT}\,\mathrm{d}\omega
\end{aligned}$$

As before, we exchange integration and summation and their limits to yield

$$\tilde{f}_{\mathrm{d}}[nT] = \frac{1}{N} \sum_{k=0}^{N-1} \int_{-\infty}^{\infty} \tilde{F}(j\omega)\delta(\omega - k\omega_0)\mathrm{e}^{j\omega nT}\,\mathrm{d}\omega$$

The integral can now be interpreted by means of the sifting property of the frequency domain impulse,

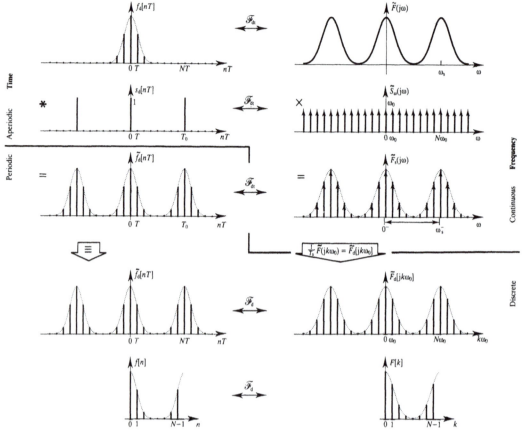

Figure 6.20   Inverse transform from discrete time.

$$\int_{-\infty}^{\infty} G(\omega)\delta(\omega - v)\,dv = G(v)$$

With $\widetilde{F}(j\omega)e^{j\omega nT} = G(v)$ and $k\omega_0 = v$, the integral takes the value $G(k\omega_0) = \widetilde{F}(jk\omega_0)e^{jk\omega_0 nT}$, so that

$$\tilde{f}_d[nT] = \frac{1}{N}\sum_{k=0}^{N-1}\widetilde{F}(jk\omega_0)e^{jk\omega_0 nT}$$

The function $\widetilde{F}(j\omega)$ is of interest only at discrete points of ω. It can be converted to the discrete-frequency variable $k\omega_0$ by the substitution

$$\boxed{\widetilde{F}_d[jk\omega_0] = \frac{1}{T_0}\,\widetilde{F}(jk\omega_0)} \tag{6.43}$$

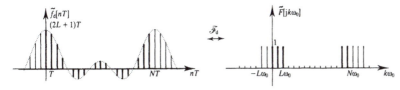

Figure 6.21   Discrete Fourier transform pair.

which is formally similar to the frequency domain conversion 6.31 of Section 6.3.1. With $T_0/N = T$, this yields the desired expression for the **inverse discrete Fourier transform**

$$\tilde{f}_d[nT] = T \sum_{k=0}^{N-1} \tilde{F}_d[jk\omega_0]e^{jk\omega_0 nT} \tag{6.44}$$

As in earlier developments, the auxiliary function $f_d[nT]$ drops out.

Example 6.4

We find the time representation $\tilde{f}_d[nT]$ of the rectangular frequency domain pulse of Figure 6.21. This is the dual and fully discrete version of Example 6.2. Using expression 6.44 with a summation interval N spanning both sides of the origin gives

$$\tilde{f}_d[nT] = T \sum_{k=-L}^{L} e^{jk\omega_0 nT} = T \sum_{l=0}^{2L} e^{j(l-L)\omega_0 nT}$$

where $l = k + L$. This is rearranged to

$$\tilde{f}_d[nT] = Te^{-jL\omega_0 nT} \sum_{l=0}^{2L} (e^{j\omega_0 nT})^l = Te^{-jL\omega_0 nT} \frac{1 - e^{j(2L+1)\omega_0 nT}}{1 - e^{j\omega_0 nT}}$$

and manipulating the numerator and denominator as in Example 6.2, yields

$$\tilde{f}_d[nT] = T \frac{\sin[(2L+1)\omega_0 nT/2]}{\sin(\omega_0 nT/2)}$$

whose value at the origin is $\tilde{f}_d[0] = (2L+1)T$.

The values used for illustration in Figure 6.21 are $N = 20$ and $L = 2$, which with $\omega_0 = 2\pi/T_0 = 2\pi/NT$, yield

$$\tilde{f}_d[nT] = T \frac{\sin(5\pi n/20)}{\sin(\pi n/20)}$$

6.4.4   Cross-verification

Expressions 6.42 and 6.44 of the discrete Fourier transform were derived along two different paths. We now show that they are complementary in that they give the forward and inverse transforms linking the discrete pair

$$\tilde{f}_d[nT] \overset{\mathscr{F}_d}{\longleftrightarrow} \tilde{F}_d[jk\omega_0]$$

We write expression 6.42 with a dummy index m,

$$\tilde{F}_d[jk\omega_0] = \frac{1}{NT} \sum_{m=0}^{N-1} \tilde{f}_d[mT]e^{-jk\omega_0 mT}$$

take the inverse transform 6.44 of both sides,

$$\mathscr{F}_d^{-1}\{\tilde{F}_d[jk\omega_0]\} = T \sum_{k=0}^{N-1} \left[\frac{1}{NT} \sum_{m=0}^{N-1} \tilde{f}_d[mT]e^{-jk\omega_0 mT} \right] e^{jk\omega_0 nT}$$

and interchange summations,

$$\mathscr{F}_d^{-1}\{\tilde{F}_d[jk\omega_0]\} = \frac{1}{N} \sum_{m=0}^{N-1} \tilde{f}_d[mT] \sum_{k=0}^{N-1} e^{jk\omega_0(n-m)T}$$

But orthogonality of the discrete exponential reduces the second sum to

$$\sum_{k=0}^{N-1} e^{jk\omega_0(n-m)T} = \begin{cases} N & \text{for } m = n \\ 0 & \text{for } m \neq n \end{cases}$$

so that all the terms of the first sum vanish, except that for $m = n$, and

$$\mathscr{F}_d^{-1}\{\tilde{F}_d[jk\omega_0]\} = \tilde{f}_d[nT]$$

which confirms that the two expressions relate domains of the same function.

6.5   Summary and interrelations

Having derived the four classes of the Fourier transform, we now take stock of results and consolidate the concepts involved. We gather the transform expressions on the framework of the signal classification chart, thus highlighting similarities and giving a deeper insight into the nature of the simplifications introduced with discrete variables.

   We also examine how the continuous exponential surface $e^{j\omega t}$ is first reduced to a set of planes $e^{j\omega nT}$ or $e^{jk\omega_0 t}$, when one variable is discrete, and then to a 'bed of nails' $e^{jk\omega_0 nT}$, when both variables are discrete. This explains how, with minor modifications, the concepts and graphical imagery developed for the continuous Fourier transform are also applicable to the discrete classes.

Time ———————————————— Aperiodic | Periodic

Continuous (Aperiodic): $\mathscr{F}$ (Section 2.1)

$$F(j\omega) = \int_{-\infty}^{\infty} f(t)\, e^{-j\omega t}\, dt \qquad (2.5)$$

$$f(t) = \frac{1}{2\pi} \int_{-\infty}^{\infty} F(j\omega)\, e^{j\omega t}\, d\omega \qquad (2.6)$$

Continuous (Periodic): $\mathscr{F}_{df}$ (Section 6.3)

$$F_d[jk\omega_0] = \frac{1}{T_0} \int_0^{T_0} \tilde{f}(t)\, e^{-jk\omega_0 t}\, dt \qquad (6.37)$$

$$\tilde{f}(t) = \sum_{k=-\infty}^{\infty} F_d[jk\omega_0]\, e^{jk\omega_0 t} \qquad (6.34)$$

Discrete (Aperiodic): $\mathscr{F}_{dt}$ (Section 6.2)

$$\tilde{F}(j\omega) = \sum_{n=-\infty}^{\infty} f_d[nT]\, e^{-j\omega nT} \qquad (6.17)$$

$$f_d[nT] = \frac{1}{\omega_s} \int_0^{\omega_s} \tilde{F}(j\omega)\, e^{j\omega nT}\, d\omega \qquad (6.22)$$

Discrete (Periodic): $\mathscr{F}_d$ Section 6.4

$$\tilde{F}_d[jk\omega_0] = \frac{1}{NT} \sum_{n=0}^{N-1} \tilde{f}_d[nT]\, e^{-jk\omega_0 nT} \qquad (6.42)$$

$$\tilde{f}_d[nT] = T \sum_{k=0}^{N-1} \tilde{F}_d[jk\omega_0]\, e^{jk\omega_0 nT} \qquad (6.44)$$

Continuous | Discrete ———————————————— **Frequency**

Figure 6.22 Signal classification chart and appropriate Fourier transforms.

6.5.1 Summary of transform expressions

The methodology used to derive the various Fourier expressions was interpreted earlier in Figure 6.18, where we used our signal classification diagram for background, and where we also included the applicable samplers.

The resulting expressions of the forward and inverse Fourier transforms are now brought together in Figure 6.22, locating each pair in the quadrant appropriate to the signals it relates. This juxtaposition highlights dualities of the discrete classes, as well as formal similarities with the continuous Fourier transform.

Examining the variables, the lower half of the diagram substitutes the discrete variable nT for each appearance of the continuous time variable t. Similarly, the right half of the diagram has the continuous frequency variable ω replaced by the related discrete variable $k\omega_0$. The fully discrete quadrant ends up with both variables replaced.

Each such change of variable causes a simplification of the integrals. In the changed domain the integral becomes a sum, while in the alternative domain the integration or summation interval is reduced from infinity to one period. These analytical simplifications are suggested in each domain of Figure 6.23, which also reveals operator symmetries between domains.

Each simplification makes it easier to evaluate a function. A Fourier series is more approachable than the related integral. The discrete Fourier transform can actually be evaluated numerically. This can be done very efficiently with the use of an FFT algorithm, to an

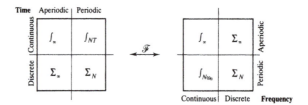

Figure 6.23 Simplifications of integrals and sums.

accuracy limited only by the word-length used. We examine the basis of that efficiency in the appendix to this chapter.

6.5.2 Simplifications to the exponential operator

In Chapter 2 we saw that the key to the continuous Fourier transform was the exponential function $e^{-j\omega t}$. We interpreted the integrand of the transform as an amplitude modulation of a surface representing the complex exponential by the function to be transformed. The related graphics called for separate surfaces for the real and imaginary parts of the exponential, the latter being slightly different for forward and inverse transforms. In the present context Figure 2.8 is typical and provides a reference for the discrete cases illustrated in Figure 6.24.

Such continuous interpretations of the sampled functions of the present chapter revealed much redundancy and wasted effort. When transforming a function of sampled variable, the whole exponential surface, with the exception of impulse-shaped planes generated by the samples, was modulated to zero, and required wasteful integration, as seen in Figures 6.7 and 6.15. Furthermore, the integration for all values yielded a function that was periodic in the transformed domain, with the consequent redundancy of information.

Conversion to discrete variable replaced the integral by a sum and reduced the integration interval of the alternative domain to one period. This is illustrated in the discrete-time and the discrete-frequency quadrants of Figure 6.24, which illustrate the transform of the appropriate unit function. The operative planes $e^{-j\omega nT}$ and $e^{-jk\omega_0 t}$ represent discrete slices of the complex exponential $e^{-j\omega t}$ and the highlighted strip represents one typical period of the alternative domain. These represent the discrete exponentials and the integration periods appearing in the expressions of Figure 6.22.

When both domains are discrete, redundancy extends to both variables. All the necessary values of the exponential $e^{-jk\omega_0 nT}$ are found on any square grid of size $N \times N$ samples, for instance that defined by the intersection of the two strips highlighted in the discrete quadrant of Figure 6.24. These values represent the intersections of the earlier dis-

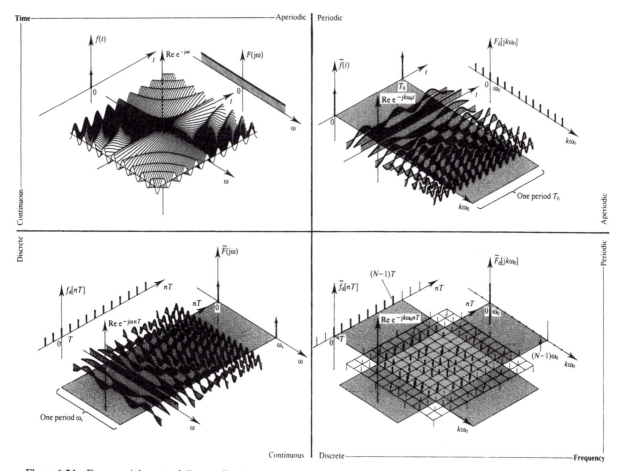

Figure 6.24 Exponential cores of discrete Fourier transforms (real parts).

crete planes. The exponential operator becomes a 'bed of nails' located on the square grid, which is redundantly repeated in both directions.

All these exponential operators of course possess real and imaginary parts. For brevity, only the real parts are shown in the figure.

6.5.3 Notation simplifications

At every stage of the foregoing developments special care was taken to identify the class of each function and each operator. This was cumbersome, but necessary to avoid mixing up fundamentally different functions that have deceptively similar notation. It is advisable to stick to such a notation whenever functions of different classes appear together.

In contrast, if all functions are of the same class, it is more expedient to simplify notation in various ways. In the case of the discrete Fourier transform, if the original function $f(t)$ from which $\tilde{f}_d[nT]$ derives is irrelevant, then the subscript and superscript are redundant. The function samples are more simply identified by the index alone as $f[n] \longleftrightarrow F[k]$. With $\omega_0 = 2\pi/T_0$ and $T_0 = NT$ the product $\omega_0 T$ in the exponent becomes a function of N alone, as $\omega_0 T = 2\pi/N$. Finally, it is customary to normalize the sampling interval to $T = 1$. With these simplifications the expressions 6.42 and 6.44 take the more compact form

$$F[k] = \frac{1}{N} \sum_{n=0}^{N-1} f[n] e^{-jnk2\pi/N}$$

$$f[n] = \sum_{k=0}^{N-1} F[k] e^{jnk2\pi/N}$$

(6.45)

In the context of the fast Fourier transform, see the appendix to this chapter, it is also convenient to write $e^{j2\pi/N} = w_N$, thus simplifying notation further to

$$F[k] = \frac{1}{N} \sum_{n=0}^{N-1} f[n] w_N^{-nk}$$

$$f[n] = \sum_{k=0}^{N-1} F[k] w_N^{nk}$$

(6.46)

In the next chapter the discrete Fourier transform will be expressed differently again, by using the notation of the z-transform.

Note that the expressions given for the discrete Fourier transform in most of the signals and systems literature differ from Equations 6.45 in that the scaling factor $1/N$ appears in the inversion formula, yielding the set

$$F[k] = \sum_{n=0}^{N-1} f[n] e^{-jnk2\pi/N}$$

$$f[n] = \frac{1}{N} \sum_{k=0}^{N-1} F[k] e^{jnk2\pi/N}$$

(6.47)

Such a set is self-consistent and is found embodied in most FFT algorithms. But extensive usage does not preclude other forms.

The set 6.45 is self-consistent too, as shown in Section 6.4.4. It has the advantage that this consistency extends to the wider set of all the discrete forms of the Fourier transform derived in this chapter, with all the symmetries and dualities highlighted in Figure 6.22.

The form of Equations 6.45 also gives consistency to certain engineering concepts, for instance d.c. value and average value. The d.c. value $F(0)$ of a continuous function $f(t)$ is its integral. The discrete-frequency equivalent is the average value $F_d[0]$ of $\tilde{f}(t)$, that is, the integral over one period T divided by the length of the period. The discrete equivalent, according to Equations 6.45, is the average value

$F[0]$ obtained by adding the samples $f[n]$ and dividing by the number of samples N. The form of Equations 6.47 transfers this averaging property to the other domain. It is all a matter of convention.

6.6   Fourier interpretations of functions

We now address the question of applying a Fourier transform of one class to a signal of another class. In this context Fourier interpretation falls into one of three categories, depending on whether the transform to be used is more, equally or less discrete than the signal.

When the signal and the transform are of the same class, the Fourier interpretation reduces to taking the transform appropriate to the class, the result being an exact alternative description of the function.

Throughout this chapter we repeatedly applied a transform that was less discrete, thereby more general, than the signal. For instance, to derive the expressions of the discrete-time Fourier transform we effectively replaced the discrete-time samples of $f[nT]$ by equivalent continuous-time impulses, as expressed in the relationship 6.13, thus making an essentially discrete function accessible to the more general continuous Fourier transform. This category is applied mainly in the analytical treatment of functions. It is also used with mixed signals, where for an overall analysis the discrete signal is replaced by an equivalent continuous signal, so that the more general transform can be applied to both signal classes.

The remaining category, where the Fourier transform is of a more discrete class than the function, is the most common in engineering practice. It is applied in the numerical evaluation of Fourier transforms and is at the heart of sampled control systems and of digital processing of continuous signals. Before the discrete transform can be applied the function needs to be sampled and windowed, and this may introduce approximation errors. We now examine the process involved in interpreting a continuous function by the discrete Fourier transform and identify the errors associated with sampling and windowing and means of reducing these errors.

6.6.1   Sampling and aliasing

The aliasing problem associated with sampling was partly addressed in the context of Section 6.2.3. We will now show that sampling a signal does not necessarily lead to loss of information and identify the conditions for which the original function can be fully recovered from the samples.

Consider the band limited function $f(t) \longleftrightarrow F(j\omega)$ with highest

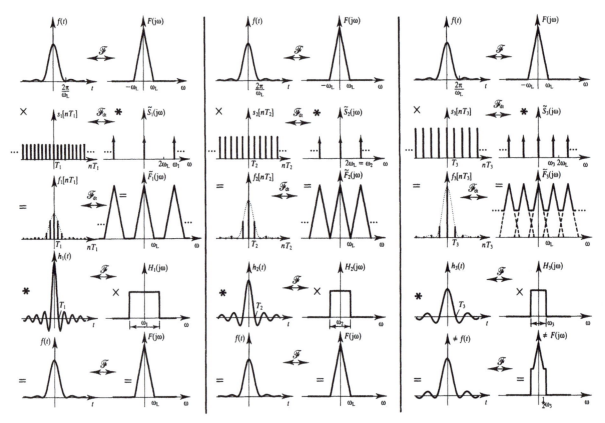

Figure 6.25 Sampling and aliasing: the sampling theorem.

frequency component ω_L, used earlier in Section 6.2.1 and repeated in the top row of Figure 6.25. We will first sample its time domain with three different sampling intervals T_1, T_2 and T_3 and then attempt to recover $f(t)$ by filtering with an appropriate ideal lowpass filter $H_L(j\omega)$, as discussed later in section 10.2.1. For brevity we use a shorthand notation for multiplication and convolution in the time domain of the figure that implies conversion between sampled-time and discrete-time values and vice versa.

A small sampling interval T_1, such that $\omega_1 > 2\omega_L$, illustrated on the left side of Figure 6.25, yields the discrete-time function $f_1[nT_1] \longleftrightarrow \widetilde{F}_1(j\omega)$, whose frequency domain becomes a periodic replication of $F(j\omega)$. If we now apply an ideal lowpass filter $H_1(j\omega)$ of a bandwidth commensurate with the sampling frequency ω_1, as shown in the figure, all periods except the fundamental are removed, leaving an exact copy of the original function $F(j\omega)$. The corresponding time domain convolution yields a function which, because of the uniqueness of the Fourier transform, must be identical to the original function $f(t)$.

Increasing the sampling interval to T_2, such that $\omega_2 = 2\omega_L$, provides the limiting case for which it is still possible, in principle, to recover the original function, as shown in the middle of Figure 6.25. This is known as the **Nyquist interval** and $2\omega_L$ is the **Nyquist frequency**.

In contrast, increasing the sampling interval beyond the limiting value $T_3 > T_2$ causes the replicas of $F(j\omega)$ to overlap, thereby corrupting their sum $\widetilde{F}_3(j\omega)$. A subsequent multiplication of $\widetilde{F}_3(j\omega)$ by the ideal filter $H_3(j\omega)$ yields a function that is clearly not the original $f(t) \longleftrightarrow F(j\omega)$. This makes it impossible to uniquely determine which function was sampled to yield $f_3[nT_3]$. The possible aliases are infinite.

The above conclusions represent the **sampling theorem**, which states that a bandlimited signal with $F(j\omega) = 0$ for $|\omega| > \omega_L$ can be fully represented by its samples, provided the sampling frequency ω_s equals or exceeds twice the highest signal frequency ω_L.

The aliasing effect is clearly seen from the identical samples of two suitably related sine waves, such as those of Figure 6.26. In the upper half of the figure the function $f(t) = \sin \omega_1 t$, illustrated by $\omega_1 = \frac{1}{8}\omega_s$, is sampled to give the pair $f_d[nT] \longleftrightarrow \widetilde{F}(j\omega)$, whose frequency domain replicates the imaginary frequency components $F(j\omega)$ at multiples of the sampling frequency. In the lower half of the figure the function $g(t) = \sin(\omega_1 + \omega_s)t$ is sampled at the same rate to give $g_d[nT] \longleftrightarrow \widetilde{G}(j\omega)$. But the replicas of the widely spaced impulses of $G(j\omega)$ adopt a configuration that is identical to $\widetilde{F}(j\omega)$, so that the two discrete functions $f_d[nT]$ and $g_d[nT]$ are identical.

To avoid gross misinterpretations it is necessary to condition the signal prior to sampling by removing all frequency components above half the sampling frequency by means of an **anti-aliasing filter**.

6.6.2   Discrete Fourier transform of a continuous function

We now examine the steps involved in interpreting a segment of length T_0 of a continuous function by means of an N-point discrete Fourier transform. For simplicity we use an even symmetric signal $f(t) \longleftrightarrow F(j\omega)$, as shown in the top row of Figure 6.27, thus avoiding phase considerations. Superimposed on the smooth signal are shown narrow noise spikes of significant amplitude but negligible energy, whose frequency representations consist of low-level disturbances extending to high frequencies (not shown).

The chosen segment length T_0 and the number N of points of the transform determine the sampling interval $T = T_0/N$ as well as the sampling frequency $\omega_s = 2\pi/T$ and the fundamental frequency component $\omega_0 = \omega_s/N$.

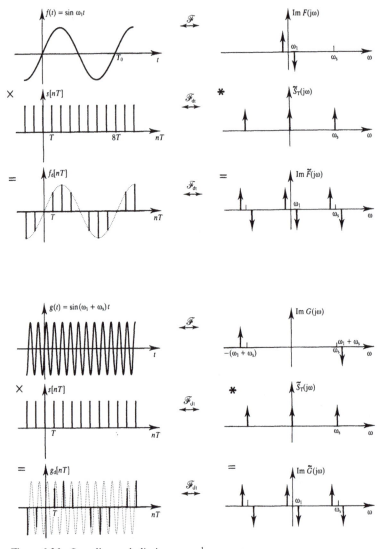

Figure 6.26   Sampling and aliasing, $\omega_1 = \frac{1}{8}\omega_s$.

If we were to sample directly, and noise spikes coincided with sampling points, their measured amplitudes would be attributed to the signal. To minimize aliasing, the signal is therefore first passed through a continuous-time anti-aliasing filter that removes all frequency components greater than half the sampling frequency. This is represented in the figure by an ideal lowpass filter $H_L(j\omega)$ of width $\pm\frac{1}{2}\omega_s$, which has a smoothing effect on time domain features and removes the noise spikes.

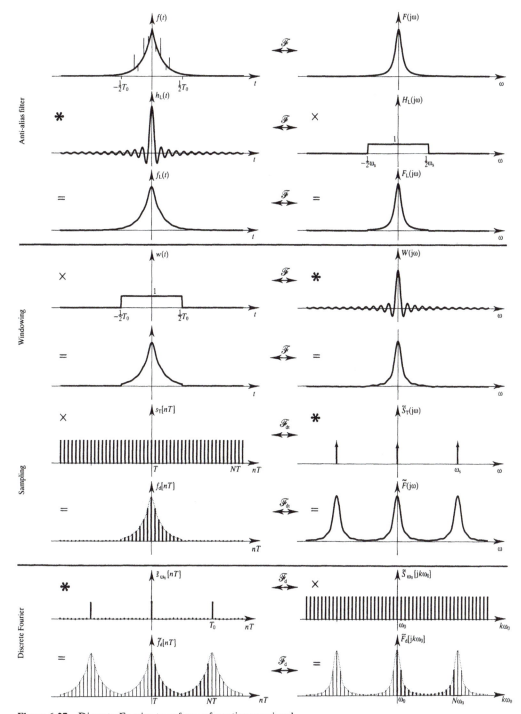

Figure 6.27   Discrete Fourier transform of continuous signal.

The next step, capturing the desired N time samples $f_d[nT]$, is represented in Figure 6.27 in two stages, namely, windowing the continuous-time signal with the window $w(t)$ of width T_0, followed by sampling with the discrete-time sampler $s_T[nT]$. When N is even, as in the case illustrated in the figure, the edges $\pm\frac{1}{2}T_0$ of the symmetric window $w(t)$ coincide with the sampling instants $\pm\frac{1}{2}NT$. In such cases, to preserve the symmetry we assign the values $f_d[\pm\frac{1}{2}NT] = \frac{1}{2}Tf_L(\pm\frac{1}{2}NT)$ to the edge samples.

The frequency domain equivalent of windowing is a convolution with $W(j\omega)$, which introduces ripples at bandedge discontinuities, once again extending the frequency representation to infinity, alas at low amplitude levels. The equivalent of sampling is a convolution with $\tilde{S}_T(j\omega)$ which yields a periodic $\widetilde{F}(j\omega)$.

The signal is now suitably conditioned for the discrete Fourier transform. Although not explicitly performed, the routine application of the discrete transform does imply sampling $\widetilde{F}(j\omega)$ with the discrete-frequency sampler $\tilde{S}_{\omega_0}[jk\omega_0]$, yielding $\widetilde{F}_d[jk\omega_0]$, and convolving the time domain with $\tilde{s}_{\omega_0}[nT]$, yielding the discrete and periodic time function $\tilde{f}_d[nT]$. In this form the samples are renumbered in the ranges $n = 0, 1, \ldots, N-1$ and $k = 0, 1, \ldots, N-1$, as highlighted in the bottom row of Figure 6.27, to agree with the discrete Fourier expression 6.42.

6.6.3   Windowing and leakage

An N-point discrete Fourier transform gives accurate results when the time function $f(t)$ is of short duration and fits one time period $T_0 = NT$. Longer signals need to be **windowed** to fit the transform, as shown above, which smears any sharp frequency domain features, thus introducing errors.

When interpreting periodic or near periodic continuous-time signals in this form, that smearing causes an effect called **leakage**. The effect can be seen in isolation by examining the discrete-frequency Fourier components (Fourier series) of a continuous and periodic signal constructed by replicating the windowed signal.

For simplicity we take a single cosine wave $f(t) = \cos \omega_c t$ of frequency ω_c and period $T_c = 2\pi/\omega_c$, represented in both domains at the top of Figure 6.28, and use a rectangular window shape. We wish to establish the effects on the discrete frequency representation when slightly different window widths are used, such that they capture either an integer or a non-integer number of cosine cycles.

The window $w_1(t)$ used in the upper half of Figure 6.28 has a width that spans precisely three cosine cycles $T_1 = 3T_c$, so that the zero crossings of its transform $W_1(j\omega)$ fall on integer multiples of

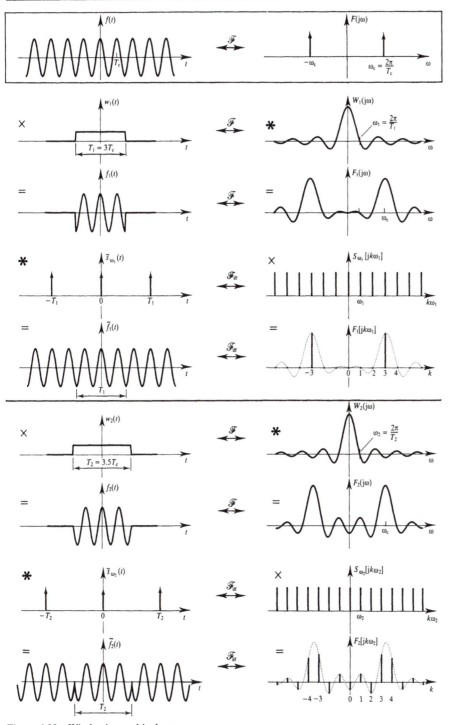

Figure 6.28   Windowing and leakage.

$\omega_1 = 2\pi/T_1 = \frac{1}{3}\omega_c$. The time domain multiplication with $w_1(t)$ corresponds to a frequency domain convolution

$$f_1(t) = f(t)w_1(t) \quad \xleftrightarrow{\mathscr{F}} \quad F_1(j\omega) = F(j\omega) * W_1(j\omega)$$

where $F_1(j\omega)$ is also zero-valued at integer multiples of ω_1 other than those coinciding with $\pm\omega_c$.

We now assume that the windowed function $f_1(t)$ represents one period of a periodic function $\tilde{f}_1(t)$ (as would be the case when applying the discrete Fourier transform) and obtain the discrete-frequency components of the latter. In terms of the discrete-frequency sampler $\tilde{s}_{\omega_1}(t) \longleftrightarrow S_{\omega_1}[jk\omega_1]$ this represents the operation

$$\tilde{f}_1(t) = f_1(t) * \tilde{s}_{\omega_1}(t) \xleftrightarrow{\mathscr{F}_{df}} F_1[jk\omega_1] = F_1(j\omega)S_{\omega_1}[jk\omega_1]$$

where we used a shorthand notation for the mixed frequency domain product.

Because the window width T_1 had an integer number of cosine periods T_c, the convolution reconstructs the original cosine $f(t)$ exactly. For the same reason all sampling points $k\omega_1$, other than $\pm 3\omega_1$, coincide with the zero-valued points of $F_1(j\omega)$, and the frequency domain multiplication gives an exact discrete-frequency equivalent of the original function $F(j\omega)$.

In contrast, if the window width is not an integer multiple of the signal period, such exact equivalence cannot be achieved. An example is given in the lower half of Figure 6.28. A window $w_2(t) \longleftrightarrow W_2(j\omega)$ of width $T_2 = 3.5T_c$ and zero-crossings at multiples of $\omega_2 = 2\pi/T_2$ applied to the same function $f(t)$ produces the windowed function $f_2(t) \longleftrightarrow F_2(j\omega)$. We now construct the periodic function

$$\tilde{f}_2(t) = f_2(t) * \tilde{s}_{\omega_2}(t) \quad \xleftrightarrow{\mathscr{F}_{df}} \quad F_2[jk\omega_2] = F_2(j\omega)S_{\omega_2}[jk\omega_2]$$

where we find that all the zero-crossings of $F_2(j\omega)$ fall half-way between integer multiples of ω_2. As a consequence, all the samples $F_2[jk\omega_2]$ are non-zero, the signal power has **leaked** into the side-lobes associated with the window.

Appendix: Fast Fourier transforms

The generic name fast Fourier transform (FFT) covers a family of algorithms developed for the fast and efficient computation of the discrete Fourier transform $\mathscr{F}_d$. The resulting values are identical to those obtainable by a direct, if lengthy, evaluation of the expressions 6.42 and 6.44, so that in terms of signal fundamentals the FFTs make no further contributions and are completely irrelevant. In terms of sheer speed their benefits are immense, making them indispensable for many applications.

We explain the principles underlying the algorithm family by examining one specific example known as the **radix 2 decimation in time FFT**. In essence, it interprets the original time sequence $f[nT]$ of length N as the sum of two interleaved sequences of length $\frac{1}{2}N$, and derives the N-point transform of $f[nT]$ from the $\frac{1}{2}N$-point transforms of the component sequences. This artifice effectively halves the number of required operations. Repeating the subdivision, or decimation, to the half-length sequences, and then to the quarter-length sequences, etc., leads to a progressive improvement in computing efficiency, without loss of accuracy.

Separation by even and odd indices n

To simplify presentation we use the real even function $f[nT] \longleftrightarrow F[jk\omega_0]$ of Example 6.4, which has an even number N of samples. This function is shown again in the top row of Figure 6.29, where $\mathcal{F}_N$ represents the N-point discrete Fourier transform operator, and we wish to interpret both domains as sums of the two functions of the lower half of the figure, where $f_0[nT]$ contains the even-indexed samples of $f[nT]$ and $f_1[nT]$ the odd-indexed samples.

To generate these two functions, especially their frequency representations, we form the auxiliary function $\bar{f}[nT] \longleftrightarrow \bar{F}[jk\omega_0]$, where $\bar{F}[jk\omega_0]$ is $F[jk\omega_0]$ frequency shifted by $\frac{1}{2}N$ samples, that is, by half the period $N\omega_0$. The corresponding time domain $\bar{f}[nT]$ can be interpreted as a fully discrete dual counterpart of Figure 1.22, in that half a revolution of the first sample $f[T]$ of Figure 6.29, taken as 'fundamental', gives a proportional number of half turns of all other samples $f[nT]$ and corresponds to a half-period displacement of the alternative domain $F[jk\omega_0]$.

The semi-sum of $f[nT]$ and $\bar{f}[nT]$ extracts the even-indexed samples of $f[nT]$ as expressed by the N-point transform pair

$$f_0[nT] = \tfrac{1}{2}(f[nT] + \bar{f}[nT])$$
$$\xleftarrow{\mathcal{F}_N} \quad F_0[jk\omega_0] = \tfrac{1}{2}(F[jk\omega_0] + \bar{F}[jk\omega_0])$$

while the semi-difference yields the complementary pair

$$f_1[nT] = \tfrac{1}{2}(f[nT] - \bar{f}[nT])$$
$$\xleftarrow{\mathcal{F}_N} \quad F_1[jk\omega_0] = \tfrac{1}{2}(F[jk\omega_0] - \bar{F}[jk\omega_0])$$

These results are shown in the bottom half of Figure 6.29 and their addition leads back to the original function,

$$f[nT] = f_0[nT] + f_1[nT]$$
$$\xleftarrow{\mathcal{F}_N} \quad F[jk\omega_0] = F_0[jk\omega_0] + F_1[jk\omega_0] \quad (6.48)$$

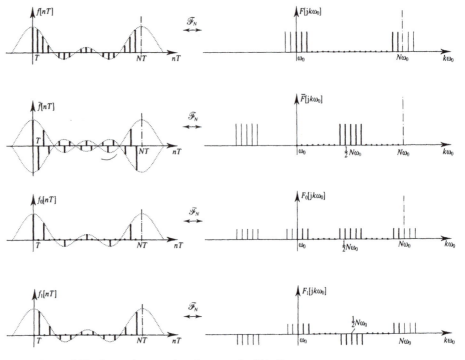

Figure 6.29   Separating samples of even and odd indices.

This addition is germane to the next argument, as expressed in the central part of Figure 6.30.

Half-length transforms

Each of the functions $f_0[nT]$ and $f_1[nT]$ has only $\frac{1}{2}N$ non-zero samples, whose frequency domain can be obtained by means of a $\frac{1}{2}N$-point transform.

The process for the even-index case is shown in the top-down sequence of Figure 6.30. The N-point transform of $f_0[nT]$ yields the N-point function $F_0[jk\omega_0]$, whose nominal period $N\omega_0$ has two identical cycles, one of which carries redundant information. All the values of $F_0[jk\omega_0]$ can be derived from the $\frac{1}{2}N$-point transform of a related sequence $g_0[mT_2] = f_0[2mT]$, whose sampling interval is $T_2 = 2T$. The transform

$$g_0[mT_2] \overset{\mathcal{F}_{N/2}}{\longleftrightarrow} G_0[jl\omega_0]$$

yields $\frac{1}{2}N$ samples of the frequency domain $G_0[jl\omega_0]$, each of which gives two samples of $F_0[jk\omega_0]$, namely

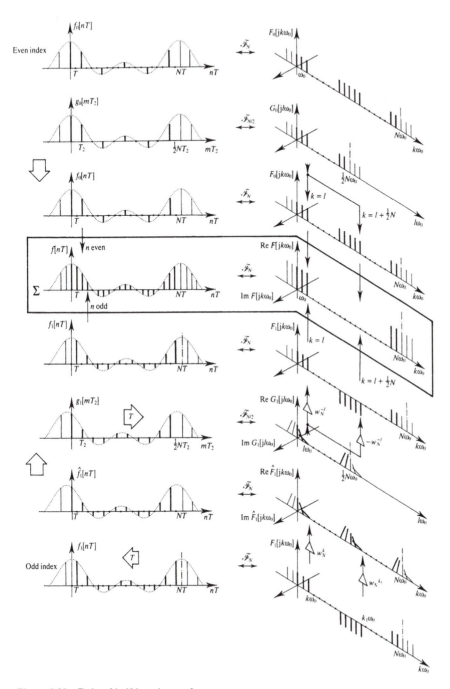

Figure 6.30 Role of half-length transforms.

$$F_0[jl\omega_0] = G_0[jl\omega_0] \quad \text{and} \quad F_0[j(l + \tfrac{1}{2}N)\omega_0] = G_0[jl\omega_0]$$
$$(6.49)$$

as shown in the top half of Figure 6.30, and provide the required N values of $F_0[jk\omega_0]$.

The odd-index function $f_1[nT]$ is ill-conditioned for such direct treatment. In the time domain its non-zero samples are positioned mid-way between sampling points mT_2, while the period of the frequency representation $F_1[jk\omega_0]$ is of full length $N\omega_0$. This ill-conditioning is overcome by the bottom-up sequence of Figure 6.30. It involves an auxiliary function $\hat{f}_1[nT]$, obtained by shifting $f_1[nT]$ one sampling interval T to the left, expressed as

$$\hat{f}_1[nT] = f_1[nT + T] \quad \overset{\mathscr{F}_N}{\longleftrightarrow} \quad \hat{F}_1[jk\omega_0] = e^{jk\omega_0 T} F_1[jk\omega_0]$$

which represents a linear-phase twisting of $F_1[jk\omega_0]$. This is consistent with the imagery of Figure 1.20, in that one full turn of the **fundamental** frequency component $F_1[j\omega_0]$ would produce N full turns of the component $F_1[jN\omega_0]$ and correspond to a time domain shift of one full period $T_0 = NT$. But the factor $e^{jk\omega_0 T}$ only produces $1/N$ turns of the fundamental, hence one full turn of $F_1[jN\omega_0]$, and corresponds to a time shift of one sampling interval T. Significantly, the mid-period sample $F_1[j\tfrac{1}{2}N\omega_0]$, which is diametrically opposite to $F_1[0]$, turns by half a revolution, thus lining it up with $F_1[0]$ and causing $F_1[jk\omega_0]$ to acquire two identical cycles in the nominal period $N\omega_0$.

The shifted function $\hat{f}_1[nT]$ is formally similar to $f_0[nT]$ regarding applicability of the $\tfrac{1}{2}N$-point transform. We associate with it a function $g_1[mT_2] = \hat{f}_1[2mT]$ with transform

$$g_1[mT_2] \quad \overset{\mathscr{F}_{N/2}}{\longleftrightarrow} \quad G_1[jl\omega_0]$$

whose frequency domain only needs to be duplicated to fill the period $N\omega_0$ of $\hat{F}_1[jl\omega_0]$. The results are back-converted to the original variables, which involves shifting the time domain to the right by one

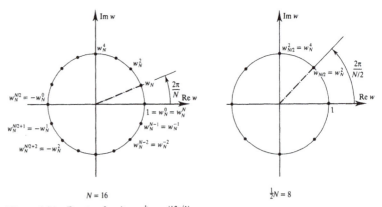

$N = 16$ $\qquad\qquad\qquad\qquad\qquad \tfrac{1}{2}N = 8$

Figure 6.31 Roots of unity $w_N^k = e^{jk2\pi/N}$.

interval T. For this, the frequency domain is multiplied by $e^{-jk\omega_0 T}$, expressed as

$$F_1[jl\omega_0] = e^{-jl\omega_0 T} G_1[jl\omega_0]$$

and   (6.50)

$$F_1[j(l + \tfrac{1}{2}N)\omega_0] = e^{-j(l+N/2)\omega_0 T} G_1[jl\omega_0]$$

The samples thus derived with the aid of half-length transforms are recombined in the middle of Figure 6.30, as mentioned earlier.

Note that in terms of the sampling interval T_2 the function $g_1[mT_2]$ is related to $f_1[nT]$ by a **half-sample time shift**, a concept to be found again in Section 11.2.4.

Roots of unity

The preceding developments explain the time domain and frequency domain implications of using half-length transforms. We now show how this concept applies to FFT algorithms in terms of the simplified expressions 6.46.

The complex number $w_N = e^{j2\pi/N}$, where the subscript N signifies the N-point transform, represents an Nth root of unity. Its integer powers $w_N^0, w_N^1, w_N^2, \ldots, w_N^{N-1}$ form the set of N distinct roots of unity, as shown in the left half of Figure 6.31 for $N = 16$. The Nth power coincides with w_N^0, that is $w_N^N = w_N^0$. Each higher power coincides with one of the distinct roots, and so do negative powers, as indicated in the figure. For even N we also have $w_N^{N/2} = -w_N^N$ and, more generally, $w_N^{l+N/2} = -w_N^l$, which we will see halves the number of multiplications in expressions 6.50.

The half-length transform uses $\tfrac{1}{2}N$ powers of the complex number $w_{N/2} = e^{j\pi/N}$ and because $w_{N/2} = w_N^2$, so that $w_{N/2}^l = w_N^{2l}$, these form a subset of the N distinct roots w_N^k, as shown in the right half of Figure 6.31. This extends to the shorter transforms, all of which take their values from the set of Nth roots.

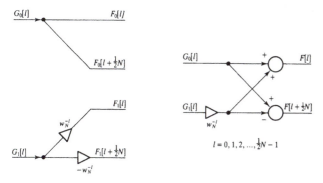

Figure 6.32   FFT butterfly.

Butterflies

With the simpler notations the result 6.49 takes the form

$$F_0[l] = G_0[l] \qquad \text{and} \qquad F_0[l + \tfrac{1}{2}N] = G_0[l]$$

and expressions 6.50 become

$$F_1[l] = w_N^{-l} G_1[l]$$

and

$$F_1[l + \tfrac{1}{2}N] = w_N^{-(l+N/2)} G_1[l] = -w_N^{-l} G_1[l]$$

which are expressed schematically by two branched diagrams in the left half of Figure 6.32. These diagrams are also found in the frequency domain of Figure 6.30, where the N individual values of $F[k]$ are obtained by combining the contributions from $F_0[k]$ and $F_1[k]$, according to expression 6.48.

This latter operation is expressed succinctly in the schematic of the right half of Figure 6.32, which combines the two branched diagrams. In FFT jargon the resulting diagram is called a **butterfly**, a reflection on its shape.

Recursive application

In the preceding discussions we built up the N-point sequence $F[k]$ from two known half-length sequences $G_0[l]$ and $G_1[l]$. This is shown in a more conventional form in the right-most stage of Figure 6.33, where four butterflies yield $N = 8$ values $F[k]$.

But the sequences $G_0[l]$ and $G_1[l]$ need not be computed directly

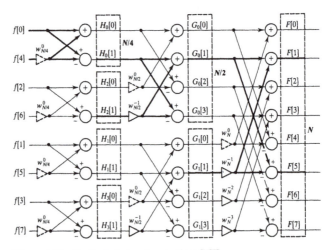

Figure 6.33 Radix 2 decimation in time FFT.

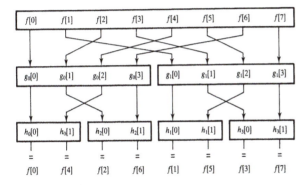

Figure 6.34 Shuffling the time sequence $f[n]$.

from the time samples $f[n]$. They can be built up from four related quarter-length sequences $H_0[m]$, $H_2[m]$, $H_1[m]$ and $H_3[m]$, as indicated in the middle stage of Figure 6.33. This involves formally identical butterflies, whose multipliers are powers of $w_{N/2}$.

This process is back-propagated until the first stage is reached. This stage involves $\frac{1}{2}N$ butterflies operating on pairs of the time samples $f[n]$, effectively representing $\frac{1}{2}N$ two-point discrete Fourier transforms. To conform with the scaling convention of Equation 6.46 it is necessary to scale either the input samples or the final results by $1/N$.

Note that the repeated time domain decimation implied in the process does not present the time samples $f[n]$ in their natural order. Recall from Figure 6.30 that $G_0[l]$ and $G_1[l]$ could have been obtained from the time sequences $g_0[m]$ and $g_1[m]$, which contained the even-index and odd-index samples of $f[n]$. This separation is clarified in Figure 6.34 for the radix 2 case $N = 8$, where the even-index and odd-index samples of $g_0[m]$ and $g_1[m]$ are similarly stripped out to give four time sequences $h_0[i]$, $h_2[i]$, $h_1[i]$ and $h_3[i]$, with the resulting equivalences $h_0[0] = f[0]$, $h_0[1] = f[4]$, etc.

Conclusions

Taking the number of multiplications as a rough guide to computing effort, the FFT algorithms are very efficient. A direct evaluation of the discrete Fourier transform expression 6.46 requires N complex multiplications for each sample $F[k]$, a total of N^2 multiplications for the set of N samples. In contrast, each stage of the FFT, as shown in Figure 6.33, requires $\frac{1}{2}N$ multiplications (including the trivial cases $w_N^0 = 1$ indicated by dashed lines in the figure). For radix 2 algorithms, where N is a power of 2, $N = 2^r$, the number of stages is $r = \log_2 N$, and the number of complex multiplications is $\frac{1}{2}rN = \frac{1}{2}N \log_2 N$.

The savings become enormous for large N. For the example of

Figure 6.33, where $N = 8$ and involves three stages, the FFT requires 12 multiplications (seven of which are by unity). Compared to $N^2 = 64$ this represents a saving by a factor of 5.3. For $N = 2^{10} = 1024$, which involves $r = 10$ stages, the FFT requires 5120 multiplications, compared to $N^2 = 1024^2$ multiplications for the direct evaluation, a ratio of approximately 200.

We illustrated the principles underlying FFTs by means of the **radix 2 decimation in time algorithm** for the forward transform. Virtually the same interpretation can be given to the inversion formula. A large variety of algorithms are found in the literature, see for instance Rabiner and Gold (1975). These differ basically in the domain chosen for decimation and in the radix. In this context the example considered in Figures 6.29 and 6.30 represents a mixed radix case, as it has $N = 20$ points and would require two stages of radix 2 and one of radix 5.

Exercises

6.1   Verify the relationships and equivalences used in Equation 6.5 to describe an impulse train $\delta_T(t)$. Discuss the usage of the Fourier series as an abstract analytical tool applied separately in both the time domain and frequency domain.

6.2   Discuss time domain and frequency domain implications when a time function $f(t)$ is sampled either with an impulse train $\delta_T(t)$ or with the related time sampler $s_T(t) = T\delta_T(t)$. If $f(t)$ is suitably band-limited to avoid aliasing, what signal is recovered in each case when an ideal lowpass filter $H(j\omega)$ of unit magnitude and width $\pm\omega_s/2$ is subsequently applied to the sampled function.

6.3   By what criterion can it be said that the sampled-time function $f_s(t)$ and the discrete-time function $f_d[nT]$ used in Equation 6.15 and in Figures 6.6 and 6.7 are equivalent. Confirm the applicability of the transform class used in each case.

6.4   Find the discrete-time Fourier transform and plot both domains of the functions

$$f_1[nT] = 0.5^n u[nT] \qquad\qquad f_2[nT] = (-0.5)^n u[nT]$$

$$f_3[nT] = \cos \omega_1 nT \qquad\qquad f_4[nT] = \sin \omega_1 nT$$

$$f_5[nT] = a^{|n|} \quad \text{with } 0 < a < 1$$

6.5   Find the inverse discrete-time Fourier transform of the periodic functions

$$\widetilde{F}_1(j\omega) = \cos \omega T \qquad\qquad \widetilde{F}_2(j\omega) = \sin \omega T$$

6.6   Derive the time-shifting property of the discrete-time Fourier transform

$$f[nT - mT] \quad \xleftrightarrow{\mathscr{F}_{dt}} \quad e^{-j\omega mT} \widetilde{F}(j\omega)$$

Apply this result to write the transform of the difference equation

$$a_2 y[nT - 2T] + a_1 y[nT - T] + a_0 y[nT] = x[nT]$$

6.7 Derive the time-convolution property of the discrete-time Fourier transform

$$f_1[nT] * f_2[nT] \quad \xleftrightarrow{\mathscr{F}_{dt}} \quad \widetilde{F}_1(j\omega) \widetilde{F}_2(j\omega)$$

Apply this result to interpret the time-shifting property of Exercise 6.6 as a time domain convolution with a shifted impulse $\delta[nT - mT]$.

6.8 Find the discrete-frequency Fourier transform of the periodic functions of Figures 1.15 and 1.16 of Chapter 1, but using the notation of Section 6.3 to express periodicity and discreteness of the relevant domains more explicitly. Re-examine the developments of Section 1.4 in this light.

6.9 Find the discrete Fourier transform of the dual case of Figure 6.21, that is, a time function of period $T_0 = NT$, whose period at the origin is described as

$$f[nT] = \begin{cases} 1 & -L \leqslant n \leqslant L \\ 0 & \text{elsewhere} \end{cases}$$

Evaluate and sketch both domains (including the frequency domain envelope) for $L = 2$ and the three cases $N = 20$, $N = 10$ and $N = 5$.

6.10 Show that the discrete Fourier transform expressions 6.42 and 6.44 are self-consistent by taking the forward transform 6.42 of the inversion formula 6.44.

6.11 Show that the properties regarding even and odd functions and real and imaginary parts of the continuous Fourier transform also apply to the discrete forms of the Fourier transform. Specifically, show that the even and odd parts of a real discrete time function $f[nT]$ transform as

$$\text{Re} \, f_e[nT] \quad \xleftrightarrow{\mathscr{F}_d} \quad \text{Re} \, F_e[jk\omega_0]$$

$$\text{Re} \, f_o[nT] \quad \xleftrightarrow{\mathscr{F}_d} \quad j \, \text{Im} \, F_o[jk\omega_0]$$

6.12 Based on Fourier transform dualities, extend the interpretation of the radix 2 decimation in time FFT to decimation in the frequency domain, by splitting the frequency representation into two 1/2-length sequences

$$F[jk\omega_0] = F_0[jk\omega_0] + F_1[jk\omega_0]$$

CHAPTER 7

Discrete-time Laplace Transform and z-transform

The z-transform is to discrete-time signals and systems what the Laplace transform is to their continuous-time counterparts. It provides a general and concise relationship between the time domain and frequency domain representations of discrete-time signals and serves to characterize systems of the applicable class and to formulate appropriate solution processes.

The similarities between these transforms are fundamental and extend to their properties and applications to systems. This is the main motivation for seeking and exploiting their relationships to give a coherent treatment of all signal processing.

For this we derive the z-transform expressions from those of the Laplace transform, introducing the discrete-time Laplace transform as the essential link. The resulting relationships are used in later chapters to give continuous-time insight into discrete-time systems and to introduce and compare various discrete-time simulations of continuous-time systems.

The similarities extend to the discrete-time Fourier transform, which is now interpreted as a subset of both the discrete-time Laplace transform and of the z-transform, in the same sense as the continuous Fourier transform is a subset of the Laplace transform. These relationships are important when formulating signal processing concepts in terms of the system's frequency response, but employing the terminology and notation of the z-transform.

All the time-frequency transforms are gathered at the end of the chapter on the framework of signal classification. The interrelationships emerging in their derivations provide the analytical basis on which to formulate simulations between continuous-time and discrete-time systems.

7.1   Discrete-time Laplace transform

In the preceding chapter we derived a discrete-time version of the Fourier transform valid for discrete-time functions $f_d[nT]$. Since the Fourier transform is a subset of the Laplace transform, we expect a similar discrete-time version of the latter. By the analogy suggested in Figure 7.1, we call this the discrete-time Laplace transform of $f_d[nT]$ and identify it by the symbol $\mathcal{L}_{dt}$.

Where practicable, we replicate the approach of Section 6.2 to derive the transform expressions, thus stressing similarities. This approach breaks down in the context of the bilateral Laplace transform of periodic functions, including the infinite impulse train $\delta_T(t)$, and we will examine the role of the unilateral Laplace transform in Section 7.5.3.

As already mentioned, our main objective here is to develop the discrete-time Laplace transform as the link to the *z*-transform. We will examine properties later in the context of the *z*-transform. This transform tends to be neglected in the literature of **signals and systems** and of **digital signal processing**, where the more compact *z*-transform takes precedence. However, it often appears in **control systems** literature, where it is unimaginatively called the **starred Laplace transform** $\mathcal{L}^*$.

7.1.1   Forward transform

We repeat the time sampling process of Section 6.2.1, as outlined in Figure 6.6, but in the wider context of the generalized frequency variable *s* of the Laplace transform. Multiplying the time domain of the

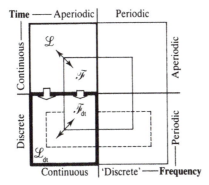

Figure 7.1   Laplace transform to discrete-time Laplace transform.

auxiliary function

$$f(t) \quad \overset{\mathcal{L}}{\longleftrightarrow} \quad F(s) \qquad \alpha < \mathrm{Re}\, s < \beta$$

by the time-sampler $s_T(t)$ yields the same sampled function $f_s(t)$,

$$f_s(t) = f(t)s_T(t) = T \sum_{m=-\infty}^{\infty} f(mT)\delta(t - mT) \tag{7.1}$$

This function is again converted to discrete time, but now specifying that the Laplace domain should remain the same.

Each continuous-time impulse $Tf(mT)\delta(t - mT)$ of Equation 7.1, with Laplace transform $Tf(mT)\mathrm{e}^{-smT}$, is replaced by an equivalent discrete-time impulse $f_d[mT]\delta[nT - mT]$, whose frequency representation $f_d[mT]\mathrm{e}^{-smT}$ is the same as that of the continuous impulse, as expressed in the scheme

$$
\begin{array}{ccc}
Tf(mT)\delta(t - mT) & \overset{\mathcal{L}}{\longleftrightarrow} & Tf(mT)\mathrm{e}^{-smT} \\
\Big\downarrow \equiv & f_d[mT] = Tf(mT) & \Big\downarrow \equiv \\
f_d[mT]\delta[nT - mT] & \overset{\mathcal{L}_{dt}}{\longleftrightarrow} & f_d[mT]\mathrm{e}^{-smT}
\end{array}
\tag{7.2}
$$

Equating frequency representations gives the conversion law $f_d[mT] = Tf(mT)$, which is identical to Equation 6.14. Applied to the time domain impulses, this law also serves to define the discrete-time Laplace transform of an impulse. Extending the latter to all the impulses of Equation 7.1 yields

$$f_d[nT] = \sum_{m=-\infty}^{\infty} f_d[mT]\delta[nT - mT]$$

$$\overset{\mathcal{L}_{dt}}{\longleftrightarrow} \quad \widetilde{F}(s) = \sum_{m=-\infty}^{\infty} f_d[mT]\mathrm{e}^{-smT}$$

which provides a convolution interpretation of the time domain and, using the dummy summation variable n for the frequency domain, defines the **forward discrete-time Laplace transform** as

$$\boxed{\; \widetilde{F}(s) = \sum_{n=-\infty}^{\infty} f_d[nT]\mathrm{e}^{-snT} \qquad \alpha < \mathrm{Re}\, s < \beta \;} \tag{7.3}$$

We will show that the frequency domain $\widetilde{F}(s)$ is a continuous function of s, that it is periodic in the imaginary component ω, whose period is the sampling frequency $\omega_s = 2\pi/T$, and that the associated region of convergence is that of the auxiliary function $F(s)$.

Example 7.1

Find the discrete-time Laplace transform of the function $f[nT] = e^{\alpha nT} u[nT]$, the second form of Example 6.1. Applying Equation 7.3 gives

$$\widetilde{F}(s) = \sum_{n=-\infty}^{\infty} e^{\alpha nT} u[nT] e^{-snT} = \sum_{n=0}^{\infty} (e^{\alpha T} e^{-sT})^n$$

and the pair

$$f[nT] = e^{\alpha nT} u[nT] \xleftrightarrow{\mathscr{L}_{dt}} \widetilde{F}(s) = \frac{1}{1 - e^{\alpha T} e^{-sT}} \qquad \text{Re } s > \alpha \qquad (7.4)$$

which is the s-plane generalization of the imaginary axis result 6.19. It will be visualized later in Figure 7.2.

Interpretation as discrete-time Fourier transform

In Section 3.1 we interpreted the Laplace transform of a function $f(t)$ as the Fourier transform of a modified function $f(t)e^{-\sigma t}$,

$$F(\sigma + j\omega) = \mathscr{F}\{f(t)e^{-\sigma t}\} \qquad (7.5)$$

where $e^{-\sigma t}$ represents a convergence factor. We based many of the Laplace transform interpretations on this relationship.

The discrete-time versions of these transforms can be similarly related. Writing the complex frequency variable s of Equation 7.3 as $s = \sigma + j\omega$, we have

$$\widetilde{F}(\sigma + j\omega) = \sum_{n=-\infty}^{\infty} \{f_d[nT]e^{-\sigma nT}\}e^{-j\omega nT} \qquad \alpha < \sigma < \beta \qquad (7.6)$$

This has the interpretation

$$\widetilde{F}(\sigma + j\omega) = \mathscr{L}_{dt}\{f_d[nT]\} = \mathscr{F}_{dt}\{f_d[nT]e^{-\sigma nT}\} \qquad (7.7)$$

which is a discrete-time version of the relationship 3.3, where time comes into play only at discrete values $t = nT$. It permits extending the results, properties and graphics of the discrete-time Fourier transform to the discrete-time Laplace transform.

Conversely, provided it exists, the discrete-time Fourier transform of a function represents a subset of the discrete-time Laplace transform, located on the imaginary frequency axis,

$$\mathscr{F}_{dt}\{f_d[nT]\} = \mathscr{L}_{dt}\{f_d[nT]\}|_{s=j\omega} \qquad (7.8)$$

This can also be expressed as

$$\widetilde{F}(j\omega) = \widetilde{F}(s)\big|_{s=j\omega} = \widetilde{F}(\sigma + j\omega)\big|_{\sigma=0}$$

For example, from expression 7.4 we derive the pair

$$f[nT] = e^{\alpha nT}u[nT] \quad \overset{\mathscr{F}_{dt}}{\longleftrightarrow} \quad \widetilde{F}(j\omega) = \frac{1}{1 - e^{\alpha T}e^{-j\omega T}} \qquad \alpha < 0$$

Graphics

Recall the constructions of Figure 3.10, which visualized the Laplace transform of a continuous function $f(t)$ on the basis of the Fourier interpretation expressed by Equation 7.5. We now extend those constructions to the discrete-time Laplace transform 7.3, on the basis of the related Fourier interpretation 7.6.

This is illustrated in Figure 7.2, using the function of Example 7.1, a discrete version of the decaying exponential of Figure 3.10. The discrete-time Fourier transform implied in Equation 7.6 brings into play only discrete planes $e^{-j\omega nT}$ of the exponential surface, and conveys to each σ-slice of $\widetilde{F}(s)$ the periodicity and aliasing associated with discreteness, as seen in the upper half of the figure.

The net result is that the whole function $\widetilde{F}(s)$ is periodic in the direction of the imaginary axis. All its features, including real and imaginary parts, magnitude and phase and the location of poles, zeros and region of convergence, are contained in the fundamental strip surrounding the origin and are redundantly repeated with period $\omega_s = 2\pi/T$, as seen in the lower half of Figure 7.2.

Applications in which both $F(s)$ and $\widetilde{F}(s)$ coexist need the two s-planes to be distinguished. In such cases we will associate the term **periodic s-plane** with $\widetilde{F}(s)$ and identify it as the $\check{s}$**-plane**, with imaginary axis $j\widetilde{\omega}$ and real axis $\check{\sigma}$, but keeping in mind that $\widetilde{F}(s)$ is not periodic in the direction of the real axis.

7.1.2   Inverse transform

The inversion formula for Equation 7.3 is now derived by the process employed in Section 3.2. Taking the inverse discrete-time Fourier transform of both sides of Equation 7.7 cancels the forward transform of the right side, and leads to the inverse interpretation

$$\boxed{f_d[nT] = e^{\sigma nT}\mathscr{F}_{dt}^{-1}\{\widetilde{F}(\sigma + j\omega)\}} \tag{7.9}$$

Applying the Fourier expression 6.22 and using a symmetrical integration interval, we write

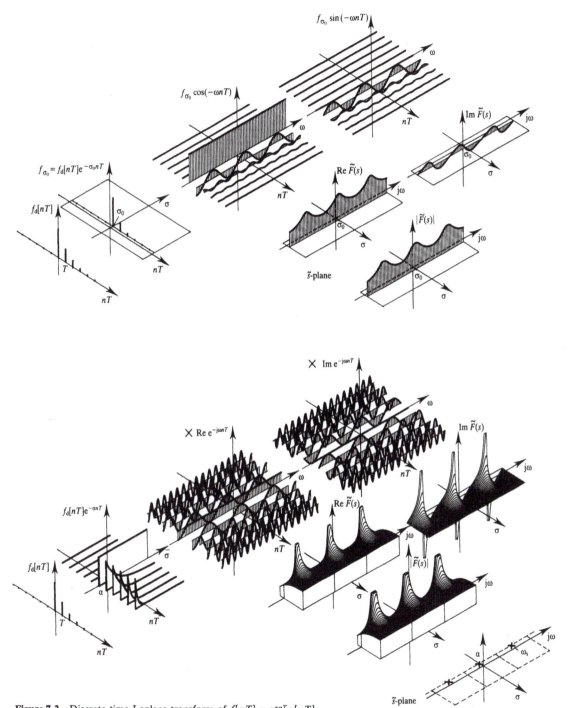

Figure 7.2   Discrete-time Laplace transform of $f[nT] = e^{\alpha nT} u[nT]$.

$$f_{\mathrm{d}}[nT] = \frac{1}{\omega_{\mathrm{s}}} \int_{-\omega_{\mathrm{s}}/2}^{\omega_{\mathrm{s}}/2} \widetilde{F}(\sigma + \mathrm{j}\omega) \mathrm{e}^{(\sigma+\mathrm{j}\omega)nT} \, \mathrm{d}\omega$$

Changing the integration variable from ω to $s = \sigma + \mathrm{j}\omega$, which implies $\mathrm{d}s = \mathrm{j}\mathrm{d}\omega$ and a change of the integration limits from $-\frac{1}{2}\omega_{\mathrm{s}} < \omega < +\frac{1}{2}\omega_{\mathrm{s}}$ to $\sigma - \frac{1}{2}\mathrm{j}\omega_{\mathrm{s}} < \sigma + \mathrm{j}\omega < \sigma + \frac{1}{2}\mathrm{j}\omega_{\mathrm{s}}$, and with $\omega_{\mathrm{s}} = 2\pi/T$ yields

$$\boxed{f_{\mathrm{d}}[nT] = \frac{T}{2\pi\mathrm{j}} \int_{\sigma-\mathrm{j}\omega_{\mathrm{s}}/2}^{\sigma+\mathrm{j}\omega_{\mathrm{s}}/2} \widetilde{F}(s) \mathrm{e}^{snT} \, \mathrm{d}s} \tag{7.10}$$

This integral has formal similarities with both the inverse Laplace transform 3.6 and the discrete-time Fourier inversion formula 6.22. It represents the area contained within one period of a σ-slice of the integrand. The slice should be interpreted in terms of the continuous Laplace transform of Figure 3.10, while the integration limits should be interpreted in terms of those of the inverse discrete-time Fourier transform of Section 6.2.2.

Referring to the upper half of Figure 7.2, the inverse discrete-time Laplace transform is interpreted according to Equation 7.9 as an inverse discrete-time Fourier transform taken on a slice parallel to the imaginary axis of the s-plane.

Compared to the case of the continuous Laplace transform, the s-plane offers even more redundancy. Any single σ-slice is still a complete and self-contained frequency representation of the original time function $f_{\mathrm{d}}[nT]$. Furthermore, as a result of periodicity in ω, a single period of any slice, such as that highlighted in the s-plane of the upper half of Figure 7.2, contains all the information of the sampled-time function, making all other periods redundant.

These results and conclusions meet our immediate requirements for turning the discrete-time Laplace transform into the link between the Laplace transform and the z-transform of the next section. For brevity we do not dwell on examples and properties of the discrete-time Laplace transform, as these are closely related to those of the z-transform, and can easily be derived from them.

7.2   The z-transform

To avoid the periodic form of redundancy of the discrete-time Laplace transform, we select a new frequency variable z, such that $\mathrm{e}^{sT} = z$, which maps the infinite cycles of the s-plane onto a single representation of the z-plane. This change of variable simplifies the notation, expressions and representation of the discrete-time Laplace transform 7.3, which is now called the z-transform. Its place on the classification chart coincides with that of the discrete-time Laplace transform, as indicated in Figure 7.3.

All the properties of the Laplace transform are thus transferred via the discrete-time Laplace transform to the z-transform. This includes the interpretation of the discrete-time Fourier transform as a subset of the z-transform, also indicated in Figure 7.3, and this last link completes our systematic study of the family of time-to-frequency transforms.

We start the section by introducing the complex frequency variable z and its associated z-plane and show the relationship of these to the complex variable s and its associated s-plane.

7.2.1 The z-variable

The proposed change of frequency variable can be interpreted in terms of mapping the complex s-plane, associated to the variable $s = \sigma + j\omega$, onto a complex z-plane, according to the mapping law

$$z = e^{sT} = e^{\sigma T}e^{j\omega T} \tag{7.11}$$

This involves a severe distortion of the s-plane, such that the fundamental cycle is stretched out to cover the entire z-plane. Points of all other identical s-plane cycles become superimposed onto the corresponding points of the z-plane.

Any features located on the s-plane, such as poles and zeros, are

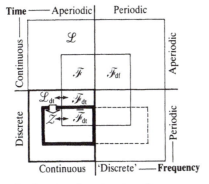

Figure 7.3 Discrete-time Laplace transform to z-transform.

moved to the corresponding z-plane locations. A function $\widetilde{F}(s)$ constructed on the s-plane is distorted accordingly, so that the function's values are read off the new z-plane coordinates.

s-plane to z-plane mapping

For mapping purposes the complex variable z is best expressed in polar form, in terms of magnitude r and angle θ, as

$$z = re^{j\theta}$$

Equated to expression 7.11 this yields the relationships

$$r = e^{\sigma T}$$
$$\theta = \omega T$$

(7.12)

which neatly decouple the mapping of the real and imaginary parts of s and are useful for visualizing the distortions of the s-plane grid.

Lines of constant σ are mapped as lines of constant radius r, that is, as the concentric circles shown at the top of Figure 7.4. The imaginary axis $j\omega$ always maps onto the unit circle $r = 1$, which becomes the main reference feature of the z-plane. Equally spaced lines of constant σ are mapped as exponentially spaced circles. This highly non-linear spacing fits the half-plane of negative σ inside the unit circle, where minus infinity maps onto the origin, $z = 0$.

In contrast, the linear relationship $\theta = \omega T$ maps lines of constant ω onto radial lines of angle θ. The real axis of the s-plane $\omega = 0$ maps onto the positive real semi-axis $\operatorname{Re} z$, and from there lines of increasing ω become radial lines of increasing θ, as seen in Figure 7.4. Those in the range $-\pi < \theta < \pi$, corresponding to $-\frac{1}{2}\omega_s < \omega < \frac{1}{2}\omega_s$, are principal values. Other lines become superimposed on these, and can not be distinguished. This provides an alternative way of looking at aliasing.

The remainder of Figure 7.4 illustrates the multi-valued correspondence between points located on such grid lines. The bottom row is a slight departure, in that it maps radial lines of the s-plane (which in Figure 4.16 represented lines of constant damping ratio, or constant Q-factor) as logarithmic spirals of the z-plane.

In conclusion, a horizontal s-plane strip of width $\omega_s = 2\pi/T$ maps onto the entire z-plane, such that the negative half of the strip falls inside the unit circle and the positive half outside it. Conversely, the corresponding portions of the z-plane can be mapped back as infinite replicas of such horizontal s-plane strips.

Inverse mapping

To map points of the z-plane onto the s-plane, it is expedient to simply invert the two independent relationships 7.12 as

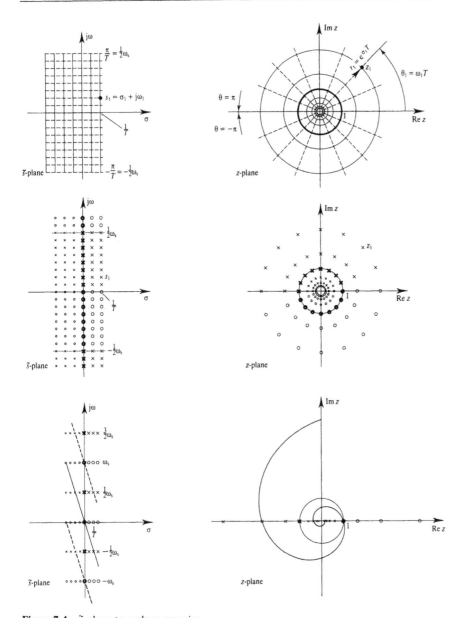

Figure 7.4 $\tilde{s}$-plane to z-plane mapping.

$$\sigma = \frac{1}{T}\ln r$$

$$\omega = \frac{1}{T}\theta = \frac{1}{T}(\Theta + n2\pi) \quad n = 0, \pm 1, \pm 2, \dots \tag{7.13}$$

The second of these is multivalued, as the angular position of a z-plane point is defined by any of the equivalent values

$$\theta = \Theta + n2\pi \qquad n = 0, \pm 1, \pm 2, \ldots$$

where Θ represents the principal value $(-\pi < \Theta \leqslant \pi)$ of θ.

These results are consistent with the definition of the natural logarithm of a complex function $w = re^{j\theta}$,

$$\ln w = \ln |w| + j\theta = \ln |w| + j(\Theta + n2\pi) \qquad n = 0, \pm 1, \pm 2, \ldots$$

which yields

$$s = \sigma + j\omega = \tfrac{1}{T} \ln r + j(\Theta + n2\pi)/T$$

But we must clarify the notation. The frequency variable s of the discrete-time Laplace transform is periodic in ω and in Chapter 11 we identify it as such by the symbol $\tilde{s} = \tilde{\sigma} + j\tilde{\omega}$. In the current chapter we deal exclusively with the discrete-time form of the Laplace transform, with the exception of Section 7.1.1 where the appropriate expressions were derived from the continuous Laplace transform. To simplify the notation of this chapter we identify the periodic frequency variable simply by $s = \sigma + j\omega$.

Effect of sampling rate

Note that both r and θ of Equations 7.12 are functions of the sampling interval T. Different sampling rates distort the continuous s-plane features to different degrees, generating different periodic $\tilde{s}$-planes and corresponding z-planes. These relationships will be explored in depth in Chapter 11, but we illustrate here with an example.

Consider for reference a continuous-time function $f(t)$, characterized in the s-plane by the location of its poles p_1 and p_1^*, as shown in the top row of Figure 7.5. Different sampling intervals T_1, T_2, T_3, replicate those poles in the periodic $\tilde{s}_i$-plane, with period $\omega_{s_i} = 2\pi/T_i$, as shown in the middle column of the figure. The relationships 7.12 map the pole coordinates to the corresponding z_i-planes of the last column of Figure 7.5. Note that the higher the sampling frequency, the closer the poles cluster in the region surrounding the point $z = 1$.

Taking the continuous-time case as the limit for $T \to 0$, we find that all the poles (the whole finite s-plane, for that matter) collapse onto the point $z = 1$. This gives one good reason for not representing continuous-time signals in the z-plane.

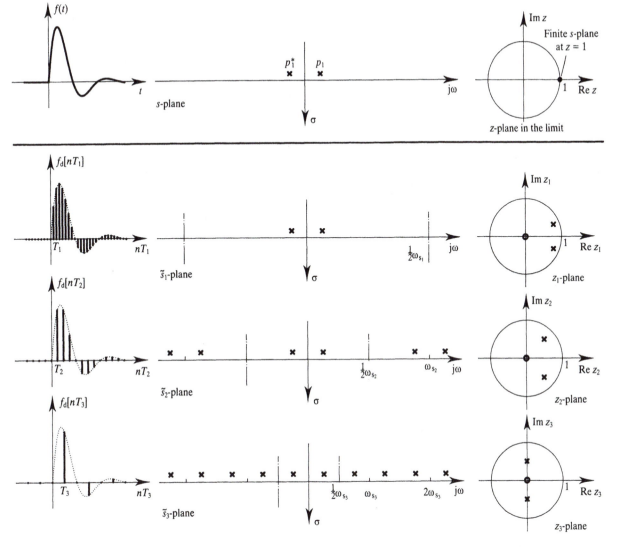

Figure 7.5 Effect of sampling rate on z-plane.

7.2.2 Forward transform

The sum defining the discrete-time Laplace transform was expressed in Equation 7.3 in terms of the frequency variable s as

$$\widetilde{F}(s) = \sum_{n=-\infty}^{\infty} f_{\mathrm{d}}[nT]\mathrm{e}^{-snT} \qquad \alpha < \mathrm{Re}\, s < \beta$$

But it can equally be interpreted as a function of the common factor e^{sT}. Taking this as a new frequency variable $e^{sT} = z$, the above sum becomes a function of z,

$$\bar{F}(z) = \sum_{n=-\infty}^{\infty} f_d[nT]z^{-n} \qquad r_\alpha < |z| < r_\beta \tag{7.14}$$

which defines the **z-transform** of a discrete-time function $f_d[nT]$ also denoted as $\mathscr{Z}\{f_d[nT]\}$.

We next relate these expressions graphically, and show that the s-plane strip of convergence of the discrete-time Laplace transform maps to a ring of convergence on the z-plane.

Graphical interpretation

Recall Figure 7.2, where we constructed the discrete-time Laplace transform $\widetilde{F}(\sigma + j\omega)$ from slices of constant σ. Each of the surfaces drawn on the periodic s-plane represented one aspect of the function $\widetilde{F}(s)$, such as real part, imaginary part, magnitude or phase. We now map those slices onto the z-plane, thus visualizing the process implied in deriving Equation 7.14.

The relevant mapping property is that of the top row of Figure 7.4, which maps lines of constant σ onto concentric circles. The associated slices $\widetilde{F}(\sigma + j\omega)$ are accordingly distorted into concentric cylinders $\bar{F}(re^{j\theta})$, as illustrated in Figure 7.6, where one period of the magnitude $|\widetilde{F}(s)|$ of Figure 7.2 is mapped onto the z-plane. The top row shows the distortion of one slice of arbitrary $\sigma = \sigma_0$, the middle row the distortion of the whole surface and the bottom row that of the slice corresponding to the imaginary axis $\sigma = 0$.

For corresponding points of the two planes the sums 7.3 and 7.14 are numerically identical, so that the function values $\widetilde{F}(s)$ and $\bar{F}(z)$ can be transferred from one plane to the other as

$$\widetilde{F}(s) = \bar{F}(z)\big|_{z=e^{sT}} \qquad \text{or} \qquad \bar{F}(z) = \widetilde{F}(s)\big|_{s=\frac{1}{T}\ln z}$$

Thus, given an arbitrary point of the z-plane (e.g. $z = 1$), we find its s-plane location $s = 1/T \ln z$ (e.g. $s = 0$) and transfer the corresponding value $\bar{F}(z) = \widetilde{F}(s)$ (e.g. $|\bar{F}(1)| = |\widetilde{F}(0)|$) to the given z-plane point. Repeated for all points of the z-plane, this process relocates values of $\widetilde{F}(s)$ on the distorted grid, thus yielding $\bar{F}(z)$.

Similar constructions apply to the surfaces $\operatorname{Re}\widetilde{F}(s)$ and $\operatorname{Im}\widetilde{F}(s)$ of Figure 7.2. All the features of the discrete-time Laplace transform, including the location of the poles and the region of convergence, are thus transferred to the z-transform. For instance, a pole located at $s = p_1$ transfers to $z = \pi_1 = e^{p_1 T}$. Similarly, the linear boundaries $\sigma = \alpha$

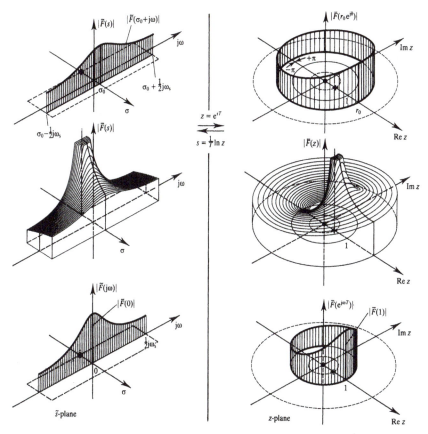

Figure 7.6 Relationship between discrete-time Laplace transform and z-transform.

and $\sigma = \beta$ of the region of convergence become the circular boundaries $r_\alpha = e^{\alpha T}$ and $r_\beta = e^{\beta T}$ of the z-plane, turning the **strip of convergence** into a **ring of convergence**.

One period of $\widetilde{F}(s)$ suffices for a full z-plane representation, as any other period would map redundantly to the same locations.

Relationship to discrete-time Fourier transform

The slice of $\widetilde{F}(s)$ containing the imaginary axis of the s-plane, see bottom row of Figure 7.6, represents the discrete-time Fourier transform $\widehat{F}(j\omega)$ of $f_d[nT]$. This slice maps onto the cylinder containing the unit circle of the z-plane, so that the associated values $\bar{F}(e^{j\omega T})$ describe the same discrete-time Fourier transform, but using z-variable notation,

$$\bar{F}(e^{j\omega T}) = \bar{F}(z)\big|_{z=e^{j\omega T}} = \bar{F}(re^{j\omega T})\big|_{r=1} \tag{7.15}$$

so that Equation 7.14 yields

$$\bar{F}(e^{j\omega T}) = \bar{\mathcal{F}}_{dt}\{f_d[nT]\} = \sum_{n=-\infty}^{\infty} f_d[nT]e^{-j\omega nT}$$

whose right side is identical in every respect to that of Equation 6.17 and the symbol $\bar{\mathcal{F}}_{dt}$ denotes z-transform notation.

Conversely, we can use this relationship to express the z-transform of a function $f_d[nT]$ as the Fourier transform of a modified function. Writing Equation 7.14 with $z = re^{j\omega T}$, we have

$$\bar{F}(re^{j\omega T}) = \sum_{n=-\infty}^{\infty} \{r^{-n}f_d[nT]\}e^{-j\omega nT} \qquad r_\alpha < r < r_\beta$$

where the sum represents the discrete-time Fourier transform of the modified function $r^{-n}f_d[nT]$, also expressed as

$$\bar{F}(re^{j\omega T}) = \mathcal{Z}\{f_d[nT]\} = \bar{\mathcal{F}}_{dt}\{r^{-n}f_d[nT]\} \tag{7.16}$$

With $r = e^{\sigma T}$, this is the z-transform equivalent of the discrete-time Laplace interpretation of Equation 7.7.

For a given radius $r_0 = e^{\sigma_0 T}$ the function 7.16 represents one cylinder of $\bar{F}(z)$, just as $\widetilde{F}(\sigma_0 + j\omega)$ represented one slice $\sigma = \sigma_0$ of $\widetilde{F}(s)$. Such a related pair is illustrated in the top row of Figure 7.6. A similar Fourier interpretation of all such cylinders of the region of convergence would build up the entire z-transform $\bar{F}(z)$.

7.2.3 Inverse transform

The inverse z-transform follows directly from the inverse discrete-time Laplace transform 7.10

$$f_d[nT] = \frac{T}{2\pi j} \int_{\sigma-j\omega_s/2}^{\sigma+j\omega_s/2} \widetilde{F}(s)e^{snT}\,ds$$

The same change of frequency variable $e^{sT} = z$, implies

$$s = \frac{1}{T}\ln z \qquad \text{and} \qquad ds = \frac{1}{T}\frac{dz}{z}$$

so that

$$f_d[nT] = \frac{1}{2\pi j} \oint_\Gamma \bar{F}(z)z^{n-1}\,dz \tag{7.17}$$

which is the expression for the **inverse z-transform** of $F(z)$. Together with Equation 7.14 it relates the two sides of the pair

$$f_d[nT] \quad \xleftrightarrow{\mathscr{Z}} \quad \bar{F}(z)$$

where $\mathscr{Z}$ represents the z-transform operator.

The analytical treatment of the integral 7.17 is outside the scope of this book, but its relationship to the corresponding integral of the discrete-time Laplace transform can be visualized intuitively in terms of the graphical construction of Figure 7.6.

Changing to the z-variable implies a change of the integration contour from a straight line of the complex s-plane to a circle of the complex z-plane. The chosen integration interval of the discrete-time Laplace transform was one period ω_s from $\omega = -\frac{1}{2}\omega_s$ to $\omega = +\frac{1}{2}\omega_s$, along a line of constant σ. This line segment maps onto the z-plane as an anticlockwise circle of radius $r = e^{\sigma T}$ and angles in the range $\theta = -\pi$ to $\theta = +\pi$, as shown in the top row of Figure 7.6, and this represents the integration contour Γ of Equation 7.17.

7.3   Poles, zeros and region of convergence

We now find the discrete-time transforms of some simple functions and build up a small table of transform pairs for later use. We pay special attention to z-plane features, such as poles and zeros, stressing the correspondence to similar s-plane features of related continuous-time pairs obtained in Section 3.3. This similarity will be exploited in Chapter 11. For simplicity we drop the subscript d in this section.

7.3.1   Examples

Some of the derivations are simplified by expressing the sum 7.14 as

$$\bar{F}(z) = \sum_{n=-\infty}^{\infty} f[nT]z^{-n} = \ldots + f[-2T]z^2 + f[-T]z^1 + f[0]z^0$$
$$+ f[T]z^{-1} + f[2T]z^{-2} + \ldots + f[mT]z^{-m} + \ldots \tag{7.18}$$

others by expressing a geometric series by its sum

$$\sum_{n=0}^{N-1} x^n = \frac{1 - x^N}{1 - x} \tag{7.19}$$

in particular the case with infinite upper limit

$$\sum_{n=0}^{\infty} x^n = \frac{1}{1 - x} \qquad |x| < 1 \tag{7.20}$$

Impulse

The time domain impulse, or unit pulse, takes the values

$$f[nT] = \delta[nT] = \begin{cases} 1 & n = 0 \\ 0 & \text{otherwise} \end{cases}$$

Taking the z-transform, all the terms of the sum 7.18 vanish, except $f[0]$, which equals 1, thus leaving

$$f[nT] = \delta[nT] \quad \overset{\mathscr{Z}}{\longleftrightarrow} \quad \bar{F}(z) = 1 \tag{7.21}$$

We conclude that the z-domain representation of the impulse is unity, that is, a function of unit magnitude and zero phase for all values of z. This applies to the unit circle of the z-plane, so that the discrete-time Fourier transform of the impulse is

$$f[nT] = \delta[nT] \quad \overset{\mathscr{F}_{dt}}{\longleftrightarrow} \quad \bar{F}(e^{j\omega T}) = 1$$

These results are the discrete-time counterparts of those provided by the Laplace and Fourier transforms for the continuous-time impulse,

$$f(t) = \delta(t) \quad \overset{\mathscr{L},\mathscr{F}}{\longleftrightarrow} \quad F(s) = F(j\omega) = 1$$

Shifted impulse

Shifting an impulse by one sampling interval T yields the function

$$f[nT] = \delta[nT - T] = \begin{cases} 1 & nT = T \\ 0 & \text{otherwise} \end{cases}$$

In the expression 7.18, only the term with $f[T]$ is different from zero, hence

$$f[nT] = \delta[nT - T] \begin{cases} \overset{\mathscr{Z}}{\longleftrightarrow} \quad \bar{F}(z) = z^{-1} = \dfrac{1}{z} & |z| > 0 \\[3mm] \overset{\mathscr{F}_{dt}}{\longleftrightarrow} \quad \bar{F}(e^{j\omega T}) = e^{-j\omega T} = \dfrac{1}{e^{j\omega T}} \end{cases} \tag{7.22}$$

As the z-plane origin $z = 0$ is approached, the magnitude $|\bar{F}(z)|$ tends to infinity, thus locating the pole of the function $\bar{F}(z)$. It also identifies the entire z-plane, except the point at the origin, with the function's region of convergence.

On the unit circle the magnitude $|\bar{F}(e^{j\omega T})| = 1$ is the same as in the case of the impulse $\delta[nT]$, but the real and imaginary parts of $\bar{F}(e^{j\omega T})$ are different, and so is the phase, as shown in Figure 7.7.

Shifting the impulse by m delays produces the function $f[nT] = \delta[nT - mT]$, whose transform is

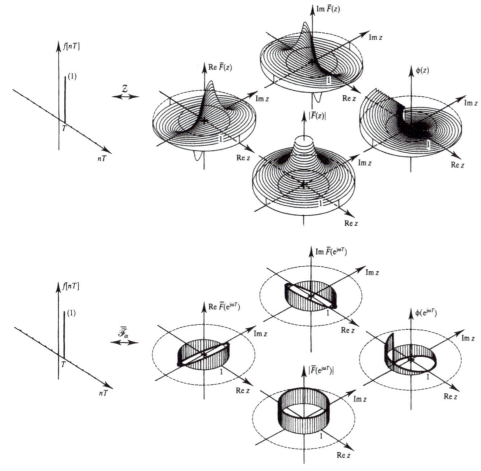

Figure 7.7   Transforms of shifted impulse.

$$f[nT] = \delta[nT - mT] \begin{cases} \xrightarrow{\;\mathscr{Z}\;} \bar{F}(z) = z^{-m} = \dfrac{1}{z^m} \quad |z| > 0 \\[3mm] \xleftarrow{\overline{\mathscr{F}}_{\mathrm{dt}}} \bar{F}(\mathrm{e}^{\mathrm{j}\omega T}) = \mathrm{e}^{-\mathrm{j}\omega mT} = \dfrac{1}{\mathrm{e}^{\mathrm{j}\omega mT}} \end{cases} \quad (7.23)$$

which has m poles superimposed at the origin, hence correspondingly larger magnitudes and higher rates of change for phase.

Unit step

The unit step function $u[nT]$ takes the values

$$f[nT] = u[nT] = \begin{cases} 1 & nT \geq 0 \\ 0 & nT < 0 \end{cases}$$

Applying Equation 7.14 gives

$$\bar{F}(z) = \sum_{n=-\infty}^{\infty} u[nT]z^{-n} = \sum_{n=0}^{\infty} (z^{-1})^n$$

which, with $z^{-1} = x$ in Equation 7.20 yields the z-transform pair

$$f[nT] = u[nT] \xleftrightarrow{\mathscr{Z}} \bar{F}(z) = \frac{1}{1 - z^{-1}} = \frac{z}{z - 1} \qquad |z| > 1 \,(7.24)$$

The function $\bar{F}(z)$ has a zero at the origin $z = 0$ and a pole at the point $z = 1$. The latter also marks the boundary of the region of convergence, which does not include the unit circle itself. Although it exists, the discrete-time Fourier transform cannot be obtained according to Equation 7.15 by evaluating expression 7.24 on the unit circle. As in the continuous-time case, the discrete-time Fourier transform has an impulse, now located at the z-plane reference point $z = 1$, to represent the d.c. value of the step,

$$f[nT] = u[nT] \xleftrightarrow{\mathscr{F}_{dt}} \bar{F}(e^{j\omega T}) = \frac{1}{1 - e^{-j\omega T}} + \pi\delta(e^{j\omega T}) \qquad (7.25)$$

The impulse $\delta(e^{j\omega T})$ is defined on the unit circle $r = 1$ of the z-plane by an equivalent of the simplified definition 1.11,

$$\oint_{r=1} \delta(e^{j\omega T}) \, dz = 1$$

$$\delta(e^{j\omega T}) = 0 \quad \text{for } \omega \neq 0 \text{ or } e^{j\omega T} \neq 1$$

The alternative form of the discrete-time Fourier transform, using discrete-time Laplace notation, is obtained by mapping expression 7.25 onto the periodic $\tilde{s}$-plane, which repeats the impulse periodically on the imaginary axis,

$$f[nT] = u[nT] \xleftrightarrow{\mathscr{F}_{dt}} \bar{F}(j\omega) = \frac{1}{1 - e^{-j\omega T}} + \sum_{k=-\infty}^{\infty} \pi\delta(\omega - k\omega_s)$$

$$(7.26)$$

Real causal exponential

A similar treatment of the real discrete-time exponential $f[nT] = a^n u[nT]$, with $az^{-1} = x$ and $|az^{-1}| < 1$ in Equation 7.20, gives the z-transform and Fourier pairs

$$f[nT] = a^n u[nT] \begin{cases} \overset{\mathscr{L}}{\longleftrightarrow} & \bar{F}(z) = \dfrac{1}{1 - az^{-1}} = \dfrac{z}{z - a} \quad |z| > |a| \\ \\ \overset{\mathscr{F}_{dt}}{\longleftrightarrow} & \bar{F}(e^{j\omega T}) = \dfrac{1}{1 - ae^{-j\omega T}} \quad |a| < 1 \end{cases} \qquad (7.27)$$

where $\bar{F}(z)$ has a pole π_1 at $z = a$ and a zero ζ_1 at $z = 0$. Setting $a = 1$, the earlier unit step becomes a special case. These two cases are shown in the top row of Figure 7.8, where they become reference functions for a chart that collects the following set of examples.

The discrete-time Laplace version of the real exponential, with the notation $a = e^{\alpha T}$, is already available from Example 7.1, where it was fully represented in Figure 7.2. It was subsequently used in Figure 7.6 to derive the relationship between the discrete-time Laplace transform and the z-transform. The corresponding version of the discrete-time Fourier transform was derived earlier in Example 6.1, and illustrated in Figure 6.8 for positive and negative values of a. A re-examination of those figures is recommended to consolidate the grasp of transform interrelationships.

Complex exponential

If the parameter a of expressions 7.27 is a complex number $\pi_1 = r_1 e^{j\theta_1}$, then the time function $f[nT]$ is a discrete-time complex exponential, with z-transform

$$f_1[nT] = r_1^n e^{jn\theta_1} u[nT] \overset{\mathscr{L}}{\longleftrightarrow} \bar{F}_1(z) = \frac{z}{z - r_1 e^{jn\theta_1}} \qquad (7.28)$$

This has a pole π_1 at the z-plane location $z = \pi_1 = r_1 e^{j\theta_1}$ and a zero ζ_1 at the origin. The complex conjugate pole $\pi_2 = \pi_1^* = r_1 e^{-j\theta_1}$ gives rise to the related pair

$$f_2[nT] = r_1^n e^{-jn\theta_1} u[nT] \overset{\mathscr{L}}{\longleftrightarrow} \bar{F}_2(z) = \frac{z}{z - r_1 e^{-jn\theta_1}} \qquad (7.29)$$

and both are found in the right half of Figure 7.8, which shows the z-plane locations of the pole and zero, as well as their mapped locations on the periodic s-plane. Corresponding special cases for $r_1 = 1$ are shown in the left half of the figure.

The pair 7.28 is expressed more concisely as

$$f[nT] = \pi_1^n u[nT] \overset{\mathscr{L}}{\longleftrightarrow} \bar{F}(z) = \frac{z}{z - \pi_1}$$

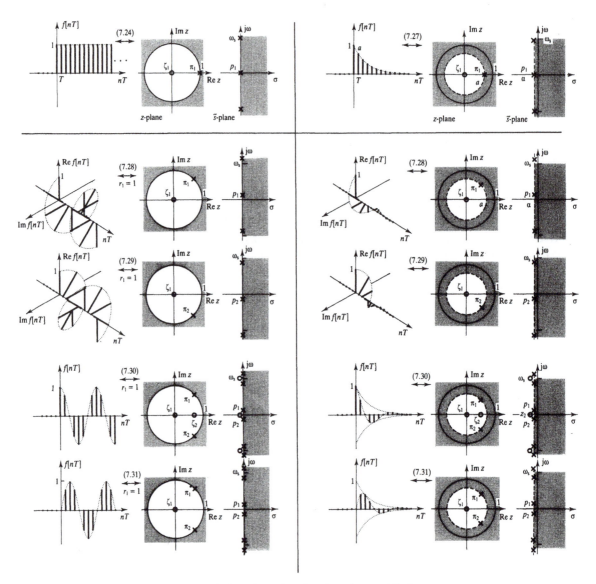

Figure 7.8 Pole and zero locations in z-plane and $\tilde{s}$-plane.

which explicitly incorporates the pole π_1 in the time function. The complex plane representation of $f[nT]$, see Figure 7.9, is particularly revealing. The sample $f[T]$ is numerically identical to π_1, in both magnitude and angle, while the values of the subsequent samples represent increasing powers of $f[T]$.

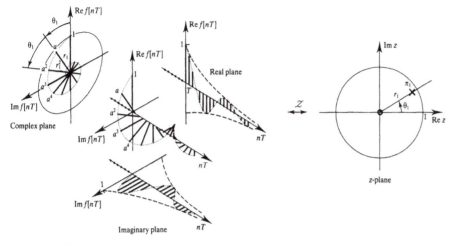

Figure 7.9 Relationship of complex exponential to pole position.

Sinusoidal functions

Writing $\theta_1 = \omega_1 T$, the semi-sum of expressions 7.28 and 7.29 yields the pair

$$r_1^n \cos \omega_1 nT\, u[nT] \quad \xleftrightarrow{\mathscr{L}} \quad \frac{z^2 - r_1 \cos \omega_1 T\, z}{z^2 - 2r_1 \cos \omega_1 T\, z + r_1^2} \qquad (7.30)$$

whose denominator is the product $(z - \pi_1)(z - \pi_2)$, so that π_1 and π_2 are also the poles of expression 7.30. The numerator has one root at the origin and another at $z = r_1 \cos \omega_1 T$, which is the projection of the pole location on the real axis of the z-plane, see Figure 7.9. The poles and zeros map onto the periodic s-plane as shown in Figure 7.8. The zero ζ_2, being on a smaller radius of the z-plane than the poles, maps further from the imaginary axis of the s-plane.

Similarly, the semi-difference of expressions 7.28 and 7.29, divided by j, gives the pair

$$r_1^n \sin \omega_1 nT\, u[nT] \quad \xleftrightarrow{\mathscr{L}} \quad \frac{r_1 \sin \omega_1 T\, z}{z^2 - 2r_1 \cos \omega_1 T\, z + r_1^2} \qquad (7.31)$$

which has the same poles as expression 7.30, but only one zero, and this is located at the z-plane origin.

These functions are shown in the right half of Figure 7.8, with the special cases for $r_1 = 1$ on their left. The entire figure is the discrete-time counterpart of Figure 3.9.

7.3.2 Convergence properties

The functions $f[nT]$ considered so far in this chapter were causal and therefore right-sided. The region of convergence of each discrete-time Laplace transform was to the right of the boundary defined by the right-most pole, while that of the z-transform was outside the corresponding boundary circle. We now examine general functions and their convergence properties.

Left-sided function

The left-sided function $f[nT] = -a^n u[-nT - T]$ of Figure 7.10 has a z-transform

$$\bar{F}(z) = -\sum_{n=-\infty}^{\infty} a^n u[-nT - T]z^{-n} = -\sum_{n=-\infty}^{-1} a^n z^{-n}$$

$$= 1 - \sum_{n=0}^{\infty} (a^{-1}z)^n$$

where unity was added and subtracted in the last step. The sum is of the form of Equation 7.20 and converges for $|a^{-1}z| < 1$, so that

$$f[nT] = -a^n u[-nT - T] \quad \overset{\mathcal{Z}}{\longleftrightarrow} \quad \bar{F}(z) = \frac{z}{z - a} \qquad |z| < |a|$$

$$(7.32)$$

The frequency representation is identical to that of expressions 7.27, except that the regions of convergence are on opposite sides of the boundary circle $r = |a|$.

A similar situation was met in Section 3.3.2 in the context of the Laplace transform, and this similarity is clearly reflected in the discrete-time Laplace transform and the associated region of convergence. Writing $a = e^{\alpha T}$ we have

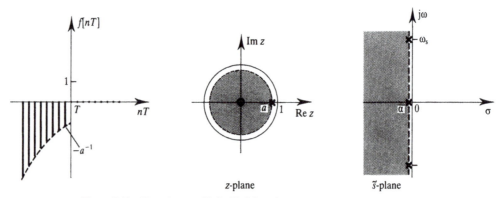

Figure 7.10   Transforms of left-sided function.

$$f[nT] = -e^{\alpha nT}u[-nT - T]$$

$$\xleftarrow{\mathscr{L}_{dt}} \quad \widetilde{F}(s) = \frac{e^{sT}}{e^{sT} - e^{\alpha T}} \qquad \mathrm{Re}\, s < \alpha$$

whose region of convergence is also shown in Figure 7.10.

Double-sided functions

By analogy to the case of the Laplace transform, a double-sided function can always be interpreted as the sum of two single-sided functions, as illustrated at the top of Figure 7.11, where the third row represents the sum of the first two,

$$f[nT] = a^n u[nT] + b^n u[-nT - T]$$

$$\xleftarrow{z} \quad \widetilde{F}(z) = \frac{(a - b)z}{(z - a)(z - b)} \qquad |a| < |z| < |b|$$

$$(7.33)$$

whose region of convergence is the intersection or overlap of the two component regions. The corresponding discrete-time Laplace version, with $a = e^{\alpha T}$ and $b = e^{\beta T}$, becomes

$$f[nT] = e^{\alpha nT}u[nT] + e^{\beta nT}u[-nT - T]$$

$$\xleftarrow{\mathscr{L}_{dt}} \quad \widetilde{F}(s) = \frac{(e^{\alpha T} - e^{\beta T})e^{sT}}{(e^{sT} - e^{\alpha T})(e^{sT} - e^{\beta T})} \qquad \alpha < \mathrm{Re}\, s < \beta$$

The other functions of Figure 7.11 give examples of regions of convergence, as well as an indication of the existence, or otherwise, of the corresponding discrete-time Fourier transform.

Convergence criteria

The arguments of Section 3.3.3 are valid for discrete-time functions. Interpreting the bilateral exponential form 7.33 as the magnitude envelope of an arbitrary function $f[nT]$ leads to a condition of convergence similar to that formulated in expression 3.16, paraphrased as follows. The z-transform of an arbitrary function converges in the region $|a| < |z| < |b|$, provided a positive scaling factor M can be found, such that the function $f[nT]$ is contained within an exponential order envelope of the form

$$|f[nT]| < \begin{cases} Ma^n u[nT] & nT \geqslant 0 \\ Mb^n u[-nT - T] & nT < 0 \end{cases} \qquad (7.34)$$

This criterion translates to the discrete-time Laplace transform as

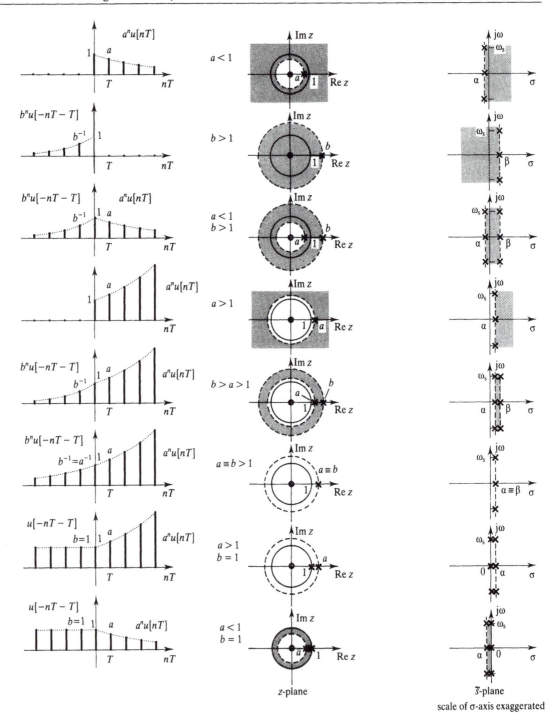

Figure 7.11 Discrete-time exponentials with z-plane and $\tilde{s}$-plane representations.

$$|f[nT]| < \begin{cases} Me^{\alpha nT} u[nT] & nT \geq 0 \\ Me^{\beta nT} u[-nT - T] & nT < 0 \end{cases} \qquad (7.35)$$

where the convergence region becomes $\alpha < \operatorname{Re} s < \beta$.

7.4 Transform properties

Since the transforms in this chapter were derived directly from the Laplace transform, the properties listed in Section 3.5 for the latter need to be modified to reflect the periodicity of the discrete-time Laplace transform and further adapted for the change of frequency variable of the z-transform. We derive some key properties stressing analogies to the continuous-time derivations of Fourier properties in Section 2.4.

Because of the great similarities of transforms and of their properties, to avoid lengthy repetitions we concentrate on properties of the z-transform required later in the book. These can easily be interpreted for the discrete-time Laplace transform by mapping to the periodic $\tilde{s}$-plane and for the discrete-time Fourier transform by evaluating on the unit circle or imaginary axis, as appropriate.

7.4.1 Linearity

By analogy to the property 3.20, two functions $f_1[nT]$ and $f_2[nT]$ with transforms of the form of Equation 7.14 yield the linearity property

$$af_1[nT] + bf_2[nT] \quad \xleftrightarrow{\mathcal{Z}} \quad a\bar{F}_1(z) + b\bar{F}_2(z) \qquad (7.36)$$

$$\max(r_{\alpha_1}, r_{\alpha_2}) < |z| < \min(r_{\beta_1}, r_{\beta_2})$$

whose region of convergence is the intersection of the individual regions, defined by the circular boundaries of radii r_{α_1}, etc. The above condition assumes that no pole-zero cancellations are involved.

7.4.2 Time scaling

The z-transform of a function $g[nT] = f[knT]$, whose time variable is a scaled version of that of $f[nT]$, is

$$\bar{G}(z) = \sum_{n=-\infty}^{\infty} f[knT]z^{-n} = \sum_{m=-\infty}^{\infty} f[mT]z^{-m/k}$$

$$= \sum_{m=-\infty}^{\infty} f[mT](z^{1/k})^{-m}$$

from which we conclude

$$f[knT] \quad \overset{\mathscr{Z}}{\longleftrightarrow} \quad \bar{F}(z^{1/k}) \tag{7.37}$$

Time-reversal represents a particular case with scaling factor $k = -1$. The transform of the function $f[-nT]$ becomes

$$f[-nT] \overset{\mathscr{Z}}{\longleftrightarrow} \bar{F}(z^{-1})$$

The function $\bar{F}(z^{-1})$ is interpreted in Section 10.6 as a 'reflection about the unit circle' of the function $\bar{F}(z)$.

7.4.3   Time shifting

We derive the discrete-time counterpart to expression 3.23 using essentially the same process as in Section 2.4.4 for the equivalent Fourier property. Given the z-transform pair $f[nT] \longleftrightarrow \bar{F}(z)$, the transform of the shifted function $g[nT] = f[nT - mT]$ is

$$\bar{G}(z) = \sum_{n=-\infty}^{\infty} f[nT - mT]z^{-n} = z^{-m} \sum_{n=-\infty}^{\infty} f[nT - mT]z^{-(n-m)}$$

$$= z^{-m} \sum_{l=-\infty}^{\infty} f[lT]z^{-l}$$

where we multiplied and divided by z^{-m} and made a change of summation variable $l = n - m$. This leads to the relationships

$$
\begin{aligned}
f[nT - mT] &\quad \overset{\mathscr{Z}}{\longleftrightarrow} \quad z^{-m}\bar{F}(z) \\
f[nT + mT] &\quad \overset{\mathscr{Z}}{\longleftrightarrow} \quad z^{m}\bar{F}(z)
\end{aligned}
\tag{7.38}
$$

for the z-transforms of right-shifted and left-shifted functions. In the system simulations of Chapter 11 these properties take the place of time domain integration and differentiation.

7.4.4   Time domain convolution

As in continuous-time functions, time domain convolution implies frequency domain multiplication. Given two functions $x[nT] \longleftrightarrow \bar{X}(z)$ and $h[nT] \longleftrightarrow \bar{H}(z)$ their time convolution is expressed according to Equation 1.6 as

$$y[nT] = x[nT] * h[nT] = \sum_{m=-\infty}^{\infty} x[mT]h[nT - mT]$$

Taking the z-transform of both sides gives

$$\bar{Y}(z) = \sum_{n=-\infty}^{\infty} \left[\sum_{m=-\infty}^{\infty} x[mT]h[nT - mT] \right] z^{-n}$$

$$= \sum_{m=-\infty}^{\infty} \left[\sum_{n=-\infty}^{\infty} x[mT]h[nT - mT] \right] z^{-n}$$

Multiplying and dividing by z^{-m} yields

$$\bar{Y}(z) = \sum_{m=-\infty}^{\infty} x[mT]z^{-m} \sum_{n=-\infty}^{\infty} h[nT - mT]z^{-(n-m)}$$

$$= \sum_{m=-\infty}^{\infty} x[mT]z^{-m} \sum_{l=-\infty}^{\infty} h[lT]z^{-l}$$

hence the convolution property

$$x[nT] * h[nT] \quad \overset{\mathcal{Z}}{\longleftrightarrow} \quad \bar{X}(z)\bar{H}(z) \tag{7.39}$$

Note the analogies with the Fourier derivation of Section 2.4.5.

7.4.5   Differentiation in frequency

Differentiating both sides of the z-transform definition 7.14 with respect to z yields

$$\frac{d\bar{F}(z)}{dz} = -z^{-1} \sum_{n=-\infty}^{\infty} nf[nT]z^{-n}$$

hence the property

$$g[nT] = nf[nT] \quad \overset{\mathcal{Z}}{\longleftrightarrow} \quad \bar{G}(z) = -z\frac{d\bar{F}(z)}{dz} \tag{7.40}$$

Example 7.2

We derive the time representation of $z/(z - a)^2$ from the known pair

$$f[nT] = a^n u[nT] \quad \overset{\mathcal{Z}}{\longleftrightarrow} \quad \bar{F}(z) = \frac{z}{z - a}$$

The derivative of $\bar{F}(z)$ is

$$\frac{d\bar{F}(z)}{dz} = \frac{1}{z - a} - \frac{z}{(z - a)^2} = \frac{-a}{(z - a)^2}$$

which with expression 7.40 yields the pair

$$na^{n-1}u[nT] \quad \overset{\mathcal{Z}}{\longleftrightarrow} \quad \frac{z}{(z - a)^2}$$

7.5 Unilateral transforms

Had we based the developments of this chapter on the **unilateral Laplace transform** (Equation 3.33), rather than on the bilateral form, an identical derivation process would have led to corresponding unilateral forms of the discrete-time Laplace transform and z-transform, with similar definitions and properties.

7.5.1 Definitions

The unilateral discrete-time Laplace transform of $f[nT]$ is thus defined on the basis of Equation 7.3 as

$$\widetilde{F}_I(s) = \mathcal{L}_{Id}\{f[nT]\} = \sum_{n=0}^{\infty} f[nT]e^{-snT}$$

whose inversion formula is formally identical to Equation 7.10,

$$f[nT]u[nT] = \frac{T}{2\pi j} \int_{\sigma-j\omega_s/2}^{\sigma+j\omega_s/2} \widetilde{F}_I(s)e^{snT}\,\mathrm{d}s$$

We inserted the unit step $u[nT]$ to signify that the function $\widetilde{F}_I(s)$ always gives rise to a causal time function.

Similarly, the z-transform definition 7.14 leads to

$$\bar{F}_I(z) = \mathcal{Z}_I\{f[nT]\} = \sum_{n=0}^{\infty} f[nT]z^{-n} \tag{7.41}$$

with inversion formula

$$f[nT]u[nT] = \frac{1}{2\pi j} \oint_{\Gamma} \bar{F}_I(z)z^{n-1}\,\mathrm{d}z$$

These transforms cover similar discrete-time situations to those covered by the unilateral Laplace transform in a continuous-time context and offer similar advantages regarding initial conditions. We will only examine the effects on the unilateral z-transform when $f[nT]$ is shifted in time. The conclusions can be extended to the discrete-time Laplace version.

7.5.2 Transform of time-shifted function

Consider a two-sided function $f[nT]$, defined for both positive and negative time, as shown on the left of Figure 7.12. The unilateral definition 7.41 represents the infinite sum

$$\bar{F}_I(z) = \sum_{n=0}^{\infty} f[nT]z^{-n} = f[0] + f[T]z^{-1} + f[2T]z^{-2} + \ldots \tag{7.42}$$

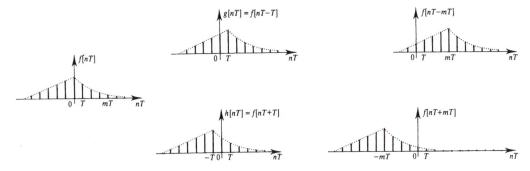

Figure 7.12 Time-shifted functions.

which effectively ignores all values $f[nT]$ of negative time $nT < 0$.

Right shifting

We construct an auxiliary function $g[nT] = f[nT - T]$, by shifting $f[nT]$ one interval T to the right, as shown in the upper row of Figure 7.12. The unilateral z-transform of $g[nT]$ is formally identical to that of $f[nT]$,

$$\bar{G}_{\mathrm{I}}(z) = \sum_{n=0}^{\infty} g[nT]z^{-n} = g[0] + g[T]z^{-1} + g[2T]z^{-2}$$
$$+ g[3T]z^{-3} + \ldots$$

Each factor $g[nT]$ of this sum represents an earlier value $f[nT - T]$ of the original function, for instance $g[3T] = f[2T]$, so that

$$\bar{G}_{\mathrm{I}}(z) = f[-T] + z^{-1}\{f[0] + f[T]z^{-1} + f[2T]z^{-2} + \ldots\}$$

which leads to the unilateral z-transform property

$$g[nT] = f[nT - T] \quad \overset{\mathscr{Z}_1}{\longleftrightarrow} \quad \bar{G}_{\mathrm{I}}(z) = z^{-1}\bar{F}_{\mathrm{I}}(z) + f[-T] \quad (7.43)$$

The first term on the right is formally similar to that obtained in expressions 7.38 for the bilateral transform. The term $f[-T] = g[0]$ represents the initial value of $g[nT]$ and gives the unilateral z-transform its special properties.

More generally, shifting $f[nT]$ by m sampling intervals to the right gives the function $f[nT - mT]$ of the upper row of Figure 7.12. A recursive application of the above process leads to the generalization

$$f[nT - mT] \quad \overset{\mathscr{Z}_1}{\longleftrightarrow} \quad z^{-m}\bar{F}_{\mathrm{I}}(z) + \{f[-T]z^{-(m-1)}$$
$$+ \ldots + f[-(m-1)T]z^{-1} + f[-mT]\}$$
$$(7.44)$$

The braces enclose a polynomial in z^{-1}, whose coefficients are the m initial values of the shifted function $g[nT]$.

Left shifting

We now construct the auxiliary function $h[nT] = f[nT + T]$ by shifting the function $f[nT]$ one interval T to the left, as shown in the bottom row of Figure 7.12. The unilateral z-transform of $h[nT]$ takes the generic form

$$\bar{H}_{\mathrm{I}}(z) = h[0] + h[T]z^{-1} + h[2T]z^{-2} + h[3T]z^{-3} + \dots$$

which can be expressed in terms of $f[nT]$ as

$$\bar{H}_{\mathrm{I}}(z) = z\{f[0] + f[T]z^{-1} + f[2T]z^{-2} + f[3T]z^{-3} + \dots\} - f[0]z$$

where we added and subtracted the term $f[0]z$. This yields the property

$$h[nT] = f[nT + T] \quad \xleftrightarrow{\mathscr{Z}_{\mathrm{I}}} \quad \bar{H}_{\mathrm{I}}(z) = z\bar{F}_{\mathrm{I}}(z) - f[0]z \qquad (7.45)$$

which accounts for the initial value $f[0]$ dropping out at the lower end. The general case of left-shifting by m intervals leads similarly to

$$f[nT + mT] \quad \xleftrightarrow{\mathscr{Z}_{\mathrm{I}}}$$

$$z^m \bar{F}_{\mathrm{I}}(z) - \{f[0]z^m + f[T]z^{m-1} + \dots + f[(m-1)T]z\} \qquad (7.46)$$

Unilateral time function

Right-shifting a causal function $f[nT]$, with $f[nT] = 0$ for $nT < 0$, or equivalently a truncated function $f[nT]u[nT]$, see Figure 7.13, yields

$$f[nT - mT]u[nT - mT] \quad \xleftrightarrow{\mathscr{Z}_{\mathrm{I}}} \quad z^{-m}\bar{F}_{\mathrm{I}}(z)$$

Similarly, left-shifting the unilateral function $g[nT]u[nT - kT]$ gives

$$g[nT + mT]u[nT - kT + mT] \quad \xleftrightarrow{\mathscr{Z}_{\mathrm{I}}} \quad z^m \bar{G}_{\mathrm{I}}(z)$$

valid for $m \le k$, as shown in Figure 7.13. The same results would have been obtained with the bilateral z-transform.

7.5.3 Laplace and z-transforms of periodic functions

The discrete forms of the Fourier transform exhibit a perfect symmetry regarding signal periodicity and discreteness in the alternative domains, as seen in Figure 6.22. This symmetry cannot be fully extended to the discrete forms of the Laplace transform, because one domain is real

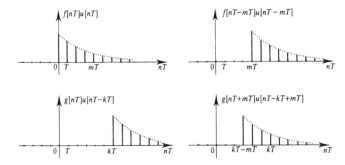

Figure 7.13 Time-shifted unilateral functions.

while the other is complex. We now examine the role of the continuous Laplace transform in relation to periodic time functions and extend the conclusions to the discrete-time Laplace transform and the z-transform.

Laplace transforms of impulse train

In the context of Chapter 6 the impulse train $\delta_T(t)$ and its Fourier transform 6.5 represented the link between a periodic domain and the discrete alternative domain. The impulse train also gives insight into the capabilities of the Laplace transform.

Applying the bilateral Laplace definition 3.1 to $\delta_T(t)$ yields

$$\mathcal{L}\{\delta_T(t)\} = \int_{-\infty}^{\infty} \sum_{m=-\infty}^{\infty} \delta(t - mT)e^{-st}\,dt$$

$$= \sum_{m=-\infty}^{\infty} \int_{-\infty}^{\infty} \delta(t - mT)e^{-st}\,dt$$

where the last integral is the Laplace transform of a displaced impulse

$$\mathcal{L}\{\delta(t - nT)\} = \int_{-\infty}^{\infty} \delta(t - mT)e^{-st}\,dt = e^{-smT}$$

so that

$$\mathcal{L}\{\delta_T(t)\} = \sum_{m=-\infty}^{\infty} e^{-smT}$$

But the sum of this infinite series does not converge for any value of s. We conclude that the impulse train $\delta_T(t)$ does not possess a bilateral Laplace transform. The corresponding Fourier transform only converged in the limit, and multiplication by any convergence factor $e^{-\sigma t}$, while assisting convergence in one direction, would only make matters worse in the other.

In contrast, the unilateral Laplace definition 3.41, whose lower limit 0^- includes the impulse at the origin, yields

$$\mathcal{L}_I\{\delta_T(t)\} = \int_{0^-}^{\infty} \sum_{m=-\infty}^{\infty} \delta(t - mT)e^{-st}\, dt = \sum_{m=0}^{\infty} e^{-smT}$$

which is a sum of the form of Equation 7.20 and gives the transform pair

$$\bar{\delta}_T(t) = \sum_{m=0}^{\infty} \delta(t - mT) \quad \overset{\mathcal{L}_I}{\longleftrightarrow} \quad \bar{F}(s) = \frac{1}{1 - e^{-sT}} \qquad \sigma > 0$$

$$(7.47)$$

whose region of convergence follows from $|e^{-sT}| = e^{-\sigma T} < 1$. The frequency domain is periodic in the imaginary component ω, as shown in Figure 7.14, by its real even and imaginary odd parts as well as by magnitude and angle for an arbitrary parameter value $T = 1$.

Unilateral Laplace transform of periodic time functions

With similar considerations, fully periodic time functions do not possess a bilateral Laplace transform. In contrast, the unilateral Laplace transform does exist in the half-plane $\sigma > 0$ and can be expressed in terms of the transform of one of its periods.

We form an auxiliary function $f_0(t)$ representing the fundamental cycle of width T_0 as

$$f_0(t) = \begin{cases} \tilde{f}(t) & 0 \leq t < T_0 \\ 0 & \text{elsewhere} \end{cases} \quad \overset{\mathcal{L}_I}{\longleftrightarrow} \quad F_0(s)$$

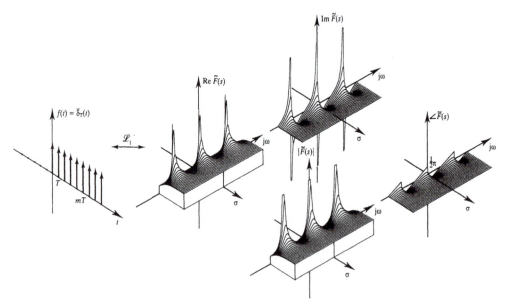

Figure 7.14   Unilateral Laplace transform of impulse train.

so that the causal periodic function $\tilde{f}(t)$ is reconstituted by the sum of shifted cycles,

$$\tilde{f}(t) = f_0(t) + f_0(t - T_0) + f_0(t - 2T_0) + \dots$$

$$= \sum_{m=0}^{\infty} f(t - mT_0)$$

as indicated in Figure 7.15. Taking the unilateral Laplace transform of all terms yields

$$F_{\mathrm{I}}(s) = F_0(s) + e^{-sT_0} F_0(s) + e^{-s2T_0} F_0(s) + \dots = F_0 \sum_{m=0}^{\infty} e^{-smT_0}$$

Evaluating the sum leads to

$$\tilde{f}(t) = f_0(t) * \bar{\delta}_{T_0}(t) \quad \overset{\mathscr{L}_{\mathrm{I}}}{\longleftrightarrow} \quad F_{\mathrm{I}}(s) = \frac{1}{1 - e^{-sT_0}} F_0(s) \qquad (7.48)$$

which interprets $\tilde{f}(t)$ as a convolution with the causal impulse train $\bar{\delta}_{T_0}(t)$ of expression 7.47 and $F_{\mathrm{I}}(s)$ as the corresponding frequency domain product. The graphical representation would exhibit a 'ribbed effect' that results from multiplying $F_0(s)$ by the surface of Figure 7.14, the 'ribs' being spaced at multiples of the fundamental frequency $\omega_0 = 2\pi/T_0$.

Discrete Laplace transform and z-transform of periodic functions

The above conclusions can be extended to the equivalent transforms of discrete-time periodic functions. Again, the bilateral form does not

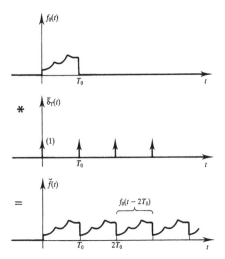

Figure 7.15   Periodic function as convolution.

exist, while the unilateral transform can be expressed in terms of that of one cycle. The same results would be obtained by taking the bilateral transform of the equivalent causal periodic function $\tilde{f}[nT]u[nT]$.

Given the fundamental causal cycle $f_0[nT]$ of length $T_0 = NT$, as shown in Figure 7.16, with unilateral or bilateral z-transform $\bar{F}_0(z)$,

$$f_0[nT] \quad \xleftrightarrow{\mathscr{Z}} \quad \bar{F}_0(z)$$

we construct the causal periodic function

$$\tilde{f}[nT]u[nT] = f_0[nT] + f_0[nT - NT] + f_0[nT - 2NT] + \ldots$$

whose terms can be transformed by the time-shifting property as

$$\bar{F}(z) = \bar{F}_0(z) + z^{-N}\bar{F}_0(z) + z^{-2N}\bar{F}_0(z) + \ldots$$

$$= \bar{F}_0(z) \sum_{m=0}^{\infty} z^{-mN}$$

hence

$$\tilde{f}[nT]u[nT] \quad \xleftrightarrow{\mathscr{Z}} \quad \bar{F}(z) = \frac{1}{1 - z^{-N}} \bar{F}_0(z) \tag{7.49}$$

valid for both the unilateral and bilateral transform.

The frequency variable substitution $z = e^{sT}$ leads to the equivalent discrete-time Laplace representation

$$\tilde{f}[nT]u[nT] \quad \xleftrightarrow{\mathscr{L}_{dt}} \quad \widetilde{F}(s) = \frac{1}{1 - e^{-sNT}} \widetilde{F}_0(s) \tag{7.50}$$

where $\widetilde{F}_0(s)$ is the corresponding transform of the fundamental cycle.

7.6 Transform interrelations

We have now derived the transforms linking the time domain and frequency domain representations of all signal classes. We used a

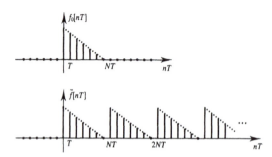

Figure 7.16 Generation of causal periodic function.

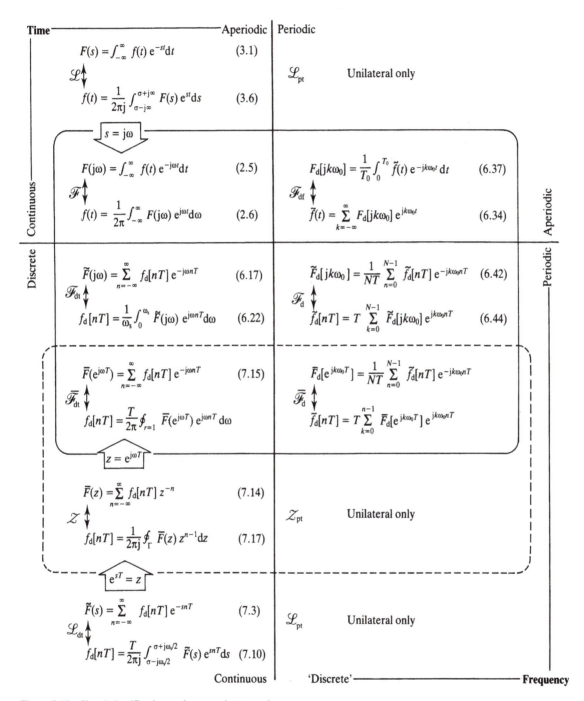

Figure 7.17 Signal classification and appropriate transforms.

strategy that interpreted each new transform as either a subset or a superset of a previously derived one, which led to a consistent set of transform expressions and a valuable body of interrelationships.

In this section we chart these results systematically on the signal classification framework, first introduced with Figure 1.2 and developed in subsequent chapters, thus revealing the close family ties of all these transforms and their ancestry in the Laplace transform.

7.6.1 Transform expressions

The forward and inverse transform expressions are collected in the chart of Figure 7.17. The Laplace transform provides the most general frequency description of the most general class of functions, continuous in both domains. All other transforms are ultimately derived from the Laplace transform and become its subsets.

We distinguish two types of subsets. One is based on signal classification, as represented by the four major quadrants of Figure 7.17. The chart clearly shows that the structure of the Laplace transform is retained, with only minor adaptations, in all other classes.

The other type relates subsets according to the generality of the complex frequency variable. Its boundary is the box at the centre of the chart, which separates Fourier subsets from the corresponding Laplace or z-transform expressions. Inside the box each general transform is evaluated on a specific contour of the applicable generalized frequency variable. The Fourier transform classification chart of Figure 6.22 forms the core of this box, which is augmented by the z-transform versions for discrete time.

The relationship linking the continuous and the discrete-time Fourier transforms inside the box is reflected in the identical relationship linking the continuous and the discrete-time Laplace transforms.

The periodic redundancy inherent in the discrete-time Laplace transform is avoided by the change of frequency variable $e^{sT} = z$, which maps points of the $\tilde{s}$-plane onto related points of the z-plane. This change of notation gives rise to the z-transform, shown enclosed by a dashed line in Figure 7.17. Evaluating the z-transform on the unit circle, the z-plane counterpart of the imaginary axis of the $\tilde{s}$-plane, brings us back to the discrete-time Fourier transform, with an alternative notation.

The Fourier classification chart in Figure 6.22 showed a complete symmetry or duality regarding discreteness and periodicity in alternative domains. This was possible because each variable had a single component, real t for time and imaginary $j\omega$ for frequency. The full symmetry cannot be extended to the corresponding periodic-time regions of the chart, firstly, because the frequency variables s and z are complex, and, secondly, because periodic time functions do not possess bilateral

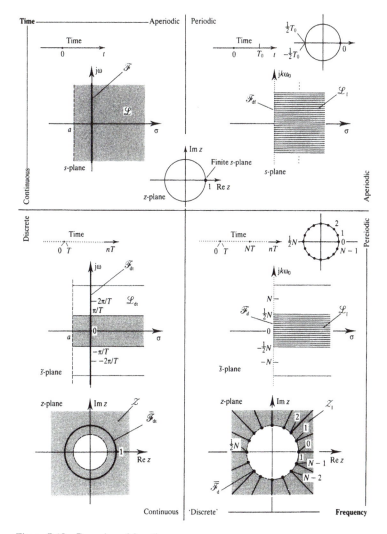

Figure 7.18   Domains of functions.

Laplace or z-transforms, as seen in Section 7.5.3. This is reflected in Figure 7.17 by writing the label 'discrete' for frequency in inverted commas and using unilateral transforms in the appropriate quadrants of the figure.

7.6.2   Domains and regions of convergence

The preceding chart of transform expressions is complemented in Figure 7.18 with summary 'floor plans' of the corresponding domains. The domain of the real time variable t or nT is represented on a line, while that of the corresponding complex frequency variable is represented on a complex plane. The latter also shows the region of convergence of the associated Laplace or z-transform and the location of the appropriate class of Fourier transform.

In the continuous quadrant the time function $f(t)$ is located on the time axis, in a plane perpendicular to the paper. The appropriate Laplace transform is located on the region of convergence of the s-plane, limited by the boundary at $\sigma = a$. If $a < 0$ the Fourier transform exists and is located on the imaginary axis $s = j\omega$, as shown.

The aperiodic discrete-time quadrant shows the discrete locations for the time function $f[nT]$ as well as the periodic $\tilde{s}$-plane, with a similar region of convergence for the discrete-time Laplace transform and the location of the discrete-time Fourier transform. The periodic redundancy is overcome by mapping the $\tilde{s}$-plane onto the corresponding z-plane, shown with its region of convergence and giving the unit circle as the location for the discrete-time Fourier transform.

The continuous and periodic time quadrant indicates the time periodicity by T_0 and the discrete locations for the Fourier transform on the imaginary axis of the s-plane. It was shown earlier that a periodic time function does not possess a bilateral Laplace or z-transform. The location of the unilateral form is indicated on the s-plane, where lines at constant values $k\omega_0$ symbolize the characteristic 'ribs' of the corresponding transforms.

The z-plane inserted in the continuous-time half of the figure is a reminder that the entire finite s-plane collapses onto the z-plane location $z = 1$.

Finally, the discrete quadrant of Figure 7.18 brings together the characteristic features of both discrete time and discrete frequency. This is indicated by the periodicity of the discrete locations on the time axis, similarly on the imaginary axis of the $\tilde{s}$-plane for the discrete Fourier transform and the 'ribs' for the discrete-time Laplace transform. The frequency domain features are also mapped onto the z-plane, where the N equidistant points on the unit circle represent the locations of the discrete Fourier transform.

7.6.3 Visual summary

To complete the summaries, Figure 7.19 generalizes the sampling processes of Figure 6.18, which led to the discrete forms of the Fourier transform, extending them to the Laplace and z-transforms. Because of size constraints functions located on complex s-planes and z-planes are indicated symbolically by their axes only.

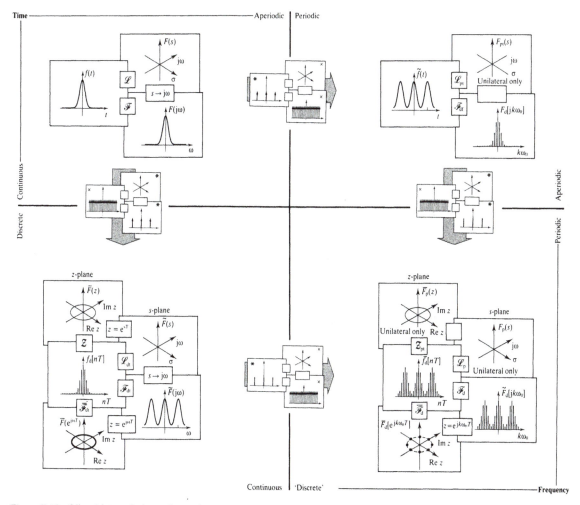

Figure 7.19 Visual interrelations of transform.

Exercises

7.1 Draw parallels between the derivation processes of the discrete-time Laplace transform in Section 7.1.1 and the discrete-time Fourier transform in Section 6.2.1, stressing the sameness of the time domain process.

7.2 Find the discrete-time Laplace transform of the function $f[nT] = e^{-\alpha|nT|}$ with $\alpha > 0$, and plot its $\tilde{s}$-plane poles and region of convergence. Determine if the discrete-time Fourier transform exists and write its expression.

7.3 Contrast the Fourier interpretation of the discrete-time Laplace transform of Figure 7.2 with the related interpretation of the continuous Laplace transform of Figure 3.10.

7.4 Discuss the correspondence between the functions $\tilde{F}(s)$ and $\bar{F}(z)$ of the two sides of Figure 7.6, and show that they describe the same frequency domain function in terms of two different sets of coordinates. Stress the correspondence between applicable poles, zeros and regions of convergence and between the function values on the imaginary axis of the $\tilde{s}$-plane and those on the unit circle of the z-plane.

7.5 Find the z-transform and plot the poles, zeros and region of convergence of the functions

$$f_1[nT] = 0.5^n \, u[nT] \qquad\qquad f_2[nT] = (-0.5)^n \, u[nT]$$

$$f_3[nT] = 0.5^{|n|} \qquad\qquad f_4[nT] = (0.4 + j0.3)^n \, u[nT]$$

$$f_5[nT] = (\tfrac{1}{2})^n \, u[nT] + (\tfrac{1}{3})^n \, u[nT]$$

$$f_6[nT] = \sin 0.8n \, u[nT] \qquad f_7[nT] = 0.7^n \cos 0.8n \, u[nT]$$

Map the poles, zeros and regions of convergence onto the periodic $\tilde{s}$-plane and write the appropriate discrete-time Laplace transform expressions.

7.6 Compare the expressions of the two forms of the discrete-time Fourier transform, when interpreted as a subset $\mathcal{F}_{dt}$ of the discrete-time Laplace transform, or a subset $\overline{\mathcal{F}}_{dt}$ of the z-transform. Use the transform of the shifted impulse $f[nT] = \delta[nT - T]$ to illustrate this graphically.

7.7 Draw the following sequence of discrete-time functions and find their z-transform using time shifting and convolution properties,

$$f_1[nT] = u[nT + 2T] - u[nT - 3T]$$

$$f_2[nT] = f_1[nT] * f_1[nT]$$

$$f_3[nT] = f_2[nT] * \delta[nT - 4T]$$

7.8 Derive the time-shifting and time domain convolution properties of the discrete-time Laplace transform and compare to the equivalent properties, Equations 7.38 and 7.39, of the z-transform.

7.9 Determine which transforms are applicable to the continuous-time and discrete-time functions

$$f(t) = e^{\alpha t} \, u(t)$$

$$f[nT] = e^{\alpha nT} \, u[nT]$$

$$f[nT] = a^n \, u[nT]$$

Find the transforms and gather them on the framework of Figure 7.17. Indicate their conditions of validity or regions of convergence, as applicable.

7.10 Derive the time-shifting property 7.44 of the unilateral z-transform, by repeatedly applying the result 7.43 to m consecutive delays.

7.11 Derive the initial value and the final value properties of the z-transform of a right-sided function $f[nT]$

$$f[0] = F(z)|_{z=\infty} \quad \text{and} \quad \lim_{N\to\infty} f[NT] = \lim_{z\to 1} (1 - z^{-1})F(z)$$

which are analogous to the properties 3.39 and 3.40 of the unilateral Laplace transform.

7.12 Discuss the transform interrelationships of Figure 7.17. Show that the Laplace transform represents the most general case and that all other transforms are essentially subsets applicable to specific signal classes.

7.13 Outline the features one would expect of a 'Fast Laplace Transform' or a 'Fast z-transform' by analogy to the 'Fast Fourier Transform'. Base the arguments on the discrete-time Fourier interpretations of the discrete-time Laplace transform and of the z-transform. Having outlined the requirements, look up the 'chirp z-transform' from the literature, in particular the concentric form, and interpret its relationship to the FFT.

CHAPTER 8

System Description

We started the continuous-time system descriptions of Chapter 4 at the physical component level, from where we proceeded at a more abstract level, characterizing those systems in terms of their time domain and frequency domain responses $h(t)$, $H(j\omega)$ and $H(s)$, which are better suited for system solution.

Discrete-time systems too employ physical devices, such as digital adders, multipliers and shift registers, interconnected according to circuit diagrams, but such specific implementation aspects fall outside the scope of this book.

We will work entirely at the systems level, interrelating functional elements by means of block diagrams and difference equations and characterizing a system by its discrete-time impulse response $h[nT]$, frequency response $\bar{H}(e^{j\omega T})$ and system function $\bar{H}(z)$ for subsequent system solution. The solution itself can then be implemented on dedicated signal processing hardware or encoded into software algorithms for general purpose computers.

The most immediate aim is a largely self-contained description of discrete-time systems. But we abbreviate and simplify the development of this chapter and the next by exploiting formal similarities with the continuous-time concepts, results and terminology of Chapters 4 and 5. In doing so we also lay the analytical foundations for the discrete-time simulations of continuous-time systems presented in Chapter 11.

8.1   System elements

The most characteristic component of discrete-time systems is the **delay**

element T, which stores a signal value for the duration of one discrete time unit T. Interpreted as an elementary system, its time domain representation is the block T shown at the top of Figure 8.1.

When an arbitrary signal $x[nT]$ is fed into the delay element, the output $y[nT]$ is a delayed version of the input signal, that is,

$$y[nT] = x[nT - T] \tag{8.1}$$

In particular, if the input is the discrete-time impulse $x[nT] = \delta[nT]$, the output $y[nT] = \delta[nT - T]$ is the same impulse, delayed by one interval T. This output is, by definition, the impulse response $h[nT]$ of the elementary system. This case is illustrated above the block diagram of Figure 8.1.

Taking the z-transform of both sides of Equation 8.1 we have

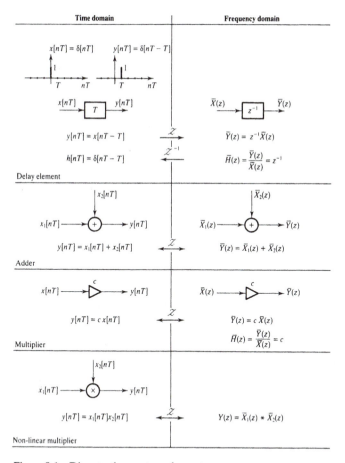

Figure 8.1 Discrete-time system elements.

$$y[nT] = x[nT - T] \quad \overset{z}{\longleftrightarrow} \quad \bar{Y}(z) = z^{-1}\bar{X}(z) \tag{8.2}$$

Defining the **discrete-time system function** $\bar{H}(z)$ as the ratio of the z-transform $\bar{Y}(z)$ of the output over the z-transform $\bar{X}(z)$ of the input, we conclude from expression 8.2 that the system function of the delay element is,

$$\bar{H}(z) = \frac{\bar{Y}(z)}{\bar{X}(z)} = z^{-1} \tag{8.3}$$

The delay element can therefore be represented by a frequency domain block identified by the symbol z^{-1}, as shown in Figure 8.1. According to expression 8.2 this block signifies multiplication by the factor z^{-1}. The function 8.3 is the z-transform 7.22 of the shifted impulse $\delta[nT - T]$, which was represented graphically in Figure 7.7.

Less frequently used, but conceptually similar, is the **advance element** $-T$, whose impulse response is the left-shifted impulse $\delta[nT + T]$, which displaces a function towards negative time. Being an acausal operator, it cannot be incorporated into a real-time system, but the concept can still be used for processing stored data, such as recorded time signals, or for spatial signals, such as graphical images, which are not directly linked to time.

Functionally, the delay element and the advance element become the discrete-time counterparts of the integrator and the differentiator of continuous-time systems. These operators reflected the capability of physical components to store energy. The functional analogy lies in the capability of the discrete-time delay element to store a function value for the duration of one interval T.

Other functional elements required to describe discrete-time systems are the **discrete-time adder**, defined by the operation

$$y[nT] = x_1[nT] + x_2[nT]$$

and the **discrete-time multiplier**,

$$y[nT] = cx[nT]$$

represented by appropriate block diagram symbols in Figure 8.1.

The multiplier factor c is a constant, which makes the multiplier a linear time-invariant operator. Note that the physical device used to multiply a signal by a constant c can also be employed to multiply two signals, giving the result $y[nT] = x_1[nT] \cdot x_2[nT]$. Used in this capacity the device is either non-linear (identical inputs) or time-variant (different inputs), and its output is represented in the frequency domain by a convolution, as indicated at the bottom of Figure 8.1.

From a functional point of view, the behaviour of discrete-time multipliers and adders is identical to that of their continuous-time counterparts.

8.2   First-order systems

An interconnection of a finite number of the above elements constitutes a discrete-time system. Recall that the order of a continuous-time system was determined by the number N of energy-storing devices or, equivalently, by the number of independent integrators and differentiators required to represent it. Similarly, the order of a discrete-time system is determined by the number N of delay elements.

A first-order system thus contains a single delay. We first examine the two most basic configurations, then a combination of these, to conclude with typical first-order responses. The presentation loosely follows that of Section 4.2.

8.2.1   Non-recursive structure

Consider the time domain block diagram of Figure 8.2. It contains one delay element T, two multipliers b_0 and b_1 and an adder. The output $y_b[nT]$ can be written by inspection as

$$y_b[nT] = b_1 x[nT - T] + b_0 x[nT] \tag{8.4}$$

This first-order difference equation describes the system output as the sum of the amplitude-scaled input $b_0 x[nT]$ and a delayed and scaled input $b_1 x[nT - T]$.

In particular, if the input is the discrete-time impulse

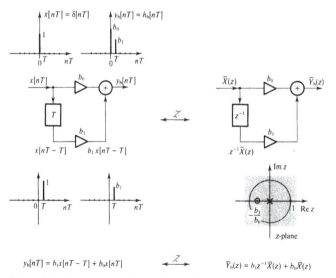

$$y_b[nT] = b_1 x[nT - T] + b_0 x[nT] \qquad \xleftarrow{\;\;\mathscr{Z}\;\;} \qquad \bar{Y}_b(z) = b_1 z^{-1} \bar{X}(z) + b_0 \bar{X}(z)$$

Figure 8.2   Non-recursive first-order system.

$x[nT] = \delta[nT]$, then the output from the delay element T is the delayed impulse $\delta[nT - T]$ and the overall output of the system is, by definition, the impulse response $h_b[nT]$ of the system

$$h_b[nT] = b_1\delta[nT - T] + b_0\delta[nT] \tag{8.5}$$

as indicated by the functions surrounding the time domain block diagram of Figure 8.2. Note that the duration of the impulse response does not extend beyond $nT = NT$ ($N = 1$ in this case). We will see that such time limitation is a characteristic of non-recursive systems, for which reason these are described as **finite impulse response systems**, FIR for short.

To find various frequency domain representations we take the z-transform of the difference equation 8.4,

$$\bar{Y}_b(z) = b_1 z^{-1} \bar{X}(z) + b_0 \bar{X}(z)$$

This can also be expressed in block diagram form, see Figure 8.2, which is identical to its time domain counterpart, except for the symbols used for the delay element and for the variables.

The system function $\bar{H}(z)$ is the output/input ratio

$$\bar{H}_b(z) = \frac{\bar{Y}_b(z)}{\bar{X}(z)} = b_1 z^{-1} + b_0 \tag{8.6}$$

a first-order polynomial in z^{-1}. Taking the inverse z-transform of $\bar{H}_b(z)$ leads back to the system's impulse response of Equation 8.5.

To give a z-plane interpretation we need an expression in z, rather than in z^{-1}. Multiplying and dividing Equation 8.6 by z yields

$$\bar{H}_b(z) = \frac{b_1 + b_0 z}{z} \tag{8.7}$$

The denominator identifies a pole at the origin $z = 0$, while the numerator shows a zero at $z = -b_1/b_0$. These are marked on the z-plane of Figure 8.2. The region of convergence is the entire z-plane, excepting the origin itself.

Evaluating $\bar{H}_b(z)$ on the unit circle gives the system's frequency response

$$\bar{H}_b(e^{j\omega T}) = \frac{\bar{Y}_b(e^{j\omega T})}{\bar{X}(e^{j\omega T})} = b_1 e^{-j\omega T} + b_0 = \frac{b_1 + b_0 e^{j\omega T}}{e^{j\omega T}}$$

where all the functions are identified with the z-transform notation of the discrete-time Fourier transform. Using discrete-time Laplace notation would give the same expression, but interpreted as belonging to the imaginary axis of the periodic $\tilde{s}$-plane,

$$\tilde{H}_b(j\omega) = \frac{b_1 + b_0 e^{j\omega T}}{e^{j\omega T}}$$

8.2.2   Recursive structure

Another basic configuration of first-order is shown in Figure 8.3, where the **delayed output** $y_a[nT - T]$ is added to the input $x[nT]$ to yield

$$y_a[nT] = \frac{1}{a_0}x[nT] - \frac{a_1}{a_0}y_a[nT - T] \tag{8.8}$$

so that the current output value depends also on previous output values. Taking the z-transform of Equation 8.8 gives the frequency domain expression

$$\bar{Y}_a(z) = \frac{1}{a_0}\bar{X}(z) - \frac{a_1}{a_0}z^{-1}\bar{Y}_a(z)$$

and the system function

$$\bar{H}_a(z) = \frac{\bar{Y}_a(z)}{\bar{X}(z)} = \frac{1}{a_1 z^{-1} + a_0} = \frac{z}{a_1 + a_0 z} \tag{8.9}$$

which has a zero at the origin $z = 0$ and a pole at the z-plane point $\pi_1 = -a_1/a_0$. The latter also defines the boundary of the region of convergence, as shown in Figure 8.3. If the pole lies inside the unit

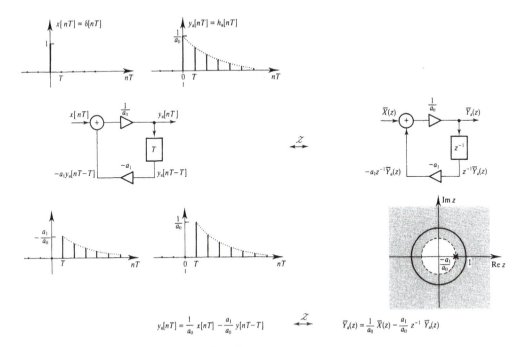

Figure 8.3   Recursive first-order system ($a_1 < 0$).

circle $|\pi_1| < 1$, the region of convergence contains the latter and the system's frequency response can be obtained from Equation 8.9 as

$$\bar{H}_\text{a}(e^{j\omega T}) = \frac{1}{a_1 e^{-j\omega T} + a_0} = \frac{e^{j\omega T}}{a_1 + a_0 e^{j\omega T}}$$

Impulse response

We express Equation 8.9 in terms of the pole π_1 as

$$\bar{H}_\text{a}(z) = \frac{1}{a_0} \frac{z}{z + a_1/a_0} = \frac{1}{a_0} \frac{z}{z - \pi_1} \tag{8.10}$$

which has the form of the z-transform pair 7.27

$$a^n u[nT] \quad \xleftrightarrow{\,z\,} \quad \frac{z}{z - a} \qquad |z| > |a|$$

and yields the impulse response

$$h_\text{a}[nT] = \frac{1}{a_0} \pi_1^n u[nT] \tag{8.11}$$

In the derivation of expression 7.27, we treated this type of function as a real and causal discrete-time exponential. This is represented in Figure 8.3 by the initial value $h_\text{a}[0] = 1/a_0$ and a rate of decay determined by the value π_1. Its relationship to the continuous-time envelope, shown by dotted lines, will be examined in Chapter 11.

This function was extensively represented in Chapters 6 and 7, where it provided visual links between real continuous-time and discrete-time exponentials and between the various discrete-time transforms, see Figures 6.8, 7.2, 7.6 and 7.8.

Recursive usage

This type of configuration describes a **recursive system**, because successive values of the output $y[nT]$ can be obtained by a recursive evaluation of the expression 8.8. The process is primed by specifying an initial output value $y[-T] = y_1$, and leads to the output sequence

$$n = -1 \qquad y[-T] = y_1$$

$$n = 0 \qquad y[0] = \frac{1}{a_0} x[0] - \frac{a_1}{a_0} y[-T]$$

$$n = 1 \qquad y[T] = \frac{1}{a_0} x[T] - \frac{a_1}{a_0} y[0]$$

$$\vdots \qquad \vdots$$

$$n = m \qquad y[mT] = \frac{1}{a_0} x[mT] - \frac{a_1}{a_0} y[mT - T]$$

If the input signal is the impulse function, $x[nT] = \delta[nT]$, and the initial value is set to $y[-T] = 0$, this sequence builds up the impulse response as

$$n = -1 \qquad h[-T] = 0$$

$$n = 0 \qquad h[0] = \frac{1}{a_0}\delta[0] - \frac{a_1}{a_0}h[-T] = \frac{1}{a_0}$$

$$n = 1 \qquad h[T] = \frac{1}{a_0}\delta[T] - \frac{a_1}{a_0}h[0] = \frac{1}{a_0}\left(-\frac{a_1}{a_0}\right)$$

$$n = 2 \qquad h[2T] = \frac{1}{a_0}\delta[2T] - \frac{a_1}{a_0}h[T] = \frac{1}{a_0}\left(-\frac{a_1}{a_0}\right)^2$$

$$\vdots \qquad \vdots$$

$$n = m \qquad h[mT] = \frac{1}{a_0}\delta[mT] - \frac{a_1}{a_0}h[mT - T] = \frac{1}{a_0}\left(-\frac{a_1}{a_0}\right)^m$$

which is the same as Equation 8.11.

Provided $|\pi_1| = |a_1/a_0| < 1$, the impulse response magnitude $|h[nT]|$ decays as time increases. This represents the time domain criterion for system stability. But because the decay is exponential, zero magnitude is never reached and the system is said to have **infinite impulse response**, IIR for short. We will see later that the mere presence of a feedback loop implies that usually, but not always, the impulse response extends to infinite time.

8.2.3 General first-order system

Cascading the two preceding systems as shown at the top of Figure 8.4 yields the general first-order system. Discrete-time block diagram manipulations are identical to those of their continuous-time counterparts of Figure 4.8. Since the delay elements have the same input, one is redundant and leads to the monolithic block diagram shown in the middle row of Figure 8.4.

The overall system function $\bar{H}(z)$ is the product of the system functions 8.9 and 8.6, that is,

$$\bar{H}(z) = \frac{\bar{Y}(z)}{\bar{X}(z)} = \frac{\bar{Y}(z)}{\bar{Y}_a(z)}\frac{\bar{Y}_a(z)}{\bar{X}(z)} = \bar{H}_a(z)\bar{H}_b(z)$$

$$\bar{H}(z) = \frac{b_1 z^{-1} + b_0}{a_1 z^{-1} + a_0} \tag{8.12}$$

This function has one pole and one zero, which are found by multiplying

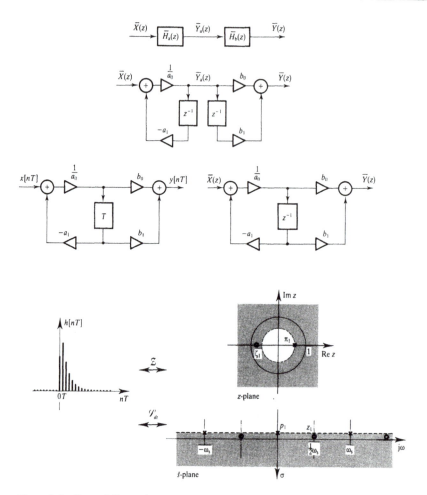

Figure 8.4 General first-order system.

both numerator and denominator by z. But it is instructive to obtain $\bar{H}(z)$ from the product of Equations 8.9 and 8.7, in the form

$$\bar{H}(z) = \bar{H}_a(z)\bar{H}_b(z) = \frac{z}{a_1 + a_0 z} \frac{b_1 + b_0 z}{z} = \frac{b_1 + b_0 z}{a_1 + a_0 z} \quad (8.13)$$

which reveals the cancellation of the zero at the origin of $\bar{H}_a(z)$ by the pole at the origin of $\bar{H}_b(z)$. This only leaves the pole of the first at the z-plane location $\pi_1 = -a_1/a_0$ and the zero of the second at $\zeta_1 = -b_1/b_0$, as expressed in

$$\bar{H}(z) = \frac{b_0}{a_0} \frac{z - \zeta_1}{z - \pi_1} \quad (8.14)$$

and seen from the corresponding z-plane diagrams. These features are mapped onto the periodic $\tilde{s}$-plane as the pole $p_1 = 1/T \ln \pi_1$ and the zero $z_1 = 1/T \ln \zeta_1$, respectively, to give the function of s

$$\widetilde{H}(s) = \frac{b_0}{a_0} \frac{e^{sT} - e^{z_1 T}}{e^{sT} - e^{p_1 T}} \tag{8.15}$$

Frequency response

The region of convergence of $\bar{H}(z)$ is determined by the location of the pole. If $|\pi_1| < 1$, this region contains the unit circle and the general first-order frequency response is the corresponding subset of Equation 8.12

$$\bar{H}(e^{j\omega T}) = \frac{b_1 e^{-j\omega T} + b_0}{a_1 e^{-j\omega T} + a_0} = \widetilde{H}(j\omega) \tag{8.16}$$

The alternative symbol $\widetilde{H}(j\omega)$ indicates that the same expression can be interpreted with $\tilde{s}$-plane notation as an evaluation of the corresponding Laplacian function $\widetilde{H}(s)$. The latter can also be expressed as a subset of Equation (8.15), that is, in terms of $\tilde{s}$-plane features, as

$$\widetilde{H}(j\omega) = \frac{b_0}{a_0} \frac{e^{j\omega T} - e^{z_1 T}}{e^{j\omega T} - e^{p_1 T}} \tag{8.17}$$

Time representation

The general system is represented in the time domain by the general first-order difference equation

$$a_1 y[nT - T] + a_0 y[nT] = b_1 x[nT - T] + b_0 x[nT] \tag{8.18}$$

Indeed, taking the z-transform of both sides,

$$a_1 z^{-1} \bar{Y}(z) + a_0 \bar{Y}(z) = b_1 z^{-1} \bar{X}(z) + b_0 \bar{X}(z)$$

leads back to the system representation of Equation 8.12.

Impulse response

Dividing the numerator polynomial of Equation 8.13 by the denominator, we write

$$\bar{H}(z) = \frac{b_1 + b_0 z}{a_1 + a_0 z}$$

$$= \frac{b_1}{a_1} + \frac{(b_0 - b_1 a_0/a_1)z}{a_1 + a_0 z} = \frac{b_1}{a_1} + \frac{(b_0/a_0 - b_1/a_1)z}{a_1/a_0 + z}$$

But $a_1/a_0 = -\pi_1$, giving the alternative system function description

$$\bar{H}(z) = \frac{b_1}{a_1} + \left(\frac{b_0}{a_0} - \frac{b_1}{a_1}\right)\frac{z}{z - \pi_1}$$

where the first term is a constant and the second term represents a system of the recursive form of Section 8.2.2. The inverse z-transform yields the general first-order impulse response,

$$h[nT] = \frac{b_1}{a_1}\delta[nT] + \left(\frac{b_0}{a_0} - \frac{b_1}{a_1}\right)\pi_1^n u[nT] \qquad (8.19)$$

The last term is a real exponential, as implied in $\pi_1^n = e^{p_1 nT}$, which for stable systems must decay with time.

Summary

The relationships between all these representations are formally similar to those expressed in the scheme 4.23 for the continuous-time first-order system, namely

$$a_1 y[nT - T] + a_0 y[nT] = b_1 x[nT - T] + b_0 x[nT]$$

$$\xleftrightarrow{\mathscr{Z}} \quad a_1 z^{-1}\bar{Y}(z) + a_0\bar{Y}(z) = b_1 z^{-1}\bar{X}(z) + b_0\bar{X}(z)$$

$$\downarrow$$

$$h[nT] = \frac{b_1}{a_1}\delta[nT] + \left(\frac{b_0}{a_0} - \frac{b_1}{a_1}\right)\pi_1^n u[nT] \left\{ \begin{array}{l} \xleftrightarrow{\mathscr{Z}} \quad \bar{H}(z) = \frac{\bar{Y}(z)}{\bar{X}(z)} = \frac{b_0}{a_0}\frac{z - \zeta_1}{z - \pi_1} \\ \\ z = e^{j\omega T} \quad \updownarrow \\ \\ \xleftrightarrow{\mathscr{F}_{dt}} \quad \bar{H}(e^{j\omega T}) = \frac{b_0}{a_0}\frac{e^{j\omega T} - \zeta_1}{e^{j\omega T} - \pi_1} \end{array} \right.$$

$$(8.20)$$

where $\pi_1 = -a_1/a_0$ and $\zeta_1 = -b_1/b_0$ are the system's pole and zero.

8.2.4 Typical responses

Apart from an overall amplitude scaling factor, the behaviour of a system is fully determined by the z-plane locations of its pole π_1 and zero ζ_1. We first consider the effect of moving the pole about when the zero is fixed at the origin $z = 0$. This represents the recursive case of Section 8.2.2, in particular the system function 8.10.

For the system to be real, the pole must be located on the real axis, while stability requires $|\pi_1| < 1$. Six cases are illustrated in Figure 8.5, each case being determined by the z-plane location of its pole π_1.

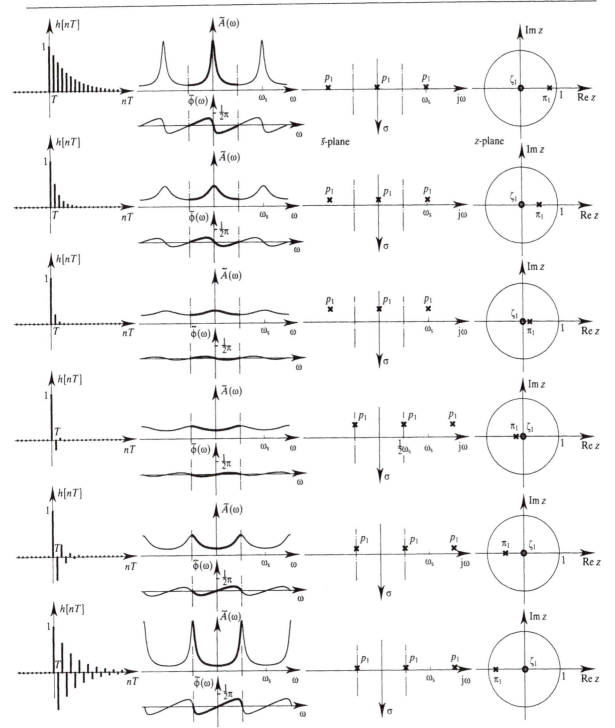

Figure 8.5   Effect of pole position with zero at origin.

The pole is mapped onto the periodic $\tilde{s}$-plane to give the repeated poles p_1, which in turn indicate the peaks of the frequency response. The impulse response of Equation 8.11 is shown on the left and all amplitudes are normalized by setting $a_0 = 1$.

A pole located close to the unit circle maps close to the imaginary axis. This implies a high magnification for the frequency response and a low rate of decay for the impulse response. The limiting case $\pi_1 = 1$ has the impulse response $h[nT] = u[nT]$, whose full frequency representation is similar to that illustrated in Figure 7.14 for the causal impulse train $\bar{\delta}_T(t)$.

As the pole approaches the z-plane origin it gradually cancels the effect of the zero, thus flattening the frequency response and making the impulse response decay sharply. The limiting case would lead to full cancellation, with unity frequency response and impulse response $h[nT] = \delta[nT]$, that is, instantaneous decay.

The negative real semiaxis of the z-plane maps onto the line $\omega = \frac{1}{2}\omega_s$ and its periodic repetitions. Poles located on that semiaxis map accordingly on the periodic $\tilde{s}$-plane, as shown in the lower half of Figure 8.5. Comparison with the symmetric cases of the upper half of the figure shows a net effect of shifting the frequency responses to half sampling frequency. The envelopes of corresponding impulse responses are the same but the sample values now alternate between positive and negative. This effectively represents a system oscillation at half the sampling frequency, $\frac{1}{2}\omega_s$, with period $2T$.

The maxima and minima of the frequency response amplitudes occur at either $z = 1$ or $z = -1$, where $H(z)$ is real and takes the values

$$H(1) = \frac{1}{1 - \pi_1} \qquad H(-1) = \frac{1}{1 + \pi_1}$$

On the $\tilde{s}$-plane imaginary axis these correspond to the points at the origin and at half sampling frequency, and at their periodic repetitions, as seen in Figure 8.5.

Effect of zero location

The location of the zero of the general first-order system function 8.14 has a comparable effect on responses. Again, for the system to be real the zero must be located on the real z-plane axis, but no such restrictions are needed for stability.

In Figure 8.6 we examine six cases with the same pole location π_1. Allowing for periodicity, the effects of displacing the zero along the real axis of the z-plane are analogous to those found in Figure 4.12 for

Figure 8.6 (*Opposite*) Effect of zero position for fixed pole.

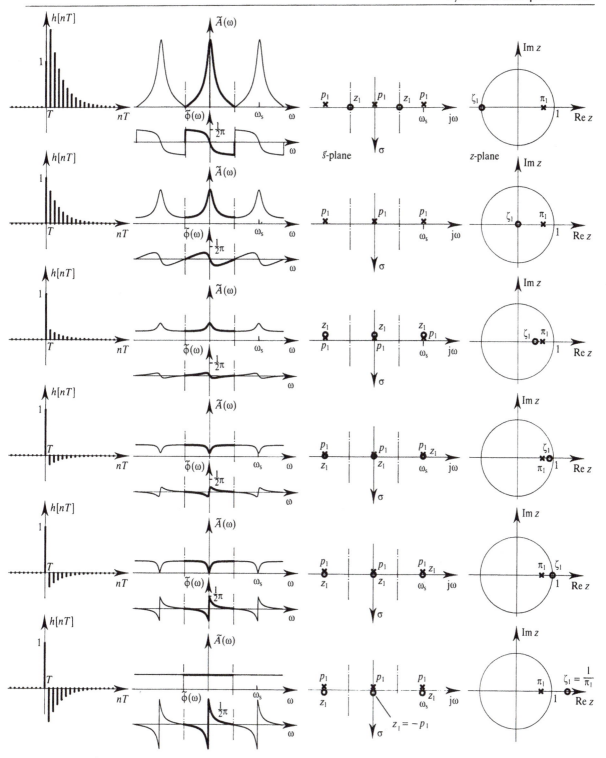

continuous-time systems. As with poles, a fundamental difference arises when the z-plane zero belongs to the negative semiaxis, as it maps onto the periodic $\tilde{s}$-plane on the line at half the sampling frequency, as illustrated in the top row of Figure 8.6.

The formal similarities with the first-order continuous-time systems of Section 4.2 will be exploited in various system simulations in Chapter 11. For instance, the **time sampling method** derives the impulse response $h[nT]$ of a discrete-time system from samples of the impulse response $h(t)$ of a related continuous-time system.

8.3   Second-order systems

Second-order sections are the basic building blocks of many Nth-order systems and are fundamental to certain interpretations of general systems and to simulations. Having examined first-order systems and established similarities between discrete-time and continuous-time structures, we take some short cuts to present second-order sections, going straight to the general case and some typical examples.

8.3.1   General expressions

A second-order system involves two delay elements. The most general case has the form shown in Figure 8.7, which represents the difference equation

$$a_2 y[nT - 2T] + a_1 y[nT - T] + a_0 y[nT]$$
$$= b_2 x[nT - 2T] + b_1 x[nT - T] + b_0 x[nT] \quad (8.21)$$

whose z-transform is

$$a_2 z^{-2} \bar{Y}(z) + a_1 z^{-1} \bar{Y}(z) + a_0 \bar{Y}(z)$$
$$= b_2 z^{-2} \bar{X}(z) + b_1 z^{-1} \bar{X}(z) + b_0 \bar{X}(z) \quad (8.22)$$

and leads to two equivalent forms of the general second-order system function

$$\bar{H}(z) = \frac{\bar{Y}(z)}{\bar{X}(z)} = \frac{b_2 z^{-2} + b_1 z^{-1} + b_0}{a_2 z^{-2} + a_1 z^{-1} + a_0} = \frac{b_0 z^2 + b_1 z + b_2}{a_0 z^2 + a_1 z + a_2} \quad (8.23)$$

The first form, a rational polynomial in z^{-1}, is representative of the delay element implementation shown in the block diagram of Figure 8.7. The second form, a rational polynomial in z, could be implemented by means of advance elements $-T$, but its main use is in the z-plane

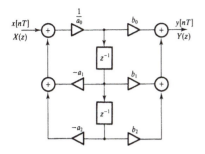

Figure 8.7   General second-order system.

representation of the system's poles and zeros. We multiply and divide by a_0 to give two further variants of the system function

$$\bar{H}(z) = \frac{d_0 z^2 + d_1 z + d_2}{z^2 + \dfrac{a_1}{a_0} z + \dfrac{a_2}{a_0}} = \frac{b_0}{a_0} \frac{(z - \zeta_1)(z - \zeta_2)}{(z - \pi_1)(z - \pi_2)} \qquad (8.24)$$

where $d_i = b_i/a_0$, and the poles π_i and zeros ζ_i represent the quadratic roots

$$\pi_1, \pi_2 = -\frac{a_1}{2a_0} \pm \sqrt{\left(\frac{a_1}{2a_0}\right)^2 - \frac{a_2}{a_0}} \qquad (8.25)$$

and

$$\zeta_1, \zeta_2 = -\frac{b_1}{2b_0} \pm \sqrt{\left(\frac{b_1}{2b_0}\right)^2 - \frac{b_2}{b_0}} \qquad (8.26)$$

These poles and zeros can be directly mapped onto poles $p_i = (1/T \ln \pi_i)$ and zeros $z_i = 1/T \ln \zeta_i$ of the periodic $\check{s}$-plane to give the corresponding discrete-time Laplace representations.

A similar set of equivalent expressions for the frequency response are obtained by the usual substitution $z = e^{j\omega T}$ in any of the forms of Equations 8.23 or 8.24, that is

$$\bar{H}(e^{j\omega T}) = \bar{H}(z)\big|_{z = e^{j\omega T}}$$

8.3.2   Characteristic properties

As with the continuous-time case of Section 4.3.3 the denominator roots determine the oscillatory character of the system, while the zeros determine the response type. The natural response associated with pole locations is most readily seen when both b_1 and b_2 are zero, that is,

$$\bar{H}(z) = \frac{b_0}{a_2 z^{-2} + a_1 z^{-1} + a_0} = \frac{d_0 z^2}{(z - \pi_1)(z - \pi_2)} \qquad (8.27)$$

which has two zeros, $\zeta_1 = \zeta_2 = 0$, at the z-plane origin. As in the continuous-time case, the relative values of the denominator coefficients yield three cases. Two of these have distinct poles and can be expressed by partial fractions (see Section 8.4.3) as

$$\bar{H}(z) = \frac{c_1}{1 - \pi_1 z^{-1}} + \frac{c_2}{1 - \pi_2 z^{-1}} = \frac{c_1 z}{z - \pi_1} + \frac{c_2 z}{z - \pi_2} \qquad (8.28)$$

with coefficients

$$c_1 = \frac{d_0 \pi_1}{\pi_1 - \pi_2} \qquad \text{and} \qquad c_2 = \frac{-d_0 \pi_2}{\pi_1 - \pi_2} \qquad (8.29)$$

which can be verified by adding the terms of Equation 8.28.

Distinct and real poles

The condition $a_1^2 > 4 a_0 a_2$ makes the radicand of Equation 8.25 real. The two poles π_1 and π_2 are real and distinct, so that c_1 and c_2 are also real and distinct and give rise to the overdamped case.

The terms of Equation 8.28 are of the form of Equation 8.10, so that the system's impulse response $h[nT]$ is the sum of terms of the form of Equation 8.11,

$$h[nT] = (c_1 \pi_1^n + c_2 \pi_2^n) u[nT] = \frac{d_0}{\pi_1 - \pi_2} (\pi_1^{n+1} - \pi_2^{n+1}) u[nT]$$

$$(8.30)$$

Written with discrete-time Laplace notation as

$$h[nT] = \frac{d_0}{\pi_1 - \pi_2} (e^{p_1(n+1)T} - e^{p_2(n+1)T}) u[nT]$$

it is clearly the sum of two decaying real exponentials. An example is shown in the top row of Figure 8.8.

Complex conjugate poles

The condition $a_1^2 < 4 a_0 a_2$ in Equation 8.25 leads to the most relevant case with complex conjugate poles, $\pi_2 = \pi_1^*$, which represents an underdamped or oscillating system. Writing the poles as

$$\pi_1 = r_1 e^{j\omega_1 T} \qquad \text{and} \qquad \pi_2 = r_1 e^{-j\omega_1 T}$$

the coefficients of Equations 8.29 become

$$c_1 = \frac{d_0 \pi_1}{\pi_1 - \pi_2} = \frac{d_0 r_1 e^{j\omega_1 T}}{r_1(e^{j\omega_1 T} - e^{-j\omega_1 T})} = \frac{d_0 e^{j\omega_1 T}}{j2 \sin \omega_1 T}$$

and

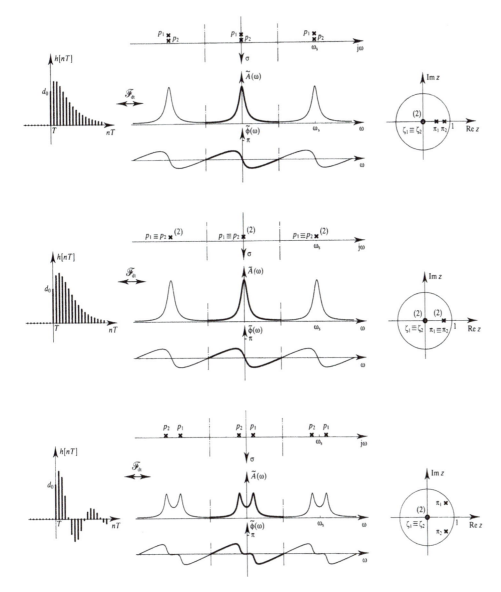

Figure 8.8 Characteristic responses.

$$c_2 = \frac{-d_0 e^{-j\omega_1 T}}{j2 \sin \omega_1 T}$$

which also form a complex conjugate pair, $c_2 = c_1^*$. The impulse response is

$$h[nT] = \frac{d_0}{\sin \omega_1 T} r_1^n \frac{e^{j\omega_1(n+1)T} - e^{-j\omega_1(n+1)T}}{2j} u[nT]$$

$$= \frac{d_0}{\sin \omega_1 T} r_1^n \sin[\omega_1(n+1)T]u[nT]$$

which is essentially a left-shifted sine function of frequency ω_1 that is amplitude modulated by a causal exponential envelope with rate of decay r_1 and initial value $d_0/\sin \omega_1 T$. An example is shown in the bottom row of Figure 8.8.

Real and coincident poles

The limiting case $a_1^2 = 4a_0a_2$ gives rise to two real coincident roots, $\pi_1 = \pi_2 = -a_1/2a_0$, for which the system function 8.27 becomes

$$\bar{H}(z) = \frac{d_0}{(1 - \pi_1 z^{-1})^2} = \frac{d_0 z^2}{(z - \pi_1)^2}$$

To find the impulse response we use the auxiliary function

$$\bar{F}(z) = \frac{z}{z - \pi_1}$$

and its derivative

$$\frac{d\bar{F}(z)}{dz} = \frac{1}{z - \pi_1} - \frac{z}{(z - \pi_1)^2}$$

These permit writing $\bar{H}(z)$ as the sum

$$\bar{H}(z) = d_0\left[\bar{F}(z) - z\frac{d\bar{F}(z)}{dz}\right]$$

which can be easily verified. Using the z-transform property 7.40 the impulse response becomes the sum

$$h[nT] = d_0\big(\pi_1^n u[nT] + n\pi_1^n u[nT]\big) = (n + 1)d_0\pi_1^n u[nT]$$

where the factor $(n + 1)u[nT]$ represents a shifted ramp that amplitude-modulates a decaying exponential $d_0 e^{p_1 nT}$. An example is found in the middle row of Figure 8.8.

8.3.3 Typical responses

We examine how the numerator coefficients b_i of Equation 8.23, or equivalently the coefficients d_i of Equation 8.24, affect the number and location of the z-plane zeros.

Three cases arise when two of the coefficients are zero, which result in either no zeros at all, or one or two zeros at the origin, as shown in the upper row of Figure 8.9. We employed the case $b_1 = b_2 = 0$ earlier to examine characteristic behaviour. Keeping the poles the same, the other two cases represent the cancellation of either one or both zeros at the origin. Each cancellation can be interpreted as a multiplication of $H(z)$ by z^{-1}, or as a placement of a pole at the origin, or by the time-shifting property 7.38. The shape of the amplitude response is not affected, only linear phase is added, which shifts the impulse response $h[nT]$ one sampling interval T to the right. The middle case, $b_0 = b_2 = 0$, corresponds to the z-transform pair 7.31.

Another special case arises when the coefficients b_0 and b_2 have the same value, for example $b_0 = b_2 = 1$. The numerator polynomial $z^2 + b_1 z + 1 = 0$ has two finite zeros, whose product is $\zeta_1 \zeta_2 = 1$ and whose sum is $\zeta_1 + \zeta_2 = -b_1$, so that they are located on either the unit circle or on the real axis, depending on the value of b_1. Some typical locations are shown in the middle row of Figure 8.9. The practical importance of conjugate zeros on the unit circle lies in that they

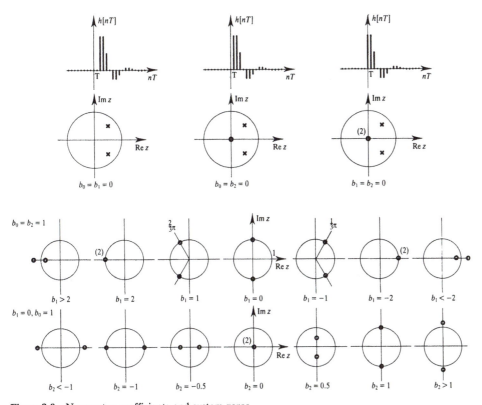

Figure 8.9 Numerator coefficients and system zeros.

represent nulls of the frequency response $H(e^{j\omega T})$, a requirement of certain systems, for instance elliptic filters.

The case when $b_1 = 0$ and $b_0 = 1$ gives the polynomial $z^2 + b_2 = 0$, which yields a symmetrical pair of zeros located either on the real axis or on the imaginary axis, as illustrated in the bottom row of Figure 8.9.

A further case, with $b_2 = 0$, corresponds to the z-transform pair 7.30, whose time domain represents a decaying cosine function.

8.4 Nth-order discrete-time systems

An Nth-order system requires N independent delay elements. The most general case is represented in Figure 8.10 by two equivalent block diagrams of the direct form and the monolithic form. The applicable expressions are listed below as generalizations of the second-order results. For analysis and implementation it is expedient to partition arbitrary systems into second-order sections connected in series or in parallel, as was done for continuous-time systems.

8.4.1 General case

The general Nth-order system responds to the general Nth-order difference equation

$$a_N y[nT - NT] + a_{N-1} y[nT - (N-1)T]$$
$$+ \ldots + a_1 y[nT - T] + a_0 y[nT] = x_f[nT]$$
$$= b_N x[nT - NT] + b_{N-1} x[nT - (N-1)T]$$
$$+ \ldots + b_1 x[nT - T] + b_0 x[nT] \tag{8.31}$$

where $x_f[nT]$ is called the system's forcing function, and represents the interim signal between the two blocks of the direct form of Figure 8.10. The z-transform of all terms yields.

$$a_N z^{-N} \bar{Y}(z) + a_{N-1} z^{-(N-1)} \bar{Y}(z) + \ldots + a_1 z^{-1} \bar{Y}(z) + a_0 \bar{Y}(z)$$
$$= \bar{X}_f(z) = b_N z^{-N} \bar{X}(z) + b_{N-1} z^{-(N-1)} \bar{X}(z) + \ldots$$
$$+ b_1 z^{-1} \bar{X}(z) + b_0 \bar{X}(z) \tag{8.32}$$

and leads directly to the general form of the Nth-order system function

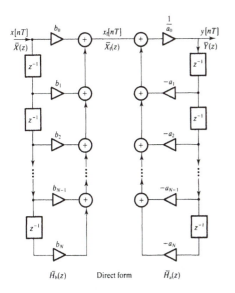 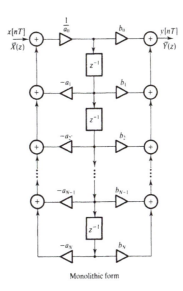

$\bar{H}_b(z)$ Direct form $\bar{H}_a(z)$ Monolithic form

Figure 8.10 General Nth-order system.

$$\bar{H}(z) = \frac{\bar{Y}(z)}{\bar{X}(z)} = \frac{b_N z^{-N} + b_{N-1} z^{-(N-1)} + \ldots + b_1 z^{-1} + b_0}{a_N z^{-N} + a_{N-1} z^{-(N-1)} + \ldots + a_1 z^{-1} + a_0}$$

(8.33)

and the frequency response

$$\bar{H}(e^{j\omega T}) = \frac{\bar{Y}(e^{j\omega T})}{\bar{X}(e^{j\omega T})}$$

$$= \frac{b_N e^{-j\omega NT} + b_{N-1} e^{-j\omega(N-1)T} + \ldots + b_1 e^{-j\omega T} + b_0}{a_N e^{-j\omega NT} + a_{N-1} e^{-j\omega(N-1)T} + \ldots + a_1 e^{-j\omega T} + a_0}$$

(8.34)

These four expressions are written more compactly as

$$\sum_{i=0}^{N} a_i y[nT - iT] = \sum_{i=0}^{N} b_i x[nT - iT]$$

$$\sum_{i=0}^{N} a_i z^{-i} \bar{Y}(z) = \sum_{i=0}^{N} b_i z^{-i} \bar{X}(z)$$

$$\bar{H}(z) = \frac{\sum_{i=0}^{N} b_i z^{-i}}{\sum_{i=0}^{N} a_i z^{-i}} \quad \text{and} \quad \bar{H}(e^{j\omega T}) = \frac{\sum_{i=0}^{N} b_i e^{-j\omega iT}}{\sum_{i=0}^{N} a_i e^{-j\omega iT}}$$

8.4.2 Cascade partitioning

Multiplying and dividing the form of Equation 8.33 by z^N, we follow the process of Section 4.4.2 and factorize the resulting function in terms of its poles and zeros as

$$\bar{H}(z) = \frac{\bar{Q}(z)}{\bar{P}(z)} = k\frac{(z - \zeta_1)(z - \zeta_2) \cdots (z - \zeta_{N-1})(z - \zeta_N)}{(z - \pi_1)(z - \pi_2) \cdots (z - \pi_{N-1})(z - \pi_N)}$$

(8.35)

where $k = b_0/a_0$ is an overall scaling factor. If N is even, this system can be replaced by a series connection of $\frac{1}{2}N$ subsystems of the general form discussed in Section 8.3,

$$\bar{H}_1(z) = \frac{(z - \zeta_1)(z - \zeta_2)}{(z - \pi_1)(z - \pi_2)}$$

such that

$$\bar{H}(z) = k\bar{H}_1(z)\bar{H}_2(z) \cdots \bar{H}_{N/2}(z) = k\prod_{i=1}^{N/2} \bar{H}_i(z)$$

For odd N a further first-order subsystem of the form discussed in Section 8.2 is required, represented by the additional factor

$$\bar{H}_0(z) = \frac{z - \zeta_0}{z - \pi_0}$$

whose roots π_0 and ζ_0 are real.

An example is given in Figure 8.11, which represents a discrete-time simulation of the continuous system of Figure 4.22, with a similar structure. The simulation process itself is explained in Section 11.6.

Geometric evaluation of system function and frequency response

By analogy to the continuous-time case of Section 4.4.2, the discrete-time functions $\bar{H}(z)$ and $\bar{H}(e^{j\omega T})$ can be evaluated graphically from the z-plane representation. The contributions from each pole and zero are clearly seen in the system function 8.35, repeated here in the compact form

$$\bar{H}(z) = k\frac{\Pi_{i=1}^{N}(z - \zeta_i)}{\Pi_{i=1}^{N}(z - \pi_i)} = |\bar{H}(z)|e^{j\theta(z)}$$

(8.36)

A typical denominator factor $(z - \pi_i)$ is a complex number representing the relative position of an arbitrary point z of the z-plane in relation to the pole π_i. It is interpreted geometrically as a directed line segment of magnitude $|z - \pi_i|$ and angle α_i, as indicated in the left half of Figure 8.12. Similarly for numerator factors and corresponding zeros.

The magnitude $|\bar{H}(z_1)|$ at an arbitrary point $z = z_1$ is thus obtained by multiplying the absolute value of the constant k by all the

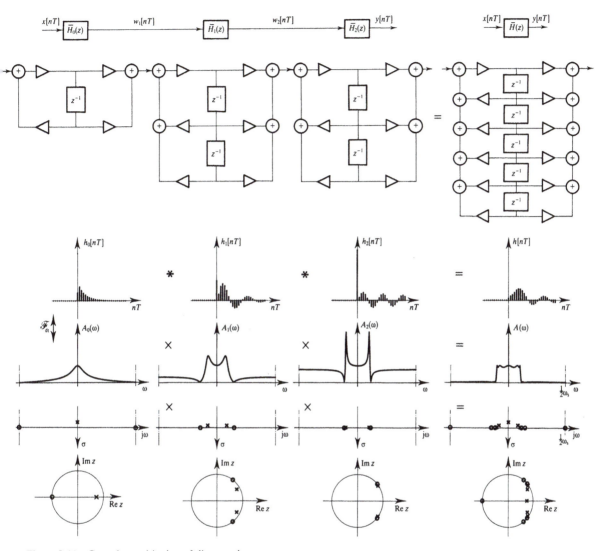

Figure 8.11 Cascade partitioning of discrete-time system.

numerator magnitudes $|z_1 - \zeta_i|$ and dividing by the product of all the denominator magnitudes $|z_1 - \pi_i|$,

$$|\bar{H}(z_1)| = |k| \frac{\Pi_{i=1}^{N}|z_1 - \zeta_i|}{\Pi_{i=1}^{N}|z_1 - \pi_i|} \tag{8.37}$$

while the argument is the corresponding sum of angles,

$$\theta(z_1) = \theta_k + \sum_{i=1}^{N} \beta_i - \sum_{i=1}^{N} \alpha_i$$

where $\theta_k = 0$ for $k > 0$ and $\theta_k = \pi$ for $k < 0$.

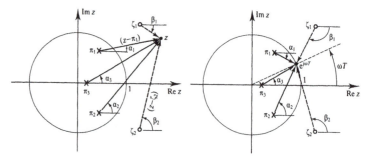

Figure 8.12 Geometric evaluation of $\bar{H}(z)$ and $\bar{H}(e^{j\omega T})$.

Evaluating Equation 8.37 for points of the unit circle, as shown in the right half of Figure 8.12, yields values of the frequency response

$$\bar{H}(e^{j\omega T}) = k\frac{\Pi_{i=1}^{N}(e^{j\omega T} - \zeta_i)}{\Pi_{i=1}^{N}(e^{j\omega T} - \pi_i)} \tag{8.38}$$

8.4.3 Parallel partitioning

By analogy to the continuous-time process of Section 4.4.3 the general system function 8.33 can also be partitioned into a sum of first-order terms

$$\bar{H}(z) = T_0(z) + T_1(z) + T_2(z) + \ldots + T_N(z) = \sum_{i=0}^{N} T_i(z) \tag{8.39}$$

where $T_0(z)$ is a constant term associated with b_N when this is non-zero. For distinct poles this represents a parallel connection of N recursive first-order systems of the form examined in Section 8.2.2,

$$\bar{H}(z) = c_0 + \frac{c_1 z}{z - \pi_1} + \frac{c_2 z}{z - \pi_2} + \ldots + \frac{c_N z}{z - \pi_N} \tag{8.40}$$

where $c_0 = T_0(z) = b_N/a_N$. Each term is associated with one of the poles of $\bar{H}(z)$, as expressed in the denominator of Equation 8.35. Writing Equation 8.40 as

$$\bar{H}(z) = c_0 + \bar{H}_1(z)$$

the coefficients c_i can be found by applying the partial fraction method of Section 4.4.3 to the auxiliary function

$$G(z) = \frac{\bar{H}_1(z)}{z} = \frac{c_1}{z - \pi_1} + \frac{c_2}{z - \pi_2} + \ldots + \frac{c_N}{z - \pi_N}$$

The coefficient c_i associated with a complex pole π_i is also complex. A conjugate pair of poles yields a conjugate pair of coefficients, which in practice are combined into a second-order section with real coefficients. Multiple poles require similar modifications to those mentioned in Section 4.4.3 for continuous time.

   We illustrate this by finding the coefficients of second-order systems.

Example 8.1

We partition the general second-order system by first dividing Equation 8.23 to give

$$\bar{H}(z) = \frac{b_2}{a_2} + \frac{(b_1 - b_2 a_1/a_2)z + (b_0 - b_2 a_0/a_2)z^2}{a_2 + a_1 z + a_0 z^2} = c_0 + \bar{H}_1(z)$$

With $d_1 = (b_1 - b_2 a_1/a_2)/a_0$ and $d_0 = (b_0 - b_2 a_0/a_2)/a_0$ we write $\bar{H}_1(z)$ as

$$\bar{H}_1(z) = \frac{d_1 z + d_0 z^2}{a_2/a_0 + (a_1/a_0)z + z^2}$$

whose denominator roots are the poles (Equation 8.25) of $\bar{H}(z)$ so that

$$G(z) = \frac{\bar{H}_1(z)}{z} = \frac{d_1 + d_0 z}{(z - \pi_1)(z - \pi_2)} = \frac{c_1}{z - \pi_1} + \frac{c_2}{z - \pi_2}$$

The coefficients c_1 and c_2 are now obtained according to Equation 4.72 as

$$c_1 = \{G(z)(z - \pi_1)\}_{z=\pi_1} = \frac{d_1 + d_0 \pi_1}{\pi_1 - \pi_2}$$

$$c_2 = \{G(z)(z - \pi_2)\}_{z=\pi_2} = \frac{d_1 + d_0 \pi_2}{\pi_2 - \pi_1}$$

and the desired general result

$$\bar{H}(z) = c_0 + \frac{c_1 z}{z - \pi_1} + \frac{c_2 z}{z - \pi_2} \tag{8.41}$$

Two particular cases with distinct poles were considered in Section 8.3.2. With $b_1 = b_2 = 0$ we have $c_0 = 0$, $d_1 = 0$ and $d_0 = b_0/a_0$, hence

$$c_1 = \frac{d_0 \pi_1}{\pi_1 - \pi_2} \quad \text{and} \quad c_2 = \frac{-d_0 \pi_2}{\pi_1 - \pi_2}$$

Additionally, if $\pi_2 = \pi_1^*$, then $c_2 = c_1^*$.

Exercises

8.1   Given the difference equation of a first-order system,

$$y[nT] = 2x[nT - T] + 3x[nT]$$

determine whether it is recursive and draw its block diagram. Find the system function and impulse response and draw the z-plane representation.

8.2   Find and plot the impulse response of the system described by

$$-2y[nT - T] + 3y[nT] = x[nT]$$

Draw the block diagram and plot the z-plane and periodic $\check{s}$-plane locations of the system's poles and zeros.

8.3   Connect the systems of Exercises 8.1 and 8.2 in series. Characterize the resulting system in full, arrange the results as in the scheme of Equation 8.20 and draw the block diagram and the z-plane representation of the system.

8.4   With the aid of Figure 8.5 discuss the effects on the frequency response and the impulse response of varying the location of the real pole π_1 of the system function

$$H(z) = \frac{z}{z - \pi_1}$$

within the range $-1 < \pi_1 < 1$. In particular, describe the transitions occurring in the vicinity of the z-plane origin.

8.5   Derive the system function $H(z)$, frequency response $H(e^{j\omega T})$ and impulse response $h[nT]$ of a system described by the general first-order difference equation

$$a_1 y[nT - T] + a_0 y[nT] = b_1 x[nT - T] + b_0 x[nT]$$

and give the locations of the system's pole and zero.

8.6   Using the recursive process suggested in Section 8.2.2, find the impulse response $h[nT]$ of the system described by the difference equation

$$y[nT] - 0.7y[nT - T] = x[nT]$$

Note that the initial value $y[-T]$ must be zero for the result to be valid.

8.7   Draw the block diagram and find the system function $H(z)$ of the second-order system

$$1.5y[nT - 2T] - 3y[nT - T] + 2y[nT]$$
$$= b_2 x[nT - 2T] + b_1 x[nT - T] + b_0 x[nT]$$

Plot the system's poles and zeros and sketch the frequency response for the three cases

$$
\begin{array}{llll}
\text{(a) } b_2 = & 1 & b_1 = -1 & b_0 = 1 \\
\text{(b) } b_2 = & 0 & b_1 = 0 & b_0 = 1 \\
\text{(c) } b_2 = & -1 & b_1 = 0 & b_0 = 1
\end{array}
$$

8.8 Draw the poles and zeros and find the impulse response of three systems represented by the system function

$$
H(z) = \frac{1}{a_2 z^{-2} - 1.5 z^{-1} + 1}
$$

when a_2 takes the values $a_2 = 0.52$, $a_2 = 0.5625$ and $a_2 = 0.8$. Then consider the cases $a_2 = 0.5$, $a_2 = 1$ and $a_2 = 2$ and state whether they lead to stable systems and if any of them possesses a frequency response.

8.9 Draw the block diagram and find the impulse response of the FIR system

$$
y[nT] = x[nT] + x[nT - T] + x[nT - 2T]
$$

Find the system's poles and zeros and sketch its frequency response.

8.10 Find and plot to scale the poles and zeros of the system function

$$
H(z) = \frac{z^2 + z + 1}{z^2 - 1.1z + 0.65}
$$

Use magnitude and angle measurements from the plot to graphically evaluate the function $H(z)$ at points of the unit circle $z = e^{j\omega T}$ and sketch the periodic frequency response $\tilde{H}(j\omega)$ in the range $-\omega_s/2 < \omega < \omega_s/2$. Discuss magnitude values and sudden phase changes as a system zero is traversed.

8.11 Partition the following system functions using partial fractions

$$
H_1(z) = \frac{12z^2 - 5z}{6z^2 - 5z + 1}
$$

$$
H_2(z) = \frac{3z^3 + 4z^2 - 19z}{(z - 1)(z + 2)(z - 3)}
$$

$$
H_3(z) = \frac{4z^{-1} - 6}{z^{-2} - 5z^{-1} + 6}
$$

Find the impulse response of each system, assuming it to be causal.

8.12 Discuss formal similarities between the time domain and frequency domain representations of continuous-time and discrete-time systems. Use the schemes 4.23 and 8.20 of first-order systems, the applicable block diagrams and s-plane or z-plane representations for illustration.

CHAPTER 9

System Solution

The solution methods applicable to discrete-time systems are virtually identical to those derived earlier for continuous-time systems. Keeping to the aim of unifying the treatment of all system classes, these methods will be presented as discrete-time counterparts of those derived in Chapter 5.

We follow the layout of Chapter 5, and this chapter should be read as an extension and complementary to it. This approach not only simplifies presentation of the discrete-time methods, but also reinforces concepts of continuous-time methods and prepares the ground for some of the simulations of Chapter 11.

As with the system descriptions of the preceding chapter, the z-transform takes the place of the Laplace transform, and the discrete-time Fourier transform that of its continuous form. We will also use the discrete-time Laplace transform, where it offers didactic advantages. Similarly, the discrete-time convolution sum takes over from the continuous-time convolution integral.

9.1 Elementary inputs and responses

The **elementary inputs** associated with discrete-time systems can be thought of as discrete-time versions of their continuous-time counterparts. The impulse $\delta[nT]$, the complex exponential $e^{j\omega nT}$ and the generalized complex exponential e^{snT}, which are the elementary signals underlying the main solution methods, are thus presented in Figure 9.1 as the discrete-time versions of the related signals of Figure 5.1.

In the preceding chapter we characterized systems in terms of

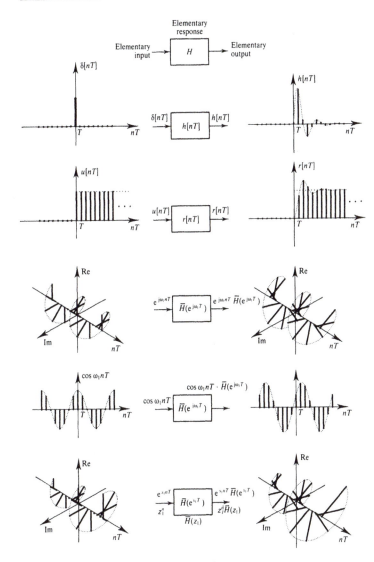

Figure 9.1   Elementary system inputs and responses.

elementary responses to elementary inputs. In this chapter we interpret arbitrary input signals as superpositions of such elementary inputs and make use of those characterizations to determine the system's output.

The continuous-time responses of Figure 5.1 are shown as dotted line envelopes of the discrete-time responses of Figure 9.1. The relationship between two actual systems is not necessarily so direct, as it depends on the method employed for system simulation. It is worth mentioning at this point, that such sampling interpretation of the impulse response $h[nT]$ forms the basis of the time sampling method of simulation, to be discussed in Section 11.3.

9.2 Time domain methods

As in the continuous-time case, two time domain methods are in common use. The elementary inputs are the impulse $\delta[nT]$ (also known as the unit pulse or the delta sequence) and the unit step $u[nT]$, which gives rise to the associated impulse response $h[nT]$ and step response $r[nT]$. There are strong similarities between the continuous and discrete classes, and the discrete-time expressions are in many ways simpler.

9.2.1 Impulse interpretation of arbitrary function

A sample at a point $nT = mT$ of an arbitrary discrete-time function $f[nT]$ can be interpreted as a scaled and shifted impulse $f[mT]\delta[nT - mT]$. The complete function can thus be written as an infinite sum of such samples,

$$f[nT] = \ldots + f[-2T]\delta[nT + 2T] + f[-T]\delta[nT + T]$$
$$+ f[0]\delta[nT - 0] + f[T]\delta[nT - T] + f[2T]\delta[nT - 2T]$$
$$+ \ldots + f[mT]\delta[nT - mT] + \ldots$$

or, more concisely as

$$f[nT] = \sum_{m=-\infty}^{\infty} f[mT]\delta[nT - mT] \tag{9.1}$$

This has the form of a discrete-time convolution, as in Equation 1.6, which **interprets a function** $f[nT]$ **as the convolution of itself with the impulse function** $\delta[nT]$

$$f[nT] = f[nT] * \delta[nT] \tag{9.2}$$

These are the discrete-time counterparts of the convolution interpretations given in Equations 5.2 and 5.3 for a continuous-time signal $f(t)$, but are conceptually more direct, as the requirement for conversion from finite to infinitesimal components does not arise.

Example 9.1

Express the function $f[nT]$ of Figure 9.2 as a sum of impulses. Since only three samples are non-zero, according to Equation 9.1 we write

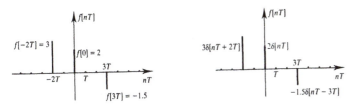

Figure 9.2   Function as sum of impulses.

$$f[nT] = f[-2T]\delta[nT + 2T] + f[0]\delta[nT] + f[3T]\delta[nT - 3T]$$
$$= 3\delta[nT + 2T] + 2\delta[nT] - 1.5\delta[nT - 3T]$$

Applied to an arbitrary input signal $x[nT]$, the convolution interpretation of Equation 9.1 resolves it into a sum of impulses, and leads directly to the impulse response method.

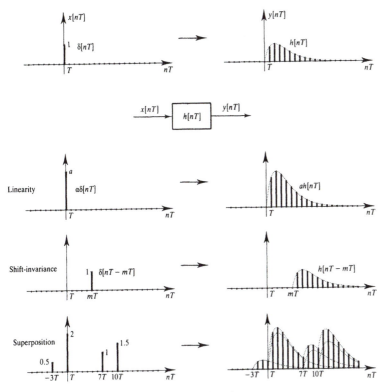

Figure 9.3   Response to scaled and shifted impulses.

9.2.2 Impulse response method

Consider the discrete-time system of Figure 9.3. By definition, when the input $x[nT]$ is the impulse $\delta[nT]$ the output is the system's impulse response $h[nT]$. For continuity, we illustrate it with a discrete-time simulation of the response $h(t)$ used earlier in Figure 5.2.

Superposition of impulses and of responses

We assume that the system is linear and time-invariant, so that a scaled impulse $a\delta[nT]$ gives rise to a similarly scaled response $ah[nT]$, and a shifted impulse $\delta[nT - mT]$ produces a similarly shifted response $h[nT - mT]$. Superposition of such inputs leads to a corresponding superposition of responses, as illustrated at the bottom of Figure 9.3 by means of an input signal

$$x[nT] = 0.5\delta[nT + 3T] + 2\delta[nT] + \delta[nT - 7T]$$
$$+ 1.5\delta[nT - 10T]$$

and the corresponding output expression

$$y[nT] = 0.5h[nT + 3T] + 2h[nT] + h[nT - 7T]$$
$$+ 1.5h[nT - 10T]$$

from which the individual values of $y[nT]$ can be computed.

Convolution sum

The impulse response method formalizes the above superposition process in the form of a convolution. We express the system's response to a weighted and shifted impulse by the input/output relationship

$$x[mT]\delta[nT - mT] \xrightarrow{\ h[nT]\ } x[mT]h[nT - mT]$$

both sides of which have the same weighting factor $x[mT]$ and the same delay mT. An arbitrary sum, as in Equation 9.1, of such elementary inputs produces a similar sum at the output,

$$x[nT] = \sum_{m=-\infty}^{\infty} x[mT]\delta[nT - mT]$$

$$\xrightarrow{\ h[nT]\ } y[nT] = \sum_{m=-\infty}^{\infty} x[mT]h[nT - mT] \quad (9.3)$$

This is to say, the system's output is the discrete convolution of that input $x[nT]$ with the impulse response $h[nT]$ of the system. This expression represents the impulse response method for discrete-time systems, which written more compactly as

$$x[nT] = x[nT] * \delta[nT] \xrightarrow{\ h[nT]\ } y[nT] = x[nT] * h[nT] \qquad (9.4)$$

can be expressed, by analogy to the continuous-time result 5.10, as: **if the input $x[nT]$ is interpreted as a convolution of itself with the impulse function $\delta[nT]$, the output $y[nT]$ can be interpreted as the convolution of the input $x[nT]$ with the impulse response $h[nT]$ of the discrete-time system.**

The result 9.3 is interpreted in Figure 9.4, based on a similar continuous-time situation illustrated in Figure 5.4. The upper half of the figure shows the contribution of an input sample located at $nT = m_1 T$. The lower half gives a full interpretation of the convolution sum, highlighting the partial result $y[m_1 T]$, up to time $nT = m_1 T$, as well as the contribution from the component $x[m_1 T]$.

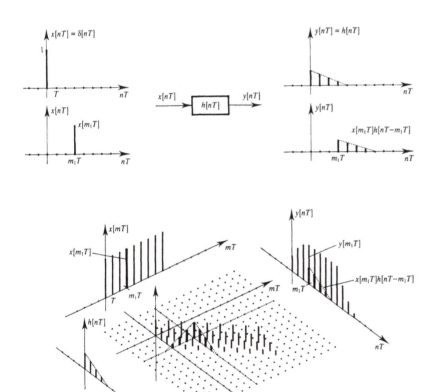

Figure 9.4 Impulse response method.

A comparison of Figures 5.4 and 9.4 helps to clarify the concepts underlying the impulse interpretation of continuous-time signals, and the visualisation of the convolution integral. The comparison is also relevant to the time sampling method for discrete-time simulation of continuous systems.

9.3   Frequency domain methods

The frequency domain methods developed in Section 5.3 for continuous-time systems find exact counterparts in discrete time. In this section we emphasize this similarity, as well as the equivalence between the various discrete-time methods.

We formulate these methods by expanding the input $x[nT]$ as superpositions of either complex exponentials $e^{j\omega nT}$ or of generalized exponentials $e^{snT} = z^n$, and finding the corresponding responses with the aid of the impulse response method just derived.

9.3.1   Response to single exponential

Given a system with known impulse response $h[nT]$, we excite it with a discrete-time complex exponential of fixed frequency $\omega = \omega_1$,

$$x[nT] = e^{j\omega_1 nT}$$

Proceeding as in Section 5.3.1, we apply the convolution method of Equation 9.4 to express the system's output $y[nT]$ as

$$y[nT] = e^{j\omega_1 nT} * h[nT] = \sum_{m=-\infty}^{\infty} h[mT]e^{j\omega_1(nT - mT)}$$

$$= e^{j\omega_1 nT} \sum_{m=-\infty}^{\infty} h[mT]e^{-j\omega_1 mT} \qquad (9.5)$$

The last sum represents one value, at $\omega = \omega_1$, of the discrete-time Fourier transform of $h[nT]$, which can be expressed either with the $\tilde{s}$-plane notation $\widetilde{H}(j\omega)$ or with the z-plane notation $\bar{H}(e^{j\omega T})$, as in the summary chart of Figure 7.17. This leads to the input/output correspondence

$$\boxed{x[nT] = e^{j\omega_1 nT} \xrightarrow{\ h[nT]\ } y[nT] = e^{j\omega_1 nT}\widetilde{H}(j\omega_1) = e^{j\omega_1 nT}\bar{H}(e^{j\omega_1 T})}$$

$$(9.6)$$

The output is a scaled version of the input signal, which makes the discrete-time complex exponential $e^{j\omega nT}$ an eigenfunction of discrete-time systems. The scaling factor $\widetilde{H}(j\omega)$ or $\bar{H}(e^{j\omega T})$ represents the associated eigenvalue.

Graphical interpretation

The result 9.6 is interpreted in Figure 9.5, showing both frequency domain alternatives. The system used for illustration is a typical second-order lowpass section of Section 8.3, as represented by $h[nT]$, $\widetilde{H}(j\omega)$ and $\bar{H}(e^{j\omega T})$.

The complex input signal $x[nT] = e^{j\omega_1 nT}$ is shown in the figure together with its real and imaginary parts. In the frequency domain, when interpreted as the discrete-time Fourier pair of Equation 6.26 and Figure 6.12, it takes the form of a shifted impulse train on the imaginary axis of the $\check{s}$-plane. In the z-plane, this train maps as a single impulse at the point $z = e^{j\omega_1 T}$ of the unit circle.

The time domain convolution process of Equation 9.5 is indicated in the time domain region of Figure 9.5 for the full complex exponential $e^{j\omega_1 nT}$ and also for its real and imaginary parts. But the result 9.6 of this

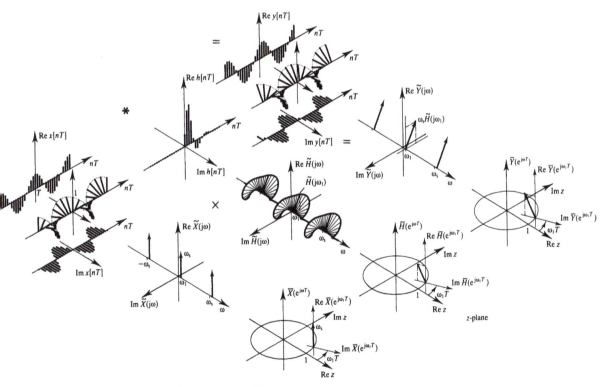

Figure 9.5 Response to single discrete exponential.

convolution is easier to visualize through its frequency domain implications, expressed as

$$x[nT] = e^{j\omega_1 nT} \quad \xrightarrow{\;h[nT]\;} \quad y[nT] = e^{j\omega_1 nT}\,\widetilde{H}(j\omega_1) = e^{j\omega_1 nT}\,\bar{H}(e^{j\omega_1 T})$$

$$\updownarrow \mathcal{F}_{dt} \qquad\qquad\qquad\qquad \updownarrow \mathcal{F}_{dt}$$

$$\widetilde{X}(j\omega) = \omega_s \sum_{k=-\infty}^{\infty} \delta(\omega - \omega_1 - k\omega_s)$$

$$\xrightarrow{\;\widetilde{H}(j\omega)\;} \quad \widetilde{Y}(j\omega) = \omega_s \sum_{k=-\infty}^{\infty} \delta(\omega - \omega_1 - k\omega_s)\widetilde{H}(j\omega_1)$$

$$s \to z \;\downarrow \qquad\qquad\qquad s \to z \;\downarrow$$

$$\bar{X}(e^{j\omega T}) = \omega_s \delta(e^{j(\omega T - \omega_1 T)}) \xrightarrow{\;\bar{H}(e^{j\omega T})\;} \bar{Y}(e^{j\omega T}) = \omega_s \delta(e^{j(\omega T - \omega_1 T)})\bar{H}(e^{j\omega_1 T})$$

$$(9.7)$$

The middle row represents the periodic $\check{s}$-plane version. It uses the notation of the discrete-time Laplace transform and is almost identical to the continuous-time case expressed earlier by the relationship 5.14 and Figure 5.5. Each impulse $\omega_s \delta(\omega - \omega_1 - k\omega_s)$ of the input train is scaled by the corresponding value of $\widetilde{H}(j\omega)$, giving a magnitude-scaled and phase-shifted version of the original impulse. Since $\widetilde{X}(j\omega)$ and $\widetilde{H}(j\omega)$ have the same periodicity, all impulses are affected by the same factor $\widetilde{H}(j\omega_1)$, and the resulting impulse train $\widetilde{Y}(j\omega) = \widetilde{X}(j\omega)\widetilde{H}(j\omega_1)$ represents a correspondingly scaled and phase-shifted time domain exponential.

This periodic $\check{s}$-plane also provides a visual link with the z-transform version of the process, represented by the bottom row of scheme 9.7 and the z-plane region of Figure 9.5. We saw that the impulse train $\widetilde{X}(j\omega)$ maps as a single impulse $\bar{X}(e^{j\omega T}) = \omega_s \delta(e^{j(\omega T - \omega_1 T)})$, located at the z-plane point $e^{j\omega T} = e^{j\omega_1 T}$ of the unit circle.

The output signal is the product $\bar{Y}(e^{j\omega T}) = \bar{X}(e^{j\omega T})\bar{H}(e^{j\omega_1 T})$. It too is a scaled and phase-shifted impulse, which maps the corresponding impulse train of the $\check{s}$-plane. To abbreviate an already overcrowded diagram the frequency response $\bar{H}(e^{j\omega T})$ is shown on the unit circle of the z-plane for the relevant value $\bar{H}(e^{j\omega_1 T})$ only.

9.3.2 Frequency response method

An arbitrary input signal $x[nT]$ can always be expanded with the aid of the inverse discrete-time Fourier transform 6.22 into an infinite sum of infinitesimal exponentials

$$x[nT] = \mathcal{F}_{dt}^{-1}\{X(j\omega)\} = \frac{1}{\omega_s}\int_0^{\omega_s} \widetilde{X}(j\omega)e^{j\omega nT}\,d\omega \qquad (9.8)$$

Proceeding as in Section 5.3.3, according to expression 9.6 one infinitesimal input exponential of frequency $\omega = v$, and amplitude and phase determined by the complex scaling factor $(1/\omega_s)\widetilde{X}(jv)\,d\omega$, elicits an infinitesimal response

$$\left[\frac{1}{\omega_s}\,\widetilde{X}(jv)\,d\omega\right]e^{jvnT} \xrightarrow{h[nT]} \left[\frac{1}{\omega_s}\widetilde{X}(jv)\,\widetilde{H}(jv)\,d\omega\right]e^{jvnT}$$

The infinite sum 9.8 of such infinitesimal components yields the output

$$y[nT] = \frac{1}{\omega_s}\int_0^{\omega_s} [\widetilde{X}(j\omega)\,\widetilde{H}(j\omega)]e^{j\omega nT}\,d\omega \qquad (9.9)$$

This output can be interpreted as the inverse Fourier transform of the product $\widetilde{X}(j\omega)\,\widetilde{H}(j\omega)$. We conclude that this product is the frequency domain output $\widetilde{Y}(j\omega)$ when the input is $\widetilde{X}(j\omega)$, and express this conclusion as the input/output relationship

$$\boxed{\widetilde{X}(j\omega) \xrightarrow{\widetilde{H}(j\omega)} \widetilde{Y}(j\omega) = \widetilde{X}(j\omega)\,\widetilde{H}(j\omega)} \qquad (9.10)$$

This represents the **frequency response method of discrete-time systems**, which is related by the discrete-time Fourier transform to the impulse response method as

$$
\begin{array}{ccc}
x[nT] = x[nT] * \delta[nT] & \xrightarrow{h[nT]} & y[nT] = x[nT] * h[nT] \\[6pt]
\Big\updownarrow \mathcal{F}_{dt} & & \Big\updownarrow \mathcal{F}_{dt} \\[6pt]
\widetilde{X}(j\omega) = \widetilde{X}(j\omega)\cdot 1 & \xrightarrow{\widetilde{H}(j\omega)} & \widetilde{Y}(j\omega) = \widetilde{X}(j\omega)\,\widetilde{H}(j\omega)
\end{array}
$$

$$(9.11)$$

The equivalence 9.11 is visualized in Figure 9.6, a discrete-time version of the earlier Figure 5.7, in which a similar continuous-time equivalence between methods was represented.

Anticipating the developments of Section 9.3.4, the lower right of Figure 9.6 also shows the alternative z-variable interpretation of the frequency response method,

$$\overline{X}(e^{j\omega T}) \xrightarrow{\overline{H}(e^{j\omega T})} \overline{Y}(e^{j\omega T}) = \overline{X}(e^{j\omega T})\,\overline{H}(e^{j\omega T}) \qquad (9.12)$$

which maps all the features of the periodic $\tilde{s}$-plane onto the z-plane. For brevity, magnitudes only are represented.

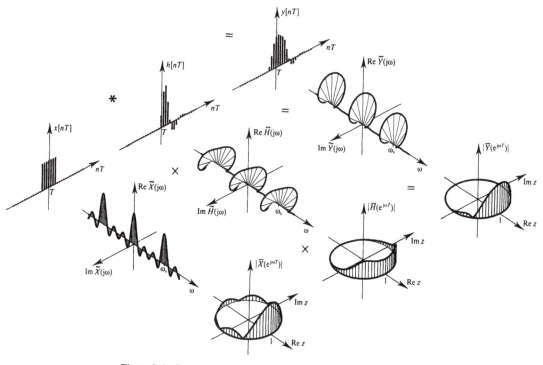

Figure 9.6 Frequency response method of discrete-time systems.

9.3.3 Response to generalized discrete-time exponential

The discrete-time exponential $x[nT] = e^{s_1 nT}$ is the counterpart of the continuous generalized exponential $x(t) = e^{s_1 t}$ of Section 5.3.5, where s_1 is again a fixed complex value. It produces a formally similar elementary response and leads to the z-transform method of solution.

Applying the impulse response method of expression 9.4 to this elementary input, we would obtain the elementary output

$$y[nT] = x[nT]*h[nT] = \sum_{m=-\infty}^{\infty} e^{s_1(nT-mT)} h[mT]$$

$$= e^{s_1 nT} \sum_{m=-\infty}^{\infty} h[mT]e^{-s_1 mT}$$

But the input exponential is more economically written using z-variable notation as

$$x[nT] = e^{s_1 nT} = z_1^n$$

where $z_1 = e^{s_1 T}$. The above convolution then takes the form

$$y[nT] = x[nT] * h[nT] = \sum_{m=-\infty}^{\infty} z_1^{n-m} h[mT]$$

$$= z_1^n \sum_{m=-\infty}^{\infty} h[mT]z_1^{-m}$$

The last summation is the bilateral z-transform $\bar{H}(z)$ of the impulse response, evaluated at $z = z_1$, and yields the input/output relationship

$$\boxed{x[nT] = z_1^n \xrightarrow{\ h[nT]\ } y[nT] = z_1^n \bar{H}(z_1)} \tag{9.13}$$

or, returning to exponential notation associated with the discrete-time Laplace transform,

$$x[nT] = e^{s_1 nT} \xrightarrow{\ h[nT]\ } y[nT] = e^{s_1 nT} \tilde{H}(s_1)$$

The generalized discrete-time exponential $z^n = e^{snT}$ is thus an eigenfunction of the discrete-time system, and $\bar{H}(z)$ or $\tilde{H}(s)$ provides the eigenvalue. The unconventional interpretation of Figure 5.11 could be extended to impulses of the periodic $\tilde{s}$-plane and the z-plane to interpret expression 9.13.

9.3.4 z-transform method

The inverse z-transform integral 7.17 is now the appropriate tool to express an arbitrary input $x[nT]$ as a superposition of generalized exponentials $z^n = e^{snT}$

$$x[nT] = \frac{1}{2\pi j} \oint_\Gamma \bar{X}(z)z^{n-1}\, dz \tag{9.14}$$

The same considerations of Section 9.3.2 applied to an infinitesimal input component yield the input/output relationship

$$\left[\frac{1}{2\pi j} \bar{X}(z)dz \right] z^{n-1} \xrightarrow{\ h[nT]\ } \left[\frac{1}{2\pi j} \bar{X}(z)\bar{H}(z)dz \right] z^{n-1}$$

so that the infinite sum 9.14 of such components leads to the output

$$y[nT] = \frac{1}{2\pi j} \oint_\Gamma \bar{X}(z)\bar{H}(z)z^{n-1}\, dz$$

and to the **discrete-time system response method** or **z-transform method**

$$\boxed{\bar{X}(z) \xrightarrow{\ \bar{H}(z)\ } \bar{Y}(z) = \bar{X}(z)\bar{H}(z)} \tag{9.15}$$

whose link to the discrete-time convolution method is

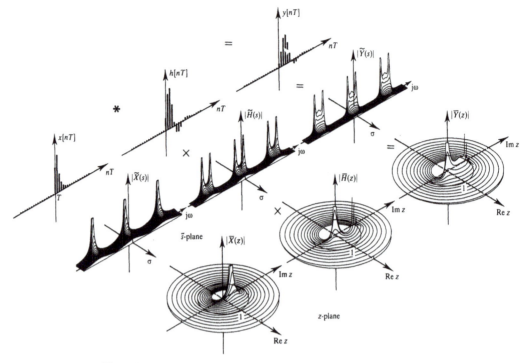

Figure 9.7 z-transform method.

$$
\begin{array}{ccc}
x[nT] & \xrightarrow{\ h[nT]\ } & y[nT] = x[nT] * h[nT] \\[4pt]
\updownarrow \mathscr{Z} & \quad \updownarrow \mathscr{Z} & \updownarrow \mathscr{Z} \\[4pt]
\bar{X}(z) & \xrightarrow{\ \bar{H}(z)\ } & \bar{Y}(z) = \bar{X}(z)\,\bar{H}(z)
\end{array}
\tag{9.16}
$$

This relationship is illustrated in Figure 9.7, which also shows the discrete-time Laplace transform version of the frequency domain, obtained by the appropriate mapping from the z-plane to the periodic $\tilde{s}$-plane.

9.4 Initial conditions and complete solution

The methods developed so far provide the complete solution of a discrete-time system only when the system is initially at rest. If initial conditions are not zero, their effect must be taken into consideration.

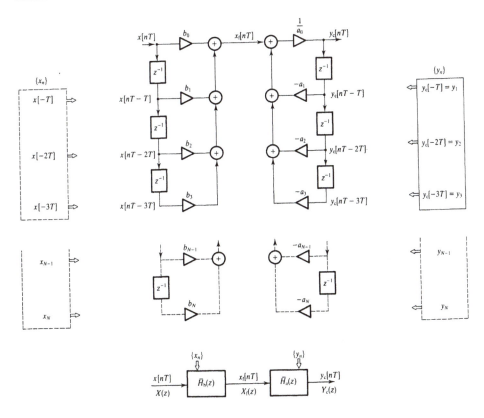

Figure 9.8 System with initial conditions, complete solution.

In continuous-time systems, the **unilateral Laplace transform** provided a routine formulation of the complete solution. It not only gave the system's response to the forcing function, but explicitly assigned places for the initial conditions. The **unilateral z-transform** provides a similar complete formulation for discrete-time systems.

To stress fundamental similarities between the two classes, we loosely follow the structure of Section 5.4, relating similar concepts, and we will find that some of these are more immediate in discrete-time. We similarly interpret the complete output $y_c[nT]$ as a sum of the forced output $y[nT]$ and the response $y_i[nT]$ due to initial conditions,

$$y_c[nT] = y[nT] + y_i[nT]$$

9.4.1   System equations and initial conditions

The general Nth-order discrete-time system is represented by the general Nth-order difference equation 8.31. We repeat it here for a third-order system, using the notation $y_c[nT]$ to signify complete output,

$$a_3 y_c[nT - 3T] + a_2 y_c[nT - 2T] + a_1 y_c[nT - T] + a_0 y_c[nT]$$

$$= x_f[nT] = b_3 x[nT - 3T] + b_2 x[nT - 2T]$$

$$+ b_1 x[nT - T] + b_0 x[nT] \tag{9.17}$$

The term $x_f[nT]$ represents the **forcing function of the difference equation**, effectively a measure of the system's internal recollection of the latest N values of the input function $x[nT]$.

In Section 8.4 we gave preference to the monolithic form of block diagram, Figure 8.10, to represent the system, as it was economical in its use of delay elements. For our present purposes the direct form, repeated in Figure 9.8, is more appropriate, as it gives a direct representation of the difference equation 9.17, and also provides explicit locations for the forcing function $x_f[nT]$ and for all initial conditions.

The expression 9.17 can be rearranged to show the output $y_c[nT]$ explicitly in terms of the forcing function and of the N latest output values as

$$y_c[nT] = \frac{1}{a_0} \{x_f[nT] - a_3 y_c[nT - 3T]$$

$$- a_2 y_c[nT - 2T] - a_1 y_c[nT - T]\} \tag{9.18}$$

This expression represents the right half of the block diagram of Figure 9.8, which has $x_f[nT]$ for input and shows clearly the significance of the delayed outputs. The forcing function $x_f[nT]$ is itself shown as the output of the left half of the block diagram, being a weighted sum of the current input value $x[nT]$ and N preceding input values.

The expression 9.18 forms the basis of a recursive algorithm often used for computing consecutive values of the output function $y_c[nT]$. Each evaluation requires the knowledge of the N latest output values. To start the process, that is, to calculate the first value, $y[0]$, it is necessary to define N initial output values. We identify these with the notation

$$y_c[-T] = y_1 \qquad y_c[-2T] = y_2 \quad \dots \quad y_c[-NT] = y_N$$

gathered in the sequence $\{y_n\}$, as shown at the right edge of Figure 9.8. For a system initially at rest all these values are zero.

Note formal similarities with the continuous-time initial conditions of Section 5.4, which were represented by the initial values of the output function $y(t)$ and its $N - 1$ derivatives, a concept that is not applicable to discrete-time signals. Anticipating the simulations of Chapter 11, a continuous function $y(t)$ is approximated by a discrete-time function $y[nT]$, while an mth-order derivative of $y(t)$ involves a set of $m + 1$ adjacent values of $y[nT]$. The N initial values $\{y_n\}$ of an Nth-order discrete-time system can thus simulate a set of derivatives up to order $N - 1$.

9.4.2   Unilateral z-transform method

In continuous-time systems the unilateral Laplace transform of the **derivative of a function** automatically incorporates initial time domain values into the frequency domain description. In discrete-time systems the unilateral z-transform achieves the same effect with the **transform of a shifted function** (expression 7.44). Applied to the output function

$$y_c[nT] \quad \overset{\mathscr{Z}_1}{\longleftrightarrow} \quad \bar{Y}_c(z)$$

it expresses the transform of the shifted function $y_c[nT - mT]$ as

$$y_c[nT - mT] \quad \overset{\mathscr{Z}_1}{\longleftrightarrow} \quad z^{-m}\bar{Y}_c(z) + z^{-(m-1)}y_1$$
$$+ \ldots + z^{-1}y_{m-1} + y_m$$

which incorporates m initial time domain values of the output.

Assuming that the input $x[nT]$ is causal, so that its own initial values $x_m = x[-mT]$ and the forcing function $x_f[nT]$ are zero for negative time, a routine application of the unilateral z-transform to the individual terms of Equation 9.17 yields the algebraic expression

$$a_3[z^{-3}\bar{Y}_c(z) + z^{-2}y_1 + z^{-1}y_2 + y_3]$$
$$+ a_2[z^{-2}\bar{Y}_c(z) + z^{-1}y_1 + y_2] + a_1[z^{-1}\bar{Y}_c(z) + y_1] + a_0\bar{Y}_c(z)$$
$$= \bar{X}_f(z) = (b_3z^{-3} + b_2z^{-2} + b_1z^{-1} + b_0)\bar{X}(z)$$

Gathering terms in $\bar{Y}_c(z)$, this can be rearranged as

$$(a_3z^{-3} + a_2z^{-2} + a_1z^{-1} + a_0)\bar{Y}_c(z)$$
$$+ [(a_3z^{-2} + a_2z^{-1} + a_1)y_1 + (a_3z^{-1} + a_2)y_2 + a_3y_3]$$
$$= \bar{X}_f(z) = (b_3z^{-3} + b_2z^{-2} + b_1z^{-1} + b_0)\bar{X}(z)$$

and dividing by the polynomial associated with $\bar{Y}_c(z)$ we can write the complete output as the sum of two distinct terms,

$$\bar{Y}_c(z) = \frac{b_3z^{-3} + b_2z^{-2} + b_1z^{-1} + b_0}{a_3z^{-3} + a_2z^{-2} + a_1z^{-1} + a_0}\bar{X}(z)$$

$$- \frac{(a_3z^{-2} + a_2z^{-1} + a_1)y_1 + (a_3z^{-1} + a_2)y_2 + a_3y_3}{a_3z^{-3} + a_2z^{-2} + a_1z^{-1} + a_0}$$

$$(9.19)$$

As the coefficients a_n and b_n are fixed parameters of the system, the first term on the right depends entirely on the input function $\bar{X}(z)$, and represents the forced response $\bar{Y}(z)$ of the system. By contrast, the second term, which depends entirely on the initial values $\{y_n\}$, represents the system's response $\bar{Y}_i(z)$ due to initial conditions, and we write

$$\bar{Y}_c(z) = \bar{Y}(z) + \bar{Y}_i(z) \tag{9.20}$$

We identify the two sides of the block diagram of Figure 9.8 with two cascaded subsystems $\bar{H}_b(z)$ and $\bar{H}_a(z)$, as indicated at the bottom of the figure. Their individual system functions are

$$\bar{H}_a(z) = \frac{1}{a_3z^{-3} + a_2z^{-2} + a_1z^{-1} + a_0}$$

and

$$\bar{H}_b(z) = b_3z^{-3} + b_2z^{-2} + b_1z^{-1} + b_0$$

so that their product yields the overall system function

$$\bar{H}(z) = \bar{H}_b(z)\bar{H}_a(z)$$

The expression 9.19 can therefore be written in the form

$$\bar{Y}_c(z) = \bar{H}(z)\bar{X}(z) - \bar{H}_a(z)\bar{Y}_0(z) \tag{9.21}$$

The function $\bar{Y}_0(z)$ represents the numerator of the second term of Equation 9.19, which can also be written as a polynomial in z^{-1}, with coefficients that are weighted sums of the initial conditions y_n,

$$\bar{Y}_0(z) = (a_3y_1)z^{-2} + (a_2y_1 + a_3y_2)z^{-1} + (a_1y_1 + a_2y_2 + a_3y_3)$$

Acausal forcing function

If the forcing function $x_f[nT]$ is not zero-valued for negative time, then the unilateral z-transform of the input signal $x[nT]$ introduces similar initial input values $\{x_n\}$ into Equation 9.19, so that the complete solution 9.21 becomes

$$\bar{Y}_c(z) = \bar{H}(z)\bar{X}(z) + \bar{H}_a(z)[\bar{X}_0(z) - \bar{Y}_0(z)] \tag{9.22}$$

where the additional term $\bar{X}_0(z)$ is a polynomial in z^{-1} containing the initial input values

$$\bar{X}_0(z) = (b_3x_1)z^{-2} + (b_2x_1 + b_3x_2)z^{-1} + (b_1x_1 + b_2x_2 + b_3x_3)$$

Generalization

The expression 9.22 is also applicable to the complete solution of a general Nth-order system, whose system function is characterized by

$$\bar{H}_a(z) = \frac{1}{\sum_{n=0}^{N} a_n z^{-n}} \qquad \bar{H}_b(z) = \sum_{n=0}^{N} b_n z^{-n}$$

$$\bar{H}(z) = \bar{H}_a(z)\bar{H}_b(z)$$

The initial conditions of the system are contained in the polynomial

$$\bar{Y}_0(z) = \sum_{n=1}^{N} c_n(z)y_n \qquad c_n(z) = \sum_{k=n}^{N} a_k z^{-(k-n)}$$

which is fully determined by the coefficients of the block $\bar{H}_a(z)$, see Figure 9.8, while those of the input signal $x[nT]$ are contained in

$$\bar{X}_0(z) = \sum_{n=1}^{N} d_n(z)x_n \qquad d_n(z) = \sum_{k=n}^{N} b_k z^{-(k-n)}$$

which is fully determined by the coefficients of the input block $\bar{H}_b(z)$.

System characterizations

If all initial conditions are zero at the time $nT = 0$, when the input signal is applied, the complete output 9.22 reduces to

$$\bar{Y}_c(z) = \bar{H}(z)\bar{X}(z) = \bar{Y}(z)$$

which is consistent with the relationship used in Chapter 8 to define the system function $\bar{H}(z)$ as an output/input ratio. It confirms the requirement that the system be initially at rest for characterization

$$\bar{H}(z) = \frac{\bar{Y}(z)}{\bar{X}(z)} = \frac{\bar{Y}_c(z)}{\bar{X}(z)}\Bigg|_{\{y_n\}=0}$$

and this requirement applies to impulse response and frequency response characterizations.

The forced response $\bar{Y}(z)$ can therefore be obtained by any of the methods presented in this chapter. But if the system has initial conditions, their effects must be added separately. For instance, the impulse response method would only give the first term of the complete solution

$$y_c[nT] = h[nT]*x[nT] + y_i[nT]$$

while $y_i[nT]$ would have to be found by other means.

9.5   Relationships between methods

We close the chapter, and Part 2 of the book, by drawing together the conclusions reached in this chapter and relating them to similar conclusions drawn for continuous-time systems at the end of Chapter 5.

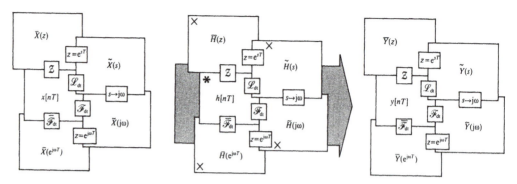

Figure 9.9 Equivalence between discrete-time methods.

9.5.1 Links between discrete-time methods

The equivalence between time domain and frequency domain methods for solving discrete-time systems was expressed in the relationships 9.11 and 9.16. These are gathered in the compact graphical statement of Figure 9.9, the discrete-time counterpart of Figure 5.15.

Using the imagery introduced with Figure 7.19, we now have five choices for describing each of the three blocks representing the input signal, the system and the output signal in Figure 9.9. In each case the time function is surrounded by the periodic $\tilde{s}$-plane representations of the discrete-time Laplace transform and its Fourier subset, and by the alternative z-plane representations of the z-transform and its version of the discrete-time Fourier transform.

This offers the choice of five paths for solving a problem. The operations appropriate to each method, convolution in time or multiplication in frequency, are indicated with the system block. A particular path would be interpreted as in the examples illustrated in Figure 5.15 for continuous time.

9.5.2 Duality of impulse and frequency response methods

The duality between the impulse response method and the frequency response method of continuous-time systems, on which we elaborated in Section 5.3.4, also extends to discrete-time.

Response to single shifted impulse

We consider the frequency implications of a system when the input $x[nT]$ is a shifted impulse. According to expression 7.23, the discrete-time Fourier transform of the input is an exponential,

$$x[nT] = \delta[nT - mT] \quad \overset{\mathcal{F}_{dt}}{\longleftrightarrow} \quad \widetilde{X}(j\omega) = e^{-j\omega mT}$$

This is interpreted in Figure 9.10 as a special case of Figure 9.6, and is the discrete-time counterpart of Figure 5.8. As in Figure 9.6, the system's frequency response $\widetilde{H}(j\omega)$ is periodic, the period being given by the sampling frequency $\omega_s = 2\pi/T$.

Since the impulse is shifted from the origin by m sampling intervals T, the exponential $\widetilde{X}(j\omega)$ takes m full turns within one period ω_s. This is also apparent in the figure from its real and imaginary parts.

The exponential is modified in magnitude and phase by the system's frequency response, giving the output $\widetilde{Y}(j\omega) = \widetilde{X}(j\omega)\widetilde{H}(j\omega)$, whose inverse Fourier transform is the equally shifted impulse response $y[nT]$, as expressed in

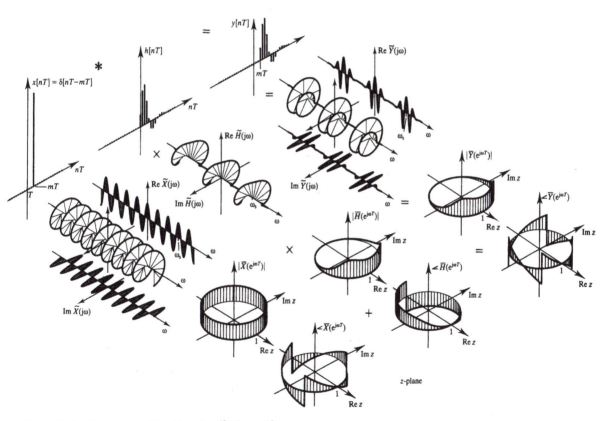

Figure 9.10 Response to shifted impulse $\delta[nT - mT]$.

$$x[nT] = \delta[nT - mT] \xrightarrow{h[nT]} y[nT] = \delta[nT - mT]*h[nT]$$
$$= h[nT - mT]$$

$$\Big\updownarrow \mathcal{F}_{dt} \qquad\qquad \Big\updownarrow \mathcal{F}_{dt}$$

$$\widetilde{X}(j\omega) = e^{-j\omega mT} \xrightarrow{\widetilde{H}(j\omega)} \widetilde{Y}(j\omega) = e^{-j\omega mT}\,\widetilde{H}(j\omega)$$

The frequency domain process is equivalently illustrated on the unit circle of the z-plane that is, as

$$\overline{X}(e^{j\omega T}) = e^{-j\omega mT} \xrightarrow{\overline{H}(e^{j\omega T})} \overline{Y}(e^{j\omega T}) = e^{-j\omega mT}\,\overline{H}(e^{j\omega T})$$

Compare this interpretation with Figure 9.5, which illustrated the system's response to a time domain exponential. That elementary input was represented in the frequency domain by a shifted impulse train on the imaginary axis of the $\tilde{s}$-plane, which mapped as a single shifted impulse on the unit circle of the z-plane. This comparison shows the role reversal between the discrete-time impulse function in one domain and the exponential function in the other. It represents the discrete-time counterpart of a similar role reversal for continuous-time systems seen in Section 5.3.4.

Duality of methods

By analogy to the continuous-time case, the impulse response method implies an interpretation of the input $x[nT]$ as a **sum of weighted time domain impulses**, each of which is seen in the frequency domain as an exponential.

Conversely, the frequency response method implies an interpretation of the discrete-time Fourier transform of the input as a **superposition of weighted frequency domain impulses**, each seen in the time domain as an exponential. According to the choice of notation, these impulses are either a single set on the unit circle of the z-plane, interpreting $\overline{X}(e^{j\omega T})$, or a periodically repeated set on the imaginary axis of the $\tilde{s}$-plane, interpreting $\widetilde{X}(j\omega)$.

The duality of continuous-time methods therefore also extends to discrete-time systems.

9.5.3 Link between continuous-time and discrete-time methods

The unified treatment of all the transforms of continuous-time and discrete-time signals, whose results were summarized in Figure 7.17, provided the base in Chapter 8 for similar characterizations of the appropriate systems. The unifying process was carried through to the present chapter, where discrete-time system solution methods were treated as extensions of the continuous-time methods of Chapter 5.

The conceptual similarities, backed up by the same notation and terminology, will be exploited further in Part 3 where synthesis of discrete-time systems will be based on related continuous systems. The heavy emphasis on graphical representations will then be justified.

Exercises

9.1 Derive the impulse response method for solving discrete-time systems

$$x[nT] \xrightarrow{h[nT]} y[nT] = x[nT] * h[nT]$$

Highlight the interpretation of the input signal $x[nT]$ as a convolution sum involving the discrete-time impulse function $\delta[nT]$.

9.2 Show that when both the impulse response $h[nT]$ and the input signal $x[nT]$ are causal the convolution sum for the output $y[nT]$ need only be evaluated in the range $0 \leq m \leq n$.

9.3 Using the impulse response method find the output $y[nT]$ from a system whose impulse response is the unit step $h[nT] = u[nT]$, for two input signals

$$x_1[nT] = 0.5^n u[nT] \quad \text{and} \quad x_2[nT] = (-0.5)^n u[nT]$$

Draw the time domain and z-plane representations of the system and signals.

9.4 With the aid of Figure 9.5 discuss the time domain and frequency domain implications of processing an acausal discrete-time complex exponential $x[nT] = e^{j\omega_1 nT}$ with a system described by its frequency response. Stress the equivalence of the two frequency domain representations associated with alternative descriptions $\widetilde{H}(j\omega)$ and $\overline{H}(e^{j\omega T})$ of the frequency response.

9.5 A discrete-time system with impulse response $h[nT] = (\frac{1}{3})^n u[nT]$ is fed the input signal $x[nT] = (\frac{1}{2})^n u[nT]$. Find the output signal $y[nT]$ using the z-transform method and plot the z-plane poles and zeros of the system and signals.

9.6 Use the impulse response method to derive the frequency response method

$$\overline{X}(e^{j\omega T}) \xrightarrow{\overline{H}(e^{j\omega T})} \overline{Y}(e^{j\omega T}) = \overline{X}(e^{j\omega T})\overline{H}(e^{j\omega T})$$

employing z-variable notation. Highlight the role of the discrete-time Fourier transform $\overline{\mathcal{F}}_{dt}$ in interpreting the input signal as a superposition of exponentials. Explain the equivalence of the two methods with the aid of Figure 9.6 and complete the z-plane representation of the figure with a sketch of the appropriate phases, indicating the applicable operation.

9.7 Derive the output $Y(e^{j\omega T})$ of a system with frequency response

$$H(e^{j\omega T}) = \frac{1}{1 - ae^{-j\omega T}}$$

for an input signal described in time by $x[nT] = b^n u[nT]$. Find the time signal $y[nT]$ for $a = 0.8$ and $b = -0.6$ and plot all z-plane poles and zeros.

9.8 Derive the discrete-time Laplace transform method,

$$\widetilde{X}(s) \xrightarrow{\widetilde{H}(s)} \widetilde{Y}(s) = \widetilde{X}(s)\widetilde{H}(s)$$

by similarity to the derivation of the z-transform method. Discuss their equivalences as shown in Figure 9.7. Relate this result to the Laplace method of continuous-time systems.

9.9 Express the function $Y(z)$ associated with the difference equation

$$y[nT] - 0.6y[nT - T] = x[nT] + 2x[nT - T]$$

in terms of the z-transform of $x[nT]$, assuming the system to be initially at rest. Then find the solution $y[nT]$ for $x[nT] = (0.8)^n u[nT]$.

9.10 Use the unilateral z-transform method to solve the difference equation

$$y_c[nT] - \tfrac{1}{6}y_c[nT - T] - \tfrac{1}{3}y_c[nT - 2T] = x[nT]$$

with initial conditions $y_c[-T] = 1$, $y_c[-2T] = 2$. Write the general expression of $Y_c(z)$ as a sum of the forced response $Y(z)$ and the response $Y_i(z)$ due to initial conditions; then find the specific output $y_c[nT]$ for the case $x[nT] = u[nT]$.

9.11 Find the numerical values of the first three samples $y[0]$, $y[T]$ and $y[2T]$ of the difference equation of Exercise 9.10 written in the form

$$y[nT] = u[nT] + \tfrac{1}{6}y[nT - T] + \tfrac{1}{3}y[nT - 2T]$$

by the recursive process suggested in Section 8.2.2, starting with the same initial conditions $y[-T] = 1$ and $y[-2T] = 2$. Verify the results against the closed form solution of Exercise 9.10.

9.12 Use the arguments applicable to the complete solution of a discrete-time system with initial conditions to show that to characterize a system by its impulse response or frequency responses, as in Chapter 8, requires the system to be initially at rest.

PART 3

System Synthesis

The first two parts of this book were devoted to the *analysis* of available signals and systems. *Signal analysis* involved the Laplace family of transforms to give a full interpretation of a signal known in one domain. *System analysis* dealt with the *characterization* of a specified system and its subsequent *solution* for a given input signal. This is typified by the formulation $Y(s) = X(s)H(s)$, where both $H(s)$ and $X(s)$ are known.

In the last part we present some fundamental aspects of system *synthesis* or *realization*. The typical formulation is the converse of the analysis problem: given an input $X(s)$ and a desired output $Y(s)$, find a realizable system that approximates the process $H(s) = Y(s)/X(s)$.

Chapter 10 covers continuous-time topics. We first describe some basic ideal systems and operators, explore inherent obstacles to realization and look at realistic system specifications. We then present a sequence of general methods that lead to realizable continuous-time approximations of a specified process.

In Chapter 11 we approximate the specified process by means of discrete-time systems. Although effective techniques are available for synthesizing directly in discrete time, for better understanding we approach the problem as a discrete-time simulation of a known continuous-time approximation.

The sections of the two chapters are structured to cover the essential steps involved in synthesizing a typical system. The interrelated design steps are summarized in the chart of Figure 11.27. This diagram also brings together systems from both classes, thus clarifying the logistics of the simulation process and indicating the roles of the applicable mapping laws and transforms and their interrelations.

CHAPTER 10

Ideal and Realizable Continuous-time Systems

This chapter starts with some strictly ideal systems, which provide a framework and a nomenclature for describing intended or desirable properties. We identify the fundamental obstacles to the realization of such systems and explore possible engineering compromises for approximating ideal specifications with realizable systems. We deal mainly with ideal filters, but the concepts are extended in Section 10.3 to other common ideal operators, such as differentiators, integrators and Hilbert transformers.

Section 10.4 forms a bridge between ideal and realizable systems. We re-examine terminal properties of lumped parameter systems, which we know from Chapter 4 to be realizable. In particular we view first-order and second-order systems as building blocks with adjustable parameters, which are combined to fit the system specification.

The remainder of the chapter introduces some important methods used for approximation. Section 10.5 examines the foundations of the most common classical methods for lowpass filter synthesis, as well as the methodology leading to realizable s-plane features. Section 10.6 presents a general method for extending the lowpass results to other characteristics, such as bandpass and highpass filters.

The primary objective of this chapter is to examine fundamental aspects of system approximation, rather than to give a detailed treatment of any particular design method. All processes are developed in both domains, including s-plane representations where applicable. This provides deeper insight and serves the further objective of preparing the ground for the discrete-time simulations of Chapter 11. To this end we will occasionally take some developments a stage further than could be justified on continuous-time objectives alone.

We will not enter into physical network considerations, which were covered in Chapter 4, where they gave sufficient assurance that realization was possible. Nor will we cover aspects of practical implementation, such as the use of specific devices and the effects of noise and component tolerances, which all fall outside the scope of this book.

10.1   System types

We briefly introduce the system types examined in this chapter, to set scope and nomenclature. Only linear and time invariant systems will be considered.

We have previously classified systems according to the class of the time domain signals that characterize them. This gave rise to continuous-time and discrete-time systems.

A second important classification distinguishes between systems that are **realizable** and those representing **idealized** mathematical abstractions. The latter provide simple models of real systems, giving insight into the performance, capabilities and fundamental limitations of the latter.

Within these large classes, systems are most commonly described by their frequency behaviour. We find it expedient for this text to divide systems according to whether the frequency response of the strictly ideal form is basically real even or imaginary odd, as indicated in Figure 10.1. This distinction will be clarified in Section 10.3, which examines the latter type.

We divide those with real even basis, which include most practical systems, into two groups. **Frequency-selective filters** offer an 'all-or-nothing' selection of specific frequency bands, as in lowpass and bandpass filters. In contrast, **frequency-shaping systems** provide a less clear-cut shaping of the frequency spectrum.

This classification is of ideal systems. Any realizable system whose response is deemed to be close to one of these, according to some error criterion, can be considered an approximation of the corresponding ideal system and be identified by the same description, as indicated at the bottom of Figure 10.1.

10.2   Ideal continuous-time filters

The descriptor **ideal** is interpreted here in contraposition to **realizable**. We will qualify this descriptor after a sequence of modifications that gradually lead us from a **strictly ideal** system, towards the requirements of a **realizable** system. For simplicity we use frequency selective filters,

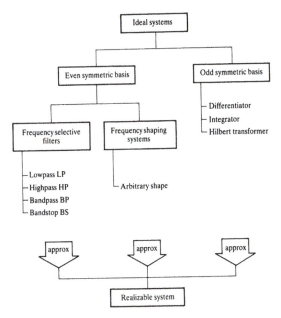

Figure 10.1   Classification of ideal and realizable systems.

of simple description and precise terminology, with the understanding that the underlying concepts also apply to other systems.

An ideal filter is specified by a passband and a stopband. It lets through all frequency components in the passband, without magnitude or phase distortions, and totally rejects the frequencies in the stopband. Such a **strictly ideal filter**, whose passband has unit magnitude and zero phase, is an unattainable mathematical abstraction. But it economically describes a specific type of behaviour, and this serves as a basis for comparison and for rating a whole family of related filters.

We introduce various types of **strictly** ideal filters in Section 10.2.1, where we interpret their characteristics in both domains. In later subsections we gradually relax some of the restrictions that make the filter strictly ideal. These include phase linearity, causality and length of the impulse response. Each restriction affects both domains, and we will see how they relate to realizability.

The fundamental concepts are first derived with the aid of the graphical imagery developed in earlier chapters, to be backed up at every stage by the relevant mathematical expressions.

In Chapter 11 we will simulate the characteristics of various ideal filters with discrete-time systems, and will find that some of the resulting limitations can be turned to advantage. For example, we will be able to design realizable linear phase FIR systems, which are impossible to achieve with the continuous-time systems of this chapter.

10.2.1 Strictly ideal filters (zero phase)

We consider four basic ideal filter types identified by the position of the passband.

Lowpass

The most familiar is the **ideal lowpass filter** shown in the upper left quadrant of Figure 10.2. Its frequency response is described by the real-valued rectangular pulse,

$$H_{\mathrm{LP}}(j\omega) = p_{\omega_1}(\omega) = \begin{cases} 1 & |\omega| < \omega_1 \\ 0 & \text{elsewhere} \end{cases}$$

The frequency ω_1, variously referred to as the cutoff frequency, corner frequency or band-edge, determines the bandwidth of the filter.

The rectangular pulse was examined in detail in Chapter 2, where its inverse Fourier transform was derived in Example 2.2. The impulse response of the filter is therefore

$$h_{\mathrm{LP}}(t) = \frac{\omega_1}{\pi} \frac{\sin \omega_1 t}{\omega_1 t} \tag{10.1}$$

Since $H_{\mathrm{LP}}(j\omega)$ is real and even, so is $h_{\mathrm{LP}}(t)$. Furthermore, since its frequency domain is strictly band-limited, the impulse response must extend to infinity, and being even this means in both positive and negative time. This presents a fundamental obstacle to the practical realization of ideal filters. A non-zero impulse response for negative time would imply that the system is capable of producing an output signal in anticipation to the input signal, which is unrealistic. And the anticipation would extend to infinite time, which is even more unrealistic.

The ideal lowpass filter is represented in the upper left quadrant of Figure 10.2, which also sets the layout of related figures. The frequency response $H_{\mathrm{LP}}(j\omega)$ is represented by its amplitude $A(\omega)$, phase $\phi(\omega) = 0$ and group delay $\tau_{\mathrm{g}}(\omega) = 0$. The time domain shows the impulse response $h_{\mathrm{LP}}(t)$ as a Fourier pair with the frequency response, and to its left we also show its time integral, the step response $r_{\mathrm{LP}}(t)$.

Highpass

The ideal highpass filter rejects all frequency components up to the band edge ω_1, letting through, free of distortions, all other frequencies. Its frequency response is

$$H_{\mathrm{HP}}(j\omega) = \begin{cases} 0 & |\omega| < \omega_1 \\ 1 & |\omega| > \omega_1 \end{cases}$$

It is complementary to the ideal lowpass filter, in the sense that their bands are reversed. This can be used to derive the impulse response of the highpass filter, as indicated in the sequence of the bottom left quadrant of Figure 10.2, where the frequency response $H_{HP}(j\omega)$ is interpreted as a lowpass filter subtracted from unity, that is,

$$H_{HP}(j\omega) = 1 - H_{LP}(j\omega)$$

The constant is interpreted as the unit function $H_{AP}(j\omega) = 1$ that represents the **strictly ideal allpass filter**, which lets through all frequency components, in other words, the ideal conductor. Its impulse response is the impulse function $\delta(t)$ itself (impulse in, impulse out). Subtracting Equation 10.1 from it yields the impulse response of the highpass filter

$$h_{HP}(t) = \delta(t) - \frac{\omega_1}{\pi} \frac{\sin \omega_1 t}{\omega_1 t} \tag{10.2}$$

Note that the impulse at the origin effectively represents the filter's behaviour at high frequency values, and gives it its highpass character.

To the left of the component impulse responses are shown their integrals. Here too the two components are added to yield the step response $r_{HP}(t)$. The sudden transition at the origin represents the step response of the allpass filter.

Bandpass and bandstop

Using a similar procedure, we construct the responses of the ideal bandpass filter. The frequency response

$$H_{BP}(j\omega) = \begin{cases} 1 & \omega_1 < |\omega| < \omega_2 \\ 0 & \text{elsewhere} \end{cases}$$

results from subtracting a narrow lowpass section, of width ω_1, from a wider lowpass section, of width ω_2, as shown in the upper right quadrant of Figure 10.2. The corresponding impulse response becomes

$$h_{BP}(t) = \frac{\omega_2}{\pi} \frac{\sin \omega_2 t}{\omega_2 t} - \frac{\omega_1}{\pi} \frac{\sin \omega_1 t}{\omega_1 t} \tag{10.3}$$

Similarly, the ideal bandstop filter

$$H_{BS}(j\omega) = \begin{cases} 0 & \omega_1 < |\omega| < \omega_2 \\ 1 & \text{elsewhere} \end{cases}$$

is the sum of a lowpass filter, of width ω_1, and a wider highpass filter, of width ω_2, as shown in the lower right of Figure 10.2. The impulse response becomes

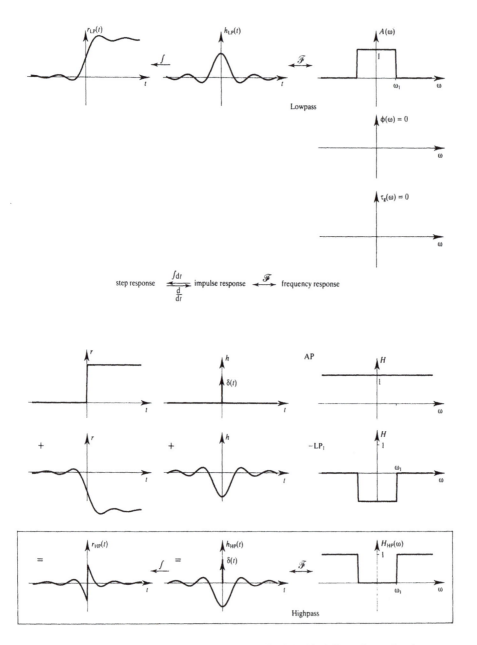

Figure 10.2 Types of strictly ideal filters (zero phase).

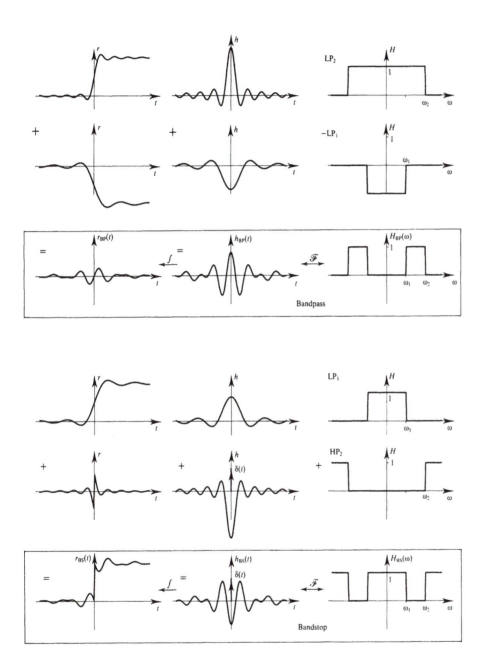

$$h_{BS}(t) = \delta(t) + \frac{\omega_1}{\pi} \frac{\sin \omega_1 t}{\omega_1 t} - \frac{\omega_2}{\pi} \frac{\sin \omega_2 t}{\omega_2 t} \tag{10.4}$$

Again, note the impulse and the sharp transition at the origin of the time responses, which represent the filter's behaviour at high frequencies.

Effects on input signals

Each of the above filter types has a characteristic effect on incoming signals, which is particularly noticeable with narrow pulses and at sharp transitions. It is revealing to compare their effects on square input pulses of different widths.

We have chosen three pulses, whose widths are respectively much smaller, comparable and much larger than the width of the main lobe of the impulse response. The outputs from the four filters are shown in Figure 10.3, which illustrates relationships of the form

$$x(t) \xrightarrow{h_{xx}(t)} y(t) = x(t) * h_{xx}(t)$$

In each case the narrow pulse, which resembles an impulse, gives rise to an output that resembles the impulse response. In contrast, the edges of the wide pulse, which can be interpreted as positive-going and negative-going steps, give rise to a similar superposition of two step responses.

Symmetries

Ideal filters preserve the symmetries of the input signal, because the distortions introduced are symmetrical too. This is clearly seen from the preceding examples, where both the input and output pulses are even symmetrical.

More generally, since any input signal can be split into a sum of even and odd parts, and the above ideal filters are real and even in both domains, it is clear that multiplication in the frequency domain preserves the symmetries of the input components. This is also true for the corresponding time domain convolutions, where the impulse responses are real and even.

10.2.2   Ideal filters with linear phase

The frequency responses of the four ideal filter types considered so far were real and even functions, with zero phase, so that their impulse responses were real and even functions of time, therefore acausal.

We now allow the frequency response to adopt a linear phase

characteristic of the form $\phi(\omega) = -t_0\omega$. This corresponds to the time-shifting property 2.31 of the Fourier transform

$$f(t - t_0) \xleftrightarrow{\mathcal{F}} e^{-j\omega t_0} F(j\omega)$$

whereby the phase $-\omega t_0$ of the exponential is added to the phase of $F(j\omega)$.

Lowpass

The time-shifting property was illustrated in the lower half of Figure 2.29. If we associate the function $F(j\omega)$ of that figure with the frequency response of the strictly ideal lowpass filter, then the function $G(j\omega)$ represents that of the ideal lowpass filter of linear phase, and the functions $f(t)$ and $g(t)$ become the corresponding impulse responses. This leads to the linear-phase system shown in Figure 10.4.

To give a clearer insight into this and later linear-phase systems we reexamine our earlier graphical interpretation in Figure 2.10 of the frequency implications of shifting a function in time. Recall that we started with the real even function of Figure 2.8 and moved back along the negative time axis. This motion was tracked in the frequency domain by twisting the frequency components into a tightening helix. The process was frozen at time $t = -\tau$, as shown in Figure 2.10, to give the frequency representation of a time-shifted function.

In the present context we interpret that twisting of the frequency components as an addition of phase $\phi(\omega) = -\tau\omega$, linear in ω and proportional to the time delay $t = \tau$. This relationship is clear from Figure 10.4, where we also show the usual modulo 2π representation of phase.

Analytically, this relationship between the zero-phase system $H_{LP}(j\omega)$ and the linear-phase system $H_L(j\omega)$ is expressed by the Fourier pair

$$h_L(t) = h_{LP}(t - \tau) = \delta(t - \tau) * h_{LP}(t)$$

$$\xleftrightarrow{\mathcal{F}} \quad H_L(j\omega) = e^{-j\omega\tau} H_{LP}(j\omega)$$

where the shifted impulse response is also interpreted as a convolution with a shifted impulse. The frequency response is interpreted at the top of Figure 10.4 as a helical segment, a portion of the complex exponential $e^{-j\omega\tau}$, truncated to the bandwidth of $H_{LP}(j\omega)$.

Note that the phase representation was continued into the stopband as a dashed line as it makes no sense to talk of the phase of a component that has zero magnitude.

Other ideal filters

Extending the same considerations and graphics to the other types of

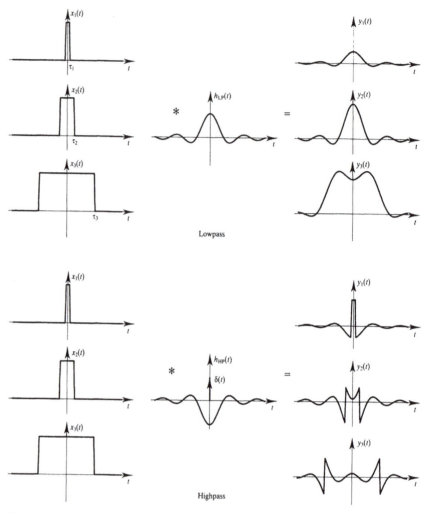

Figure 10.3 Responses from strictly ideal filters.

strictly ideal filters leads to similar results, namely, a linear-phase twisting of the frequency response and a time delay to $t = \tau$ of the impulse and step responses, as summarized in Figure 10.5 for all four types.

Linear phase implies constant group delay. There are two commonly used measures of delay, namely phase delay $\tau_\phi(\omega)$ and group delay $\tau_g(\omega)$, defined respectively by a ratio and a derivative, as

$$\tau_\phi(\omega) = -\frac{\phi(\omega)}{\omega} \quad \text{and} \quad \tau_g(\omega) = -\frac{d\phi(\omega)}{d\omega}$$

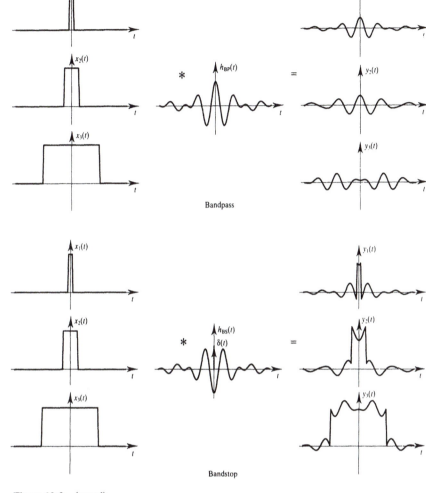

Figure 10.3   (contd)

For systems whose phase is linear with frequency and zero at the origin the two values coincide, $\tau_\phi(\omega) = \tau_g(\omega) = \tau$, and result in a pure time-shift to $t = \tau$ of the related zero-phase response.

Effects on signals

The net effect of adding linear phase to a zero-phase system is to time-shift the system's output. Indeed, if

$$x(t) * h(t) = y(t)$$

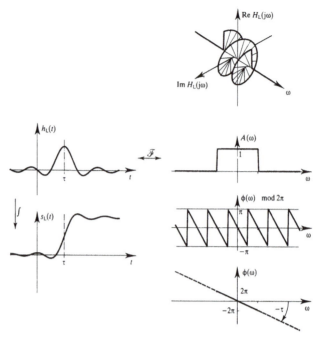

Figure 10.4 Ideal lowpass filter with linear phase.

represents the input/output relationship of a zero-phase system $h(t)$, then the output from the related linear-phase system $h(t - \tau)$ becomes

$$x(t) * h(t - \tau) = x(t) * h(t) * \delta(t - \tau)$$
$$= y(t) * \delta(t - \tau) = y(t - \tau)$$

Thus, shifting the impulse responses of the filters of Figure 10.3 would simply shift their outputs by the same amount, as illustrated in Figure 10.6 for the lowpass case. Simply delaying the output, without further shape distortions, can often be tolerated in practical applications, so that in many systems a linear phase characteristic is a desirable feature.

Causality

Adding linear phase to strictly ideal filters shifts the most prominent features of the impulse response from the time origin to positive time. But the systems of Figure 10.5 are still acausal. Increasing the delay τ to reduce the effects of the acausal tail has adverse affects on the delay of the output signal.

Another solution for making the system causal is to multiply the impulse response by some function that is zero for negative time. We next investigate symmetric time windows.

10.2.3　Linear phase and finite impulse response

The progression from strictly ideal to linear-phase filters is now taken one step further, making the impulse response causal and of finite duration. This response can still not be realized as a lumped parameter continuous-time system, but it forms the basis of related discrete-time and sampled-time systems, which **can** be implemented.

Rectangular window

Consider the zero-phase ideal lowpass filter $H_0(j\omega)$ shown at the top of Figure 10.7. We first truncate its impulse response $h_0(t)$ by multiplying with a symmetric rectangular window $w(t)$ of width 2τ, as indicated in the figure, and this implies a frequency domain convolution,

$$h_0(t)w(t) = h_F(t) \quad \overset{\mathcal{F}}{\longleftrightarrow} \quad H_0(j\omega)*W(j\omega) = H_F(j\omega)$$

Both functions are real and even and yield a real and even response $H_F(j\omega)$, which is no longer frequency limited, because the narrow $\sin x/x$ pulse representing the window causes typical Gibbs oscillations at discontinuities, which are characteristic of rectangular time windows.

We next shift the finite impulse response $h_F(t)$ by convolving with the shifted impulse $\delta(t - \tau)$, as indicated in the time domain of Figure 10.7. In the frequency domain this attaches the phase $\phi(\omega) = -\tau\omega$ of the exponential to the amplitude $A_F(\omega)$ of the filter,

$$h_F(t)*\delta(t - \tau) = h_{FL}(t) \quad \overset{\mathcal{F}}{\longleftrightarrow} \quad H_F(j\omega)e^{-j\omega\tau} = A_F(\omega)e^{-j\omega\tau}$$

The result is a system whose impulse response $h_{FL}(t)$ is causal, of finite length 2τ and symmetrical about its midpoint $t = \tau$, and whose frequency response $H_{FL}(j\omega)$ has an amplitude $A_F(\omega)$ that approximates that of the original zero-phase system $H_0(j\omega)$ and a phase characteristic $\phi(\omega)$ that is linear with frequency.

The mid-point symmetry is the result of the original system being real and even in both domains. The causality requirement sets a lower limit, while mid-point symmetry implies a symmetrical upper limit, so that the impulse response is finite. The mid-point of the impulse response also represents the constant group delay $\tau_g(\omega) = \tau$ of the resulting system.

The time-shifting operation is further interpreted at the bottom of Figure 10.7 with the imagery of Figure 2.10. The frequency response $H_F(j\omega)$ is wound back in time until its phase reaches a value appropriate to half the window width, that is, until the time value coincides with the left edge of the window, $t = -\tau$. The time reference is then changed to this value, yielding the causal impulse response $h_{FL}(t)$ and its near-helical frequency response $H_{FL}(j\omega)$, as shown in the figure.

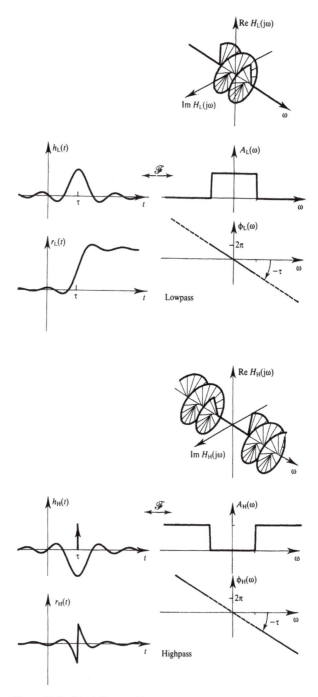

Figure 10.5 Ideal filters with linear phase.

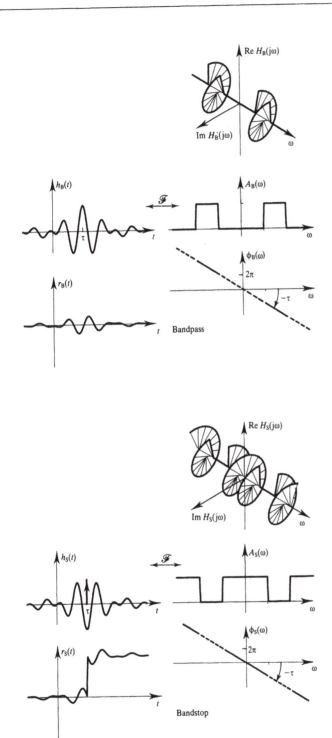

Bandpass

Bandstop

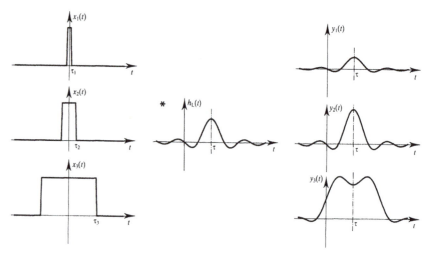

Figure 10.6 Responses from ideal lowpass filter with linear phase.

Generalization

Other zero-phase systems, such as the ideal highpass, bandpass and bandstop filters, can be similarly windowed and time shifted to yield corresponding causal filters with finite impulse response and linear phase. The above development implies that any real-valued impulse response that is causal, of finite length and symmetrical about its mid-point must represent a system with linear phase, whose frequency response must have originated in a real and even function.

Unfortunately, none of these filters can be faithfully implemented as a lumped parameter system of the form seen in Chapter 4. We will discuss this in some detail in Section 10.4. But the developments of this section are highly relevant to the design of linear-phase discrete-time finite impulse response (FIR) filters, and will be taken up again in Section 11.2.

Amplitude, magnitude and phase terminology

To stress phase linearity we refrained from using **magnitude** components, as expressed by the absolute value $|H(j\omega)|$, to represent the frequency domain. Instead we used real-valued **amplitude** functions $A(\omega)$ capable of taking positive and negative values. A function $H(j\omega)$ is thus expressed in the equivalent forms

$$H(j\omega) = |H(j\omega)|e^{j\theta(\omega)} = A(\omega)e^{j\phi(\omega)}$$

When the amplitude $A(\omega)$ is a real positive function, free of polarity

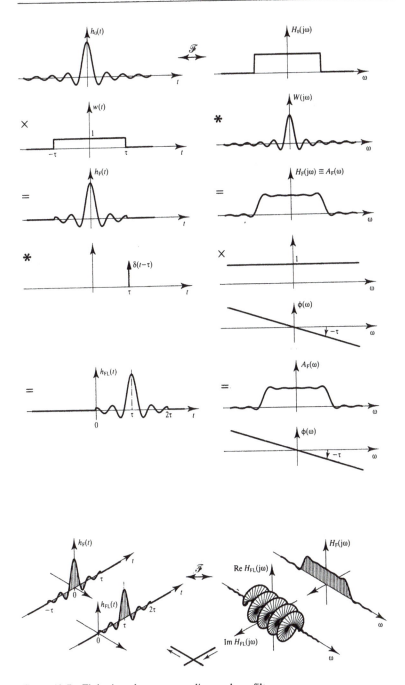

Figure 10.7   Finite impulse response, linear phase filter.

reversals, there is no conflict. Otherwise we need to distinguish between **amplitude** and **magnitude**. We illustrate this with the frequency domain sequence of Figure 10.7.

The frequency response $H_0(j\omega)$ is a real positive function which does not need the distinction. The window $W(j\omega)$ is shown as an **amplitude**, with zero angle, and so is the convolution

$$H_F(j\omega) = H_0(j\omega) * W(j\omega) = A_F(\omega)$$

The subsequent multiplication by the complex exponential $e^{-j\omega\tau}$, of unit magnitude and linear phase $\phi(\omega) = -\omega\tau$, transfers this phase to $H_{FL}(j\omega)$. The latter is illustrated by its **amplitude** $A_F(\omega)$ and the obviously linear phase $\phi(\omega)$.

The same function is represented in Figure 10.8 by its **magnitude** $|H_{FL}(j\omega)|$ and phase $\theta(\omega)$, where polarity reversal of $A(\omega)$ become phase discontinuities $\Delta\theta = \pm\pi$ of $\theta(\omega)$, thus obscuring linearity.

Furthermore, most algorithms compute phase values in the range $(-\pi, \pi)$, so that it is customary to represent phase in the form of **modulo 2π** functions, as shown in the lower part of Figure 10.8. Continuous phase representations can be obtained from **modulo 2π** phase values by using so-called phase unwinding, or phase unwrapping algorithms.

Alternative windows

The amplitude response $A_F(\omega)$ attained in Figure 10.7 is only an approximation of the ideal lowpass filter $H_0(j\omega)$, and the approximation error depends entirely on the shape and width of the window $w(t)$.

The rectangular window used for illustration in Figure 10.7 introduced ripples and overshoots at band-edge. These may be desirable when a steep band-edge transition is the prime consideration. Increasing the window width 2τ would improve this characteristic.

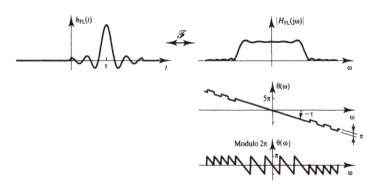

Figure 10.8   Alternative magnitude and phase representations of Figure 10.7.

But it is often preferable to trade off some of this steepness for a flatter amplitude response. Keeping constant the window width 2τ, and therefore the group delay $\tau_g(\omega)$, less severe window shapes give smoother amplitude responses. This is illustrated in Figure 10.9, where two alternative windows, triangular and raised cosine, are applied to the same ideal lowpass filter. The shape of the impulse response $h_F(t)$ is slightly modified, mainly by flattening its skirts, and this may give a better compromise for the shape of the amplitude response $A_F(\omega)$.

10.3 Ideal operators

The concepts introduced with ideal filters are equally applicable to other frequency-shaping systems. We now examine how they apply to ideal

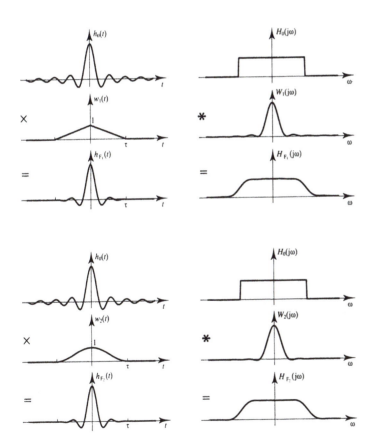

Figure 10.9 Alternative windows.

systems from the other main group of the classification of Figure 10.1, those identified as of odd-symmetric basis.

The ideal filters of Section 10.2 originated in functions that were real and even-symmetric. The addition of linear phase simply shifted the point of symmetry along with the impulse response. A second class of symmetric systems results when the original function is odd-symmetric about the origin. To give real-valued impulse response the corresponding frequency response must then be imaginary and odd. The integrator, differentiator and Hilbert transformer of this section fall into this class. Adding linear phase, as before, similarly shifts the point of odd-symmetry along with the impulse response.

We first review the representations of strictly ideal operators, then band-limit and add linear phase, as with ideal filters. We also introduce the identity operator as a link between the differentiator and the integrator, and this serves to clarify the above basis for classification.

This examination of ideal operators distinguishes between fundamental operator properties and considerations specific to continuous-time or discrete-time implementation.

10.3.1   Strictly ideal operators

We now give a **systems interpretation** of some analytical operators, treated earlier as abstract functions.

Differentiator

A strictly ideal differentiator is an elementary system whose output $y(t)$ is the derivative of the input $x(t)$, that is

$$y(t) = \frac{dx(t)}{dt} \tag{10.5}$$

as represented in Figure 10.10. When the input signal is an impulse $\delta(t)$, the output is its derivative, the doublet $d(t) = \delta'(t)$. By definition this represents the impulse response $h(t) = d(t)$ of the system. We write this statement as the input/output relationship

$$x(t) = \delta(t) \xrightarrow{\;d(t)\;} y(t) = d(t) = \frac{d\delta(t)}{dt} = \delta'(t)$$

The differential of an arbitrary input signal $x(t)$ can then be expressed in terms of the impulse response method of expression 5.10 as a convolution with the doublet $d(t)$,

$$x(t) \xrightarrow{\;d(t)\;} y(t) = \frac{dx(t)}{dt} = x(t) * d(t) \tag{10.6}$$

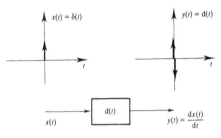

Figure 10.10 Differentiator as elementary system.

The relationship 10.5 also represents the elementary system's differential equation, and, as seen in Chapter 4, all the implied responses can be obtained by means of the Laplace transform. Thus, the transform

$$Y(s) = sX(s)$$

yields the system function $D(s)$ of the ideal differentiator as

$$D(s) = \frac{Y(s)}{X(s)} = s \qquad (10.7)$$

The frequency response $D(j\omega)$, obtained by evaluating Equation 10.7 on the imaginary axis $s = j\omega$, is thus a pure imaginary function,

$$D(j\omega) = j\omega$$

The impulse response $d(t)$ is of course related to the last two results by the Laplace transform and the Fourier transform, respectively, as

$$d(t) = \mathcal{L}^{-1}\{s\} = \mathcal{F}^{-1}\{j\omega\}$$

The relationship between the impulse response and the frequency response is expressed graphically by the Fourier pair at the top of Figure 10.11, where the frequency response is shown by magnitude and phase as well as in full complex form.

Integrator

The integral $y(t)$ of a function $x(t)$ can be interpreted as a convolution with the unit step. Indeed, by similarity to Example 5.1, we have

$$y(t) = x(t)*u(t) = \int_{-\infty}^{\infty} u(t - \tau)x(\tau)\,d\tau = \int_{-\infty}^{t} x(\tau)\,d\tau$$

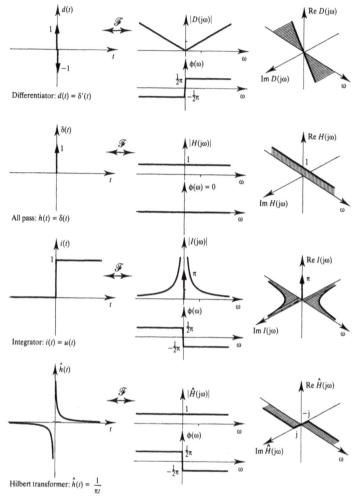

Figure 10.11 Strictly ideal operators.

In terms of the impulse response method of expression 5.10, this convolution identifies the ideal integrator as an elementary system whose impulse response $i(t)$ is the unit step

$$i(t) = u(t) \tag{10.8}$$

The integrator's system function $I(s)$ is the Laplace transform of $i(t)$. From the transform of the unit step, Equation 3.8, we have

$$I(s) = \frac{1}{s} \qquad \text{Re } s > 0 \tag{10.9}$$

Since this function is not defined on the imaginary axis, the frequency response $I(j\omega)$ must be found from the Fourier transform definition. In the present context it is instructive to interpret the step function as the integral of the impulse function, and to find its transform from the time domain integration property 2.41

$$\int_{-\infty}^{t} f(\tau)\,d\tau \quad \overset{\mathcal{F}}{\longleftrightarrow} \quad \frac{1}{j\omega} F(j\omega) + \pi\delta(\omega)F(0)$$

where $f(t) = \delta(t)$ and $F(j\omega) = F(0) = 1$, so that

$$I(j\omega) = \frac{1}{j\omega} + \pi\delta(\omega) \tag{10.10}$$

The ideal integrator is thus represented in the third row of Figure 10.11.

Identity operator

We inserted the **ideal allpass filter** into the second row of Figure 10.11, as a link between the differentiator and the integrator. Its impulse response is the impulse function itself, giving the input/output relationship

$$x(t) \quad \overset{\delta(t)}{\longrightarrow} \quad y(t) = x(t) * \delta(t) = x(t)$$

This filter, whose impulse response is the identity element of convolution, and whose frequency response, the unit function, is the identity element of multiplication, can thus be called the **identity operator**.

The time derivative of its impulse response becomes the impulse response of the differentiator. Its time integral is the unit step, that is, the impulse response of the integrator. With the graphical interpretations given in Section 2.4.6 for the Fourier transform of the derivative and of the integral of a function, this somewhat trivial identity operator provides a clear link between the other two.

In later developments, when we band limit and phase shift the strictly ideal responses, this link takes a less trivial role.

Real even and imaginary odd bases

In Figure 10.11 the identity operator represents a real even function in both domains, thus typifying the filters of Section 10.2. Its time derivative is real odd, and, if the d.c. component is ignored, so is its integral. The frequency responses of the differentiator and integrator are thus essentially imaginary odd.

This is the basis of the classification of Figure 10.1. We next introduce another important operator of imaginary odd basis, the time domain Hilbert transformer.

Time domain Hilbert transformer

The Hilbert transformer $\mathcal{H}$ is an operator that relates the real and imaginary parts of a causal function. It is most frequently defined in terms of a convolution in the chosen domain. In the alternative domain, this is seen as a multiplication of the corresponding Fourier transforms.

We will here examine the time domain form, which we identify with the subscript t, and define by the time domain convolution

$$\hat{f}(t) = \mathcal{H}_t\{f(t)\} = f(t)*1/(\pi t) \tag{10.11}$$

This relationship is expressed symbolically as the Hilbert transform pair

$$f(t) \quad \overset{\mathcal{H}_t}{\longleftrightarrow} \quad \hat{f}(t)$$

The operator definition 10.11 has the convolution form of the impulse response method, in the sense that the output function $\hat{f}(t)$ is the result of convolving the input $f(t)$ with $1/\pi t$. The latter represents the impulse response $\hat{h}(t)$ of the Hilbert transformer, related by the Fourier transform to the frequency response $\hat{H}(j\omega)$ as

$$\hat{h}(t) = 1/(\pi t) \quad \overset{\mathcal{F}}{\longleftrightarrow} \quad \hat{H}(j\omega) = -j\,\mathrm{sgn}\,\omega$$

and shown at the bottom of Figure 10.11. Having odd imaginary frequency response, this function falls in the same classification scheme as the differentiator and integrator.

The definition 10.11 can therefore be interpreted to represent an elementary system with input/output relationship

$$x(t) = f(t) \quad \overset{\hat{h}(t)}{\longrightarrow} \quad y(t) = \hat{f}(t) = f(t)*\hat{h}(t) \tag{10.12}$$

whose frequency domain equivalent is

$$X(j\omega) = F(j\omega) \quad \overset{\hat{H}(j\omega)}{\longrightarrow} \quad Y(j\omega) = \hat{F}(j\omega) = -j\,\mathrm{sgn}\,\omega\hat{H}(j\omega)$$

Effects on signals

The effects of the ideal differentiator, integrator and Hilbert transformer on rectangular pulses of different width are shown in Figure 10.12.

For this type of input signal the differentiator highlights the sharp transitions of the pulse, thus acting as an edge detector, while the integrator yields the area of the pulse. The Hilbert transformer produces a function which, despite its very different shape, has the same frequency components as the original function, but with their phases rotated by $+\frac{1}{2}\pi$ or $-\frac{1}{2}\pi$.

As in the case of the ideal filters of Figure 10.3, the output associated with the narrow pulse resembles the impulse response, while that caused by the wide pulse represents the sum of two step responses.

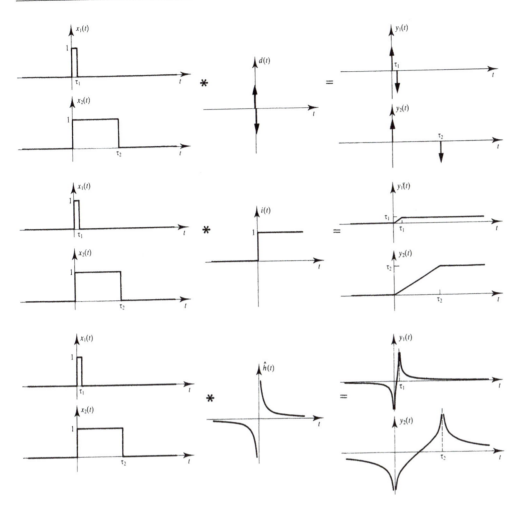

Figure 10.12   Outputs from strictly ideal operators.

10.3.2   Band-limited and causal differentiator

Two fundamental obstacles to the realization of these ideal operators are causality and bandwidth. We will attempt to overcome the first by adding linear phase and windowing, as we did with filters. Regarding the second, in real life applications it is valid to limit the width of the frequency response, provided the resulting bandwidth is broader than that of the input signal.

These limitations in frequency and in time affect the alternative domain and cannot be pursued in isolation. We now investigate the mutual effects, taking the differentiator as a typical example, and later extend the findings to other operators.

The increasing magnitude of the ideal differentiator's frequency response implies large magnification of high frequency components. The response can be band-limited by multiplying with the ideal lowpass filter $H_{LP}(j\omega)$, as indicated in the sequence of the left half of Figure 10.13. The corresponding time domain convolution

$$d(t) * h_{LP}(t) = d_b(t)$$

can be given alternative system interpretations. As a system, the lowpass filter $h_{LP}(t)$ would distort the input doublet $d(t)$, spreading it out from the origin. Alternatively, considering the filter's impulse response as an input to the differentiator $d(t)$, the output becomes the derivative of $h_{LP}(t)$, which is easier to visualize.

Windowing the impulse response $d_b(t)$, as indicated at the bottom left of Figure 10.13, introduces the usual ripples and overshoots at frequency domain discontinuities. As in the case of the earlier filters, a smoother window shape would reduce these distortions.

Finally, by adding linear phase the impulse response is shifted into the causal region, as shown at the bottom right of Figure 10.13 (convolution with shifted impulse is not indicated in the figure).

Alternative sequence

The same results are achieved by an alternative path, indicated by the clockwise arrow sequence of Figure 10.13, where the doublet $d(t)$ is first displaced into the causal region, thus adding linear phase, followed by lowpass filtering and finally by windowing.

10.3.3 Extension to other operators

We extend the preceding results to the other operators of Figure 10.11, taking them together through a band-limiting, time-shifting and time-windowing sequence.

Band-limiting

For reference we first band-limit the allpass filter, or identity operator of the second row of Figure 10.11. Multiplying its frequency response by the ideal lowpass filter $H_{LP}(j\omega)$ replicates the latter in both time and frequency, as shown in the second row of Figure 10.14.

Multiplying the frequency representations of the other operators by the ideal filter truncates their magnitudes, without affecting phase, as shown earlier for the differentiator. The corresponding convolutions in

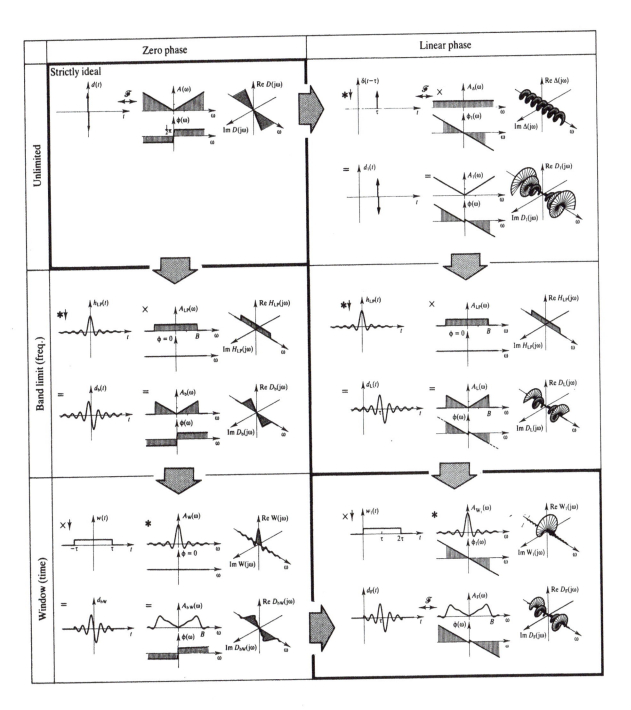

Figure 10.13 Finite impulse response, linear phase differentiator.

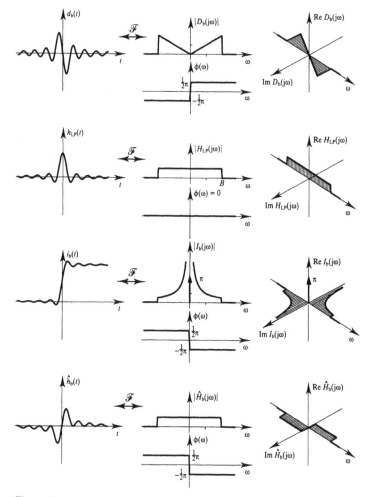

Figure 10.14 Band-limited ideal operators.

the time domain turn the impulse response of the differentiator into the derivative of that of the ideal filter

$$d_b(t) = \frac{dh_{LP}(t)}{dt} = h_{LP}(t) * d(t)$$

and that of the integrator into the integral of $h_{LP}(t)$,

$$i_b(t) = \int_{-\infty}^{t} h_{LP}(\tau)\,d\tau = h_{LP}(t) * i(t)$$

The corresponding graphics of Figure 10.14 are consistent with the graphical rules given in Section 2.4.6 for the transforms of a derivative and an integral.

The impulse response of the band-limited Hilbert transformer can be similarly interpreted as the ideal Hilbert transform of the lowpass impulse response $h_{LP}(t)$,

$$\hat{h}_b(t) = h_{LP}(t) * \hat{h}(t)$$

Time shifting

Convolving the impulse responses of Figure 10.14 with the impulse $\delta(t - \tau)$ shifts them by an amount $t = \tau$, as shown in Figure 10.15. This does not affect the magnitude responses, but adds the appropriate amounts of linear phase, $\Delta\phi(\omega) = -\omega\tau$, effectively causing the phase responses to pivot about values $\pm\frac{1}{2}\pi$ at the origin, and twisting the complex representations as shown.

Windowing

Finally, windowing the time domain with a causal pulse of width 2τ turns these ideal operators into causal, finite impulse response, linear-phase operators. The results of using a rectangular window are illustrated in Figure 10.16. In the case of the integrator, the real even impulse associated with an otherwise imaginary odd frequency response makes the interpretation of windowing less intuitive, so that this case was excluded from the figure.

Conclusions

We see that desirable system-conditioning operations in one domain tend to produce undesirable effects in the alternative domain. Although the results derived in this section are still ideal, in the sense that they cannot be realized as lumped parameter systems, they give insight into the overall effects of such conditioning processes. In the next section we examine the fundamental limitations of lumped parameter systems and consider ways of specifying system characteristics that **can** be realized.

The reason we proceeded so far towards a seemingly dead end is that the responses of Figure 10.16 only need to be time-sampled to produce eminently realizable discrete-time systems.

10.4   Realizable systems

In the preceding sections we looked at some idealized models of filters and other operators, which had simple and precise mathematical definitions, but were physically unattainable.

The models provided simple terminology and revealed fundamental constraints between time and frequency representations, which

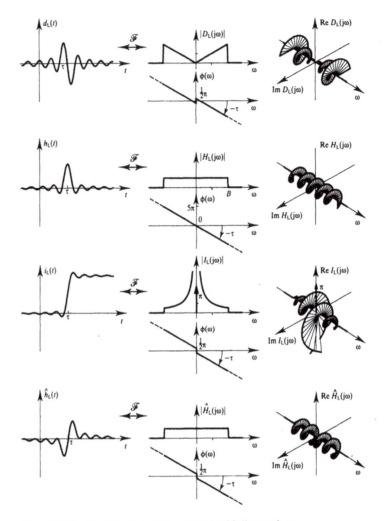

Figure 10.15 Band-limited ideal operators with linear phase.

warned against arbitrary changes in one domain, without considering their effects in the other. But that path did not lead to practical methods for designing realizable continuous-time systems.

In Chapter 4 we studied the properties and limitations of lumped parameter systems, which are obviously realizable. We now make use of the terminal properties derived there to synthesize realizable systems that fit desired specifications.

In this section we first examine what form the specifications of realizable systems take, and then interpret some of the simplest building

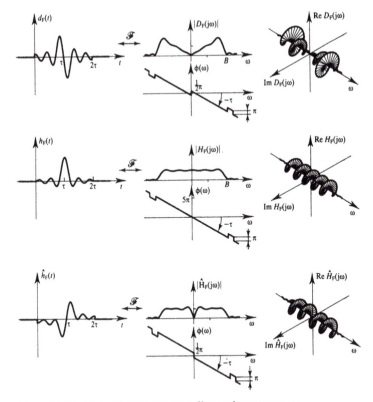

Figure 10.16 Finite impulse response, linear phase operators.

blocks of Chapter 4 as crude approximations of ideal filters or other operators. More elaborate approximations of lowpass filters are presented in Section 10.5 and other filter types in Section 10.6.

10.4.1 General and terminology

Before proceeding with realizable systems we must clarify terminology in the more general context of real-life physical systems and of the phases involved in system design.

In this book the term **realizable** applies to a system whose characteristics can be implemented by the class of linear, time-invariant, lumped parameter systems, with idealized elements, as presented in Chapter 4. That is to say, we will be dealing with mathematical models, which are linearized idealizations of real-life systems.

Realizable systems and real-life physical systems

Physical processes are basically smooth and do not allow for infinitely sharp changes of value. They are effectively limited in both time and frequency.

If we were pedantic, we would state that the Fourier transform cannot be applied exactly to any physical process. For instance, recalling the definition 2.5 of the Fourier transform, to compute one frequency value $F(j\omega)$ would involve an infinite integration in time, hence a full knowledge of the past and future of the time function $f(t)$. Similarly for the inverse Fourier transform.

To take advantage of transform methods we need a simplified mathematical model of the physical system that behaves according to the rules of the Fourier or Laplace transform, making it amenable to analysis by those tools. The results may not represent the real-life system exactly, but give a good likeness of its behaviour to allow realistic predictions.

In this sense the term **realizable** does not imply **real life**.

Design phases of a system

It is customary to distinguish three phases in the design of a system, namely specification, approximation and realization. These are not entirely independent and the sequence is not rigid. For instance, specification and approximation may depend on the technologies available for realization, and vice versa, and the whole process may require several iterations.

The **specification phase** consists of defining a mathematical model with the desired characteristics, typically in terms of the frequency response. Since ideal responses cannot be implemented, a realistic specification must allow deviations from the ideal. This takes the form of permissible error specifications, or tolerance bands, to be seen in Section 10.4.2.

The purpose of the **approximation phase** is to find a realizable system function that fits the specification tolerance bands in a manner deemed to be optimal by some criterion. The realizability assumption implies an earlier choice of the class of system, and possibly of a structure.

The **realization phase** consists of synthesizing a physical system capable of realizing the system function derived in the approximation phase. It involves the choice of technology and of the system's structure, if these were not chosen at the outset, and the detailed calculation of all system elements, down to component values.

The remainder of this book deals exclusively with the approximation stage, in the above interpretation, of both continuous-time and discrete-time systems. The term **approximation** will be used in the sense of finding a realizable response that fits the specification. When the approximation bridges class boundaries, for instance, a discrete-time approximation of a continuous-time specification, we call it a **system simulation**. We will not use the term **realizable** in the sense of the **realization phase** of system design.

10.4.2   Limitations of realizable systems

The lumped parameter systems of Chapter 4 involve a finite number of interconnected linear time-invariant elements and are realizable. They are also amenable to analysis and solution by transform techniques, which makes them attractive for approximating realizable systems. We now re-examine their capabilities and limitations regarding ideal specifications.

In Chapter 4 we concluded that any lumped parameter system containing N independent energy storage elements can be described by an Nth-order differential equation. Further, in Section 4.4.2 we showed that the system can be partitioned into a cascade of second-order sections, plus one first-order section when N is odd.

It is customary to specify systems, in particular filters, in terms of the frequency response $H(j\omega)$, most commonly by the magnitude $|H(j\omega)|$ alone. Since the overall magnitude response of a cascaded system is the product of the magnitudes and the overall phase response is the sum of the phases, the frequency response capabilities of the constituent first-order and second-order sections reveal the capabilities of the general system.

Magnitude and phase limitations

Examining the typical responses of Figures 4.12 and 4.19 suggests that the frequency responses of first- and second-order sections are bounded in both magnitude and phase.

The system function magnitude $|H(s)|$ is only zero-valued in correspondence to the system's zeros. If these fall on the imaginary axis, then $|H(j\omega)|$ is zero-valued at those points only, as illustrated in Figure 4.19 for second-order sections that have two zeros. Consequently, the frequency response $H(j\omega)$ of an Nth-order system cannot be exactly zero at more than N points of the frequency axis, and ideal stopbands are unrealizable.

An Nth-order polynomial only admits $N - 1$ non-zero derivatives, and this sets a limit to the flatness attainable at any particular point. An ideal passband characteristic that is truly constant over any finite frequency band is therefore unattainable. Allpass characteristics with unity magnitude response are the exception.

The values of a second-order phase response fall within a band of magnitude 2π, as seen in Figure 4.19. The phase range of an Nth-order system is $N\pi$, so that it can never represent a truly linear phase characteristic. This can only be approximated over a finite frequency range.

Specification parameters

As a result of the above limitations, a realizable system is specified by a **tolerance envelope**, which admits controlled deviations from the desired response. For filters, the ideal magnitude response $|H(j\omega)|$ is replaced by an envelope $E(\omega)$, such that the constant passband magnitude is replaced by a **passband tolerance band**, the zero-valued stopband is replaced by an **upper tolerance boundary** and, since realizable systems do not allow sudden magnitude transitions, bandedges are replaced by **frequency transition bands**. The specification parameters of the four filter types of Section 10.2 take the forms shown in Figure 10.17.

Although not often required in practice, a system's phase response can be similarly specified by means of a **phase envelope**, or by a **group delay envelope**. But even when the phase response is of prime importance, it is customary to specify the magnitude response and to choose a polynomial type known to give desirable phase characteristics, as discussed later.

10.4.3 Associating operators with basic sections

The developments of Chapter 4 established a relationship between elementary physical components, such as resistors and capacitors, and the more abstract system response $H(s)$. Little reference was made to functional operators, such as filters and integrators, that could be associated with the system. In this section we interpret some elementary systems as crude approximations of ideal operators.

Filters

The six examples shown earlier in Figure 4.19 were associated there with lowpass, bandpass and other filter responses, in anticipation of the

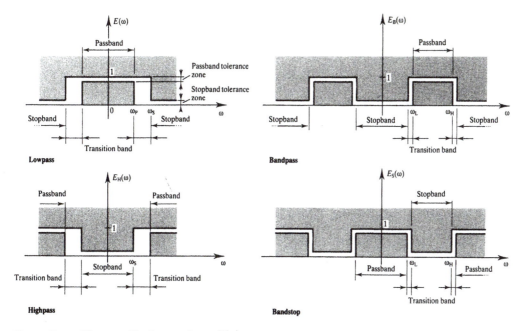

Figure 10.17   Filter specification envelopes $E(\omega)$.

present treatment. Surrounding those magnitude responses with suitable tolerance bands, as shown in Figure 10.18, we now interpret them in terms of the specifications of Figure 10.17.

The three responses of the left half of Figure 10.18 justify the descriptions of Figure 4.19, and so does the bandstop response on the right. If the zeros of the highpass notch are sufficiently close to the origin, magnitudes near the origin are small and the response fits a highpass filter envelope. Similarly for the lowpass notch, when the zeros are sufficiently removed from the pole frequency.

Thus, modifying the numerator coefficients c_i of the system function

$$H(s) = \frac{c_2 s^2 + c_1 s + c_0}{s^2 + (\omega_r/Q)s + \omega_r^2}$$

permits changing the filter type. Similarly, modifying the pole frequency ω_r and the Q-factor gives control over the peak shapes.

A single second-order section can only offer a crude approximation to ideal filters. In filter jargon this is described as low frequency selectivity, characterized by relatively large magnitude deviations in the passband, large stopband values and a wide transition band with slow rolloff. First-order sections offer even more primitive approximations.

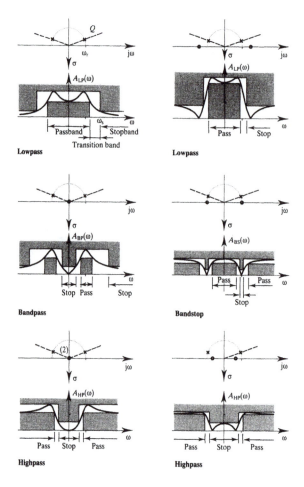

Figure 10.18 Second-order sections as primitive filters.

It is easy to see, however, that flatter and more elaborate specifications can be achieved by combining a number of such sections that are individually tuned for pole-frequency and pole-Q. This is illustrated in Figure 10.19, where series or parallel combinations of two second-order sections lead to systems with higher overall selectivity. This represents the system synthesis counterpart of partitioning a system in Section 4.4 into cascade and parallel connections for analysis purposes. In Section 10.5 we formalize the search for optimal combinations.

Integrator

The ideal integrator was introduced in Section 10.3.1, where its impulse

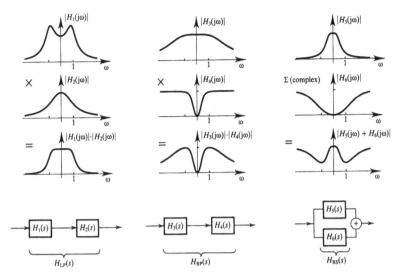

Figure 10.19   Higher-order filters from series and parallel connections.

response $i(t)$, system response $I(s)$ and frequency response $I(j\omega)$ were expressed by Equations 10.8, 10.9 and 10.10 respectively as

$$i(t) = u(t) \qquad I(s) = \frac{1}{s} \qquad \text{and} \qquad I(j\omega) = \frac{1}{j\omega} + \pi\delta(\omega)$$

These responses can be interpreted as the limiting case of the progression indicated in Figure 10.20.

Recall the first-order RC network examined in detail in Section 4.2, see Figure 4.5. For moderate values of R and C, the pole position $p_1 = -1/RC$ lies far from the s-plane origin and produces the responses shown in the top row of Figure 10.20. These give a poor approximation to the ideal integrator responses shown in the bottom row of the figure. Only at high frequencies, $\omega \gg \omega_c$, where ω_c is the cut-off frequency defined by $\omega_c = 1/RC$, does the function $H(j\omega)$

$$H(j\omega) = \frac{1/RC}{j\omega + 1/RC} \simeq \frac{1}{j\omega RC}$$

resemble the shape of the ideal response $I(j\omega)$ in both amplitude and phase.

Higher component values bring the pole closer to the origin, thus improving the shapes of the approximation in all its representations, as shown in the middle row of Figure 10.20. The resulting reduction of the cut-off frequency permits using the integrator for lower frequency values. Note that the frequency components in the vicinity of the origin are predominantly real, thus representing an approximation to the impulse at the origin.

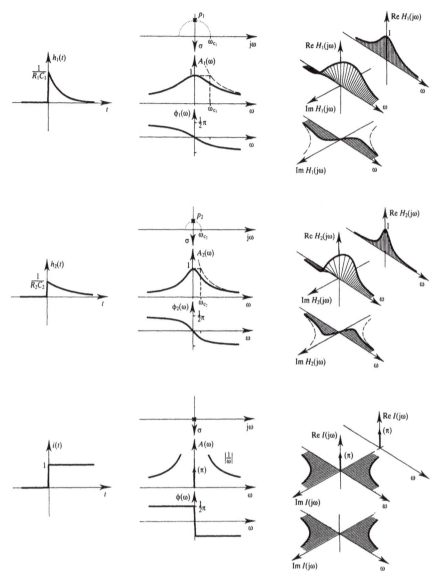

Figure 10.20   Approximation of integrator.

This trend suggests very large component values for a perfect approximation. The penalty is a proportional reduction of the output level, which calls for an engineering compromise to keep the signal above the noise level. In practice the problem is overcome by means of active circuits, which permit bringing the pole close to the origin without losing signal level.

Differentiator

The ideal differentiator was characterized in Section 10.3.1 by the responses

$$d(t) = \delta'(t) \qquad D(s) = s \qquad \text{and} \qquad D(j\omega) = j\omega$$

whose s-plane representation is a zero at the origin. It is approximated by a first-order system with its zero at the origin, see the chart of Figure 4.12, with the pole located far from the origin.

The series RC network realization of the differentiator exchanges the resistor and the capacitor of the integrator circuit. Taking the voltage across the resistor for output, this yields the system function

$$H(j\omega) = \frac{j\omega}{j\omega + 1/RC}$$

whose cut-off frequency is defined as $\omega_c = 1/RC$. For frequency values $\omega \ll \omega_c$ we have

$$H(j\omega) \simeq j\omega RC$$

The responses for moderate and for small values of the product RC are illustrated in the progression of Figure 10.21, which shows the ideal differentiator as the limiting case. As with the integrator, improvements to the response shapes are obtained at the cost of signal level, and call for an engineering compromise to stay above the noise threshold, or for the use of active circuits.

10.5 Classical filter approximations

Several methods are available for approximating filters in which the frequency response is described in terms of well-known polynomial functions taken from the vast repertoire of mathematical functions. These responses are usually named after those polynomials or, equivalently, after the implied optimal approximation criterion, and are extensively tabulated and widely accepted in engineering practice.

Our main objective is to give an insight into the common foundations of these methods. We will also introduce some of the better-known types, such as Butterworth, Chebyshev, elliptical and Bessel, and examine the optimal criteria they represent. It is not our aim to develop these methods in detail, or to provide filter design data, merely to illustrate the process leading from a system specification to a possible realization.

All the developments of this section involve lowpass filter characteristics. In Section 10.6 we present a family of frequency transformations, which treat those lowpass filters as prototypes and convert them into related filters with highpass, bandpass, bandstop and other characteristics.

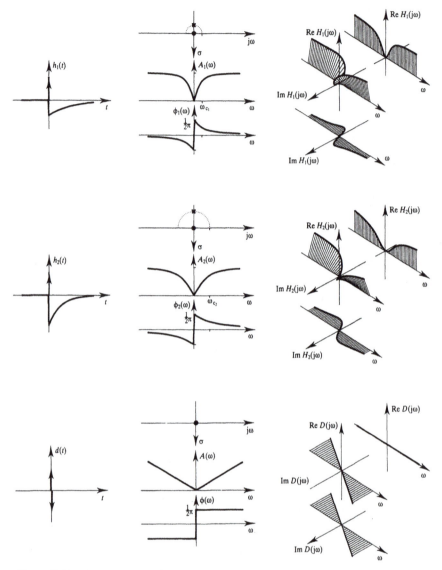

Figure 10.21   Approximation of differentiator.

10.5.1   General methodology

The classical methods presented here fit a rational polynomial $A(\omega)$ of the general form

$$A(\omega) = |H(j\omega)| = \frac{1}{\sqrt{1 + \varepsilon^2 F_N^2(\omega)}} \tag{10.13}$$

into the specified lowpass filter envelope $E(\omega)$. The function $F_N(\omega)$ is a polynomial of degree N, which determines the filter type, and ε is an associated scaling factor. Note that the functions $E(\omega)$, $A(\omega)$ and $F_N(\omega)$ are all real functions of real variable ω, whereas the complex function $H(j\omega)$ associates the same real variable ω with the imaginary axis of the s-plane.

We examine three aspects of the problem. Firstly, we relate the initial specification $E(\omega)$ to the polynomial $F_N(\omega)$. This involves manipulating the amplitude of $E(\omega)$ to allow the selection of a specific function $F_N(\omega)$ of standard form and then fitting the latter back into the envelope $E(\omega)$, such that the function 10.13 is satisfied.

Secondly, we need a mechanism that unambiguously assigns a complex system function $H(s)$ to the real magnitude function $A(\omega)$, such that the resulting system is stable.

Finally, we want to relate the special characteristics of some common types of polynomials $F_N(\omega)$ to the characteristics of the resulting frequency response. That is, we want to establish in what sense the choice of polynomial type makes the response optimal.

10.5.2   Link between $E(\omega)$ and $F_N(\omega)$

The mechanism by which a specification envelope $E(\omega)$ is linked to the polynomial $F_N(\omega)$ takes essentially three stages, namely scaling the envelope to the normalized format of the chosen polynomial family $F_N(\omega)$, finding the required degree N, so that the polynomial fits the scaled envelope, and manipulating the resulting polynomial back into the original specification. The required steps are illustrated in Figure 10.22.

We assume that the specification envelope $E(\omega)$ is available in normalized form, such that the passband frequency limit ω_p is unity and the maximum passband value is also unity. The envelope $E(\omega)$ is thus fully defined by three values: the stopband edge ω_s, the lower passband limit k_p and the stopband limit k_s. Critical points of the envelope and their subsequent locations are indicated in Figure 10.22 by small circles.

Envelope manipulation

This stage consists of a sequence of amplitude distortions that map the passband region of the envelope $E(\omega)$ onto the standard reference region of the polynomial $F_N(\omega)$, normally taken as the square limited to plus and minus unity in amplitude and frequency. The steps involved in this mapping, indicated on the left side of Figure 10.22, anticipate the sequence indicated on the right, which will take the polynomial $F_N(\omega)$ back to the denominator of $A(\omega)$.

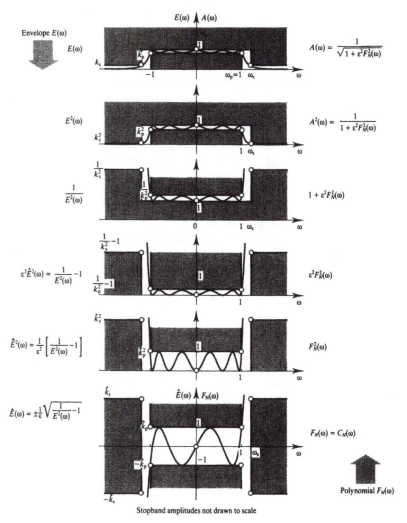

Figure 10.22 Transformations of envelope and polynomial.

The first three steps, leading to the function $\varepsilon^2 \hat{E}^2(\omega)$ are obvious. The next step requires division by ε^2 which is not yet available. The objective is to map the tolerance strip of the passband to a reference strip of unit height. This implies mapping the line $E(\omega) = k_p$ onto the line $\hat{E}^2(\omega) = 1$, as

$$\hat{E}^2(\omega)\big|_{E(\omega) = k_p} = \frac{1}{\varepsilon^2}\left(\frac{1}{k_p^2} - 1\right) = 1$$

which links the design parameter ε^2 to the passband boundary k_p as

$$\varepsilon^2 = \frac{1}{k_p^2} - 1 \tag{10.14}$$

The final step takes a square root, which expresses the mapped envelope $\hat{E}(\omega)$ as

$$\hat{E}(\omega) = \pm \frac{1}{\varepsilon} \sqrt{\frac{1}{E^2(\omega)} - 1}$$

The net result is that the line representing the upper value $E(\omega) = 1$ of the passband maps onto the frequency axis $\hat{E}(\omega) = 0$, the lower limit $E(\omega) = k_p$ maps onto the line pair $\hat{k}_p = \pm 1$ and the stopband limit $E(\omega) = k_s$ maps onto the line pair

$$\hat{k}_s = \pm \frac{1}{\varepsilon} \sqrt{\frac{1}{k_s^2} - 1}$$

Note that $\hat{k}_s$ depends on both k_s and k_p.

Polynomial fitting

The second stage involves fitting a real-valued polynomial $F_N(\omega)$ into the mapped envelope $\hat{E}(\omega)$, as illustrated at the bottom of Figure 10.22. Assuming that the polynomial family was given as part of the specification (e.g. Chebyshev), this stage reduces to finding the degree N that satisfies stopband requirements. Any member of the family that does not violate the stopband boundary $\hat{k}_s$ is valid, but the higher N, the greater the number of poles and the cost of the system. The solution is the polynomial of lowest degree that still clears the stopband envelope.

The choice of polynomial family affects the shape of the response contained in the magnitude tolerance bands, and also the phase response. This is considered in Section 10.5.4.

Return to specification

The third stage, manipulating $F_N(\omega)$ back into $E(\omega)$, simply backtracks the amplitude distortion sequence employed to map the envelope. Since the same operations apply to the envelope and to the function inside it, the magnitude $A(\omega)$ of the resulting rational polynomial $H(j\omega)$ is bound to be contained within the envelope $E(\omega)$.

10.5.3   Link between $A(\omega)$ and $H(s)$

To derive a realizable system function $H(s)$ from the real-valued function $A(\omega)$ we first form an auxiliary system function $G(s)$, defined as the product

$$G(s) = H(s)\bar{H}(s) \tag{10.15}$$

where $H(s)$ represents the yet unknown system and $\bar{H}(s)$ represents a **virtual system** related to $H(s)$. The latter is a fictitious system, a concept analogous to **virtual images** of optical systems.

The realizable system $H(s)$ has an input $X(s)$ and an output $Y(s)$, as usual. Its output $Y(s)$ is taken as the input to $\bar{H}(s)$, which would have $Z(s)$ for output, as shown at the bottom of Figure 10.23. The auxiliary system $G(s)$ therefore relates the input $X(s)$ with the output $Z(s)$, as indicated at the top of the figure.

The purpose of introducing $\bar{H}(s)$ and $G(s)$ is to generate a **real-valued** function $G(j\omega)$ that can be equated to the **real-valued** function $A(\omega)$. This can be achieved by choosing an $\bar{H}(s)$ that is symmetrical to $H(s)$ about the s-plane origin.

The strategy is to write the function $G(s)$ from the knowledge of $A(\omega)$, find its poles and zeros and then allocate the stable half of the poles to the realizable system $H(s)$ and the unstable half to the virtual system $\bar{H}(s)$, as indicated in the centre of Figure 10.23.

Auxiliary system $G(s)$

Recall the graphical interpretation of values of $H(s)$ in terms of the s-plane location of its poles and zeros (Equation 4.68),

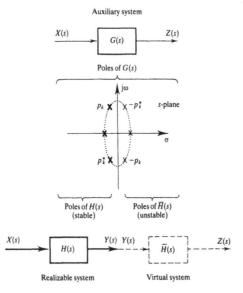

Figure 10.23 Assigning stable poles of $G(s)$ to $H(s)$.

$$H(s) = k \frac{\Pi_{n=1}^{N}(s - z_n)}{\Pi_{n=1}^{N}(s - p_n)}$$

We extend that interpretation to the product 10.15, writing it as

$$G(s) = k \frac{\Pi_{n=1}^{N}(s - z_n)}{\Pi_{n=1}^{N}(s - p_n)} \, k \frac{\Pi_{n=1}^{N}(s + z_n)}{\Pi_{n=1}^{N}(s + p_n)} \tag{10.16}$$

The contributions to $G(s)$ from one conjugate pair of poles p_k and p_k^* of $H(s)$, and from the exact counterparts $-p_k$ and $-p_k^*$ of $\bar{H}(s)$ are represented in the upper half of Figure 10.24 for an arbitrary point of the s-plane. The lower half shows the contributions of the same poles when s is located on the imaginary axis $s = j\omega$.

The latter case is highly relevant, because the frequency response $G(j\omega)$ is **real-valued** for all ω. Because of the double s-plane symmetry, about the real axis for real-valued impulse response and about the origin for the construction of $\bar{H}(s)$, the poles shown in Figure 10.24 are also symmetrical about the imaginary axis. As a result, the contribution to $G(j\omega)$ from each factor of the form $(j\omega - p_k)(j\omega - p_k^*)$ of $H(j\omega)$ is

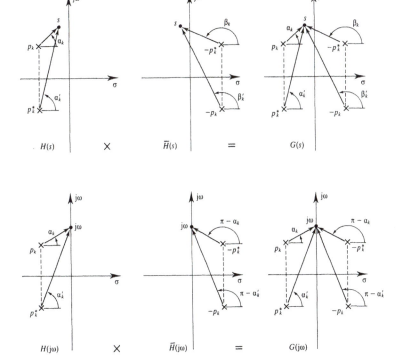

Figure 10.24   Relationship of $H(s)$, $\bar{H}(s)$ and $G(s)$.

matched by a symmetrical factor $(j\omega + p_k^*)(j\omega + p_k)$ of $\bar{H}(j\omega)$. Their magnitudes are identical and the sum of their angles is 2π, so that their product is a real number. The same applies to the system zeros.

Collectively, the numerator and denominator products of Equation 10.16 yield a real function $G(j\omega)$

$$G(j\omega) = H(j\omega)\bar{H}(j\omega) = |H(j\omega)|e^{j\phi(\omega)}|H(j\omega)|e^{-j\phi(\omega)} = |H(j\omega)|^2$$

where $\phi(\omega)$ represents the phase of $H(j\omega)$.

Since $G(j\omega)$ represents $G(s)$ evaluated on the imaginary axis, we write

$$G(j\omega) = G(s)|_{s=j\omega} = A^2(\omega) \tag{10.17}$$

where we replaced $|H(j\omega)|$ by $A(\omega)$, as defined in Equation 10.13. Conversely, making the change of variable $j\omega = s$, the expression 10.17 becomes

$$G(s) = A^2(\omega)|_{\omega=s/j} = A^2\left(\frac{s}{j}\right) \tag{10.18}$$

which gives the desired link between the real function of real variable $A(\omega)$ and the complex function of complex variable $G(s)$.

Allocation of poles

When the polynomial $F_N(\omega)$ is of degree N, the denominator of $G(s)$ is a polynomial of degree $2N$, which gives it $2N$ poles with a double symmetry, about the real axis and about the imaginary axis, as seen earlier.

We now allocate the N stable poles of $G(s)$ to the realizable system function $H(s)$, as planned, leaving the remaining N unstable poles to the virtual system $\bar{H}(s)$ as shown in Figure 10.23. That the poles of the latter are in the unstable half of the s-plane need not give cause for concern, since there will never be a need to implement such a system. It has already served its purpose.

This procedure will be illustrated and clarified in Section 10.5.4, in the context of the Butterworth approximation.

10.5.4   Polynomial families

The procedure described in Section 10.5.2 fits an arbitrary polynomial $F_N(\omega)$ into the specification envelope $E(\omega)$. It ensures that the magnitude error remains within the prescribed tolerance bands but does not determine the magnitude variations within those bands. These depend entirely on the choice of the polynomial family $F_N(\omega)$, which also determines the general character of the phase response.

We now introduce the most common polynomials used in filter design, and examine the criteria by which they are optimal.

Butterworth

The simplest polynomial family has a single term of the form

$$F_N(\omega) = \omega^N$$

which are alternatively even and odd functions. The first five are shown in the top row of Figure 10.25 against the reference square of values ± 1 in both coordinates.

Setting $\varepsilon = 1$, the magnitude $A_N(\omega)$ takes the form

$$A_N(\omega) = \frac{1}{\sqrt{1 + \omega^{2N}}} \qquad (10.19)$$

known as the Butterworth response. At bandedge $\omega = \pm 1$ all responses take the value

$$A_N(1) = \frac{1}{\sqrt{2}} \simeq 0.707$$

which also represents the value k_p of Figure 10.22.

Since the function values $F_N(\omega)$ increase monotonically with

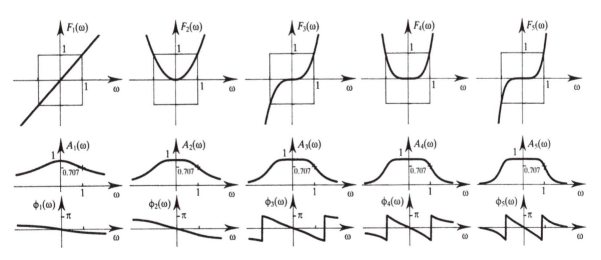

Figure 10.25 Butterworth responses.

frequency, the associated magnitudes $A_N(\omega)$ decrease monotonically. A Taylor expansion of $F_N(\omega)$ about the origin,

$$F_N(\omega) = F_N(0) + \omega F'_N(0) + \frac{\omega^2}{2!}F''_N(0) + \ldots + \frac{\omega^N}{N!}F_N^{(N)}(0) + \ldots$$

shows that all the derivatives at the origin, up to order $N - 1$ are zero, so that $F_N(\omega)$ is the flattest function attainable with a polynomial of degree N. This property also extends to $A_N(\omega)$, see lower half of Figure 10.25, so that this family is described as having **maximally flat frequency response**. The larger N, the flatter the response at the origin.

Poles and zeros

We use the Butterworth response to illustrate the procedure outlined in Section 10.5.3. Introducing the Butterworth magnitude from Equation 10.19 into the general relationship 10.18 gives

$$G(s) = A_N^2(\omega)\big|_{\omega=s/j} = \frac{1}{1 + (s/j)^{2N}} = \frac{1}{1 + (-1)^N s^{2N}}$$

whose denominator roots yield the $2N$ poles p_k of the auxiliary system $G(s)$. For odd N the roots of $1 - s^{2N} = 0$ are the $2N$ roots of $s = 1$,

$$p_k = e^{jk\pi/N}$$

illustrated in Figure 10.26 for $N = 3$. For even N the roots of $1 + s^{2N} = 0$ are those of $s = -1$,

$$p_k = e^{j(\pi/2N + k\pi/N)}$$

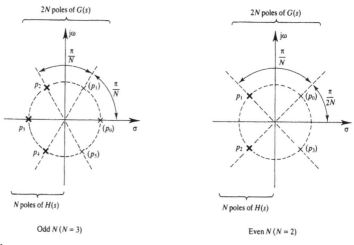

Figure 10.26 Poles of $G(s)$ and $H(s)$.

illustrated for $N = 2$ in Figure 10.26. In all cases the roots lie equidistant on the unit circle. Those of the left half of the s-plane are the poles of $H(s)$, as discussed earlier. These come in complex conjugate pairs, except for odd N, which also has one real pole at $s = -1$. All zeros are at infinity.

Selectivity increases with N. The ideal lowpass filter represents the limiting case which would require an infinite number of poles on the unit circle, with the consequent infinite group delay.

Chebyshev

Forcing the magnitude to oscillate within the tolerance band gives a more efficient utilization of the passband specification. The Chebyshev polynomials, defined by functions of the form

$$C_N(\omega) = \cos Nz \qquad z = \cos^{-1} \omega \tag{10.20}$$

produce the most typical equi-ripple characteristics, and these were used in Section 10.5.2 to formulate the general envelope-fitting process.

The functions $C_N(\omega)$ are written explicitly as polynomials in the frequency variable ω by expressing the cosines of multiples of an angle z as

$$\cos 1z = \cos z$$

$$\cos 2z = 2\cos^2 z - 1$$

$$\cos 3z = 4\cos^3 z - 3\cos z$$

Writing the second expression of Equations 10.20 as $\cos z = \omega$, the first expression leads to the Chebyshev polynomial sequence

$$C_1(\omega) = \omega$$

$$C_2(\omega) = 2\omega^2 - 1$$

$$C_3(\omega) = 4\omega^3 - 3\omega$$

$$C_4(\omega) = 8\omega^4 - 8\omega^2 + 1$$

$$C_5(\omega) = 16\omega^5 - 20\omega^3 + 5\omega$$

illustrated in the top row of Figure 10.27. These functions could also be drawn by using Equations 10.20 directly, but to be valid for all ω the cosines must be interpreted as complex cosine functions of complex variable.

The sequence can be continued recursively as

$$C_{N+1}(\omega) = 2\omega C_N(\omega) - C_{N-1}(\omega)$$

Inserted in the general form of Equation 10.13 the polynomials $C_N(\omega)$ yield corresponding Chebyshev responses

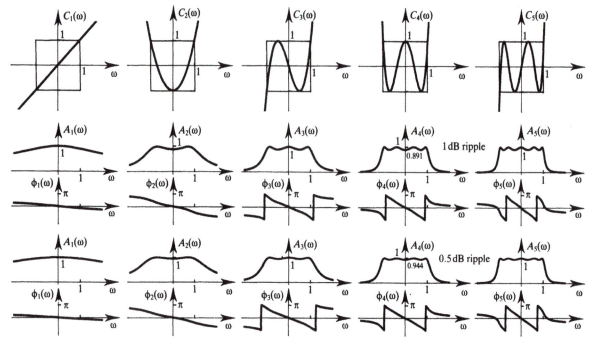

Figure 10.27   Chebyshev responses.

$$A_N(\omega) = \frac{1}{\sqrt{1 + \varepsilon^2 C_N^2(\omega)}} \tag{10.21}$$

The step-by-step link between the two functions was illustrated in Figure 10.22. The resulting amplitude distortions depend on the desired ripple height, which also determines the rate of decay in the transition band. The trade-off between passband ripple and attainable stopband performance is illustrated in Figure 10.27.

Example 10.1

We find the poles and system function of a third-order Chebyshev filter with 1 dB bandpass ripple. Taking the upper limit as $H(0) = 1$, the specified ripple implies a lower limit $k_p = 0.8913$ and, according to Equation 10.14, a design factor $\varepsilon^2 = 0.2589$.

The third-order Chebyshev polynomial squared becomes the sixth-order function

$$C_3^2(\omega) = 16\omega^6 - 24\omega^4 + 9\omega^2$$

which, inserted in Equation 10.21 yields

$$A_3^2(\omega) = \frac{1}{1 + \varepsilon^2 C_3^2(\omega)} = \frac{1}{4.143\omega^6 - 6.214\omega^4 + 2.330\omega^2 + 1}$$

The substitution in Equation 10.18 gives the auxiliary system function $G(s)$,

$$G(s) = \frac{1}{-4.143s^6 - 6.214s^4 - 2.330s^2 + 1}$$

The six complex roots of the denominator of $G(s)$ can be found by standard numerical methods. The three stable roots become the poles of $H(s)$

$$p_0 = -0.4942$$

$$p_1 = -0.2471 + j0.9660 \qquad\qquad (10.22)$$

$$p_1^* = -0.2471 - j0.9660$$

while the three unstable roots $\bar{p}_0 = -p_0$, $\bar{p}_1 = -p_1$ and $\bar{p}_1^* = -p_1^*$ are discarded.

Combining the two complex poles as

$$(s - p_1)(s - p_1^*) = s^2 - (p_1 + p_1^*)s + p_1 p_1^*$$

$$= s^2 + 0.4942s + 0.9941$$

the desired system function can be expressed in the factorized form

$$H(s) = \frac{0.4913}{(s + 0.4942)(s^2 + 0.4941s + 0.9942)} \qquad\qquad (10.23)$$

or be combined further into the polynomial form

$$H(s) = \frac{0.4913}{s^3 + 0.9883s^2 + 1.2384s + 0.4913} \qquad\qquad (10.24)$$

The purpose of the above example was to illustrate the link between a Chebyshev response and its generating polynomial, as discussed in earlier sections. The algorithms used in engineering practice obtain the Chebyshev poles by effectively scaling the circle of corresponding Butterworth poles and projecting the latter on an ellipse,

$$\frac{\sigma^2}{\sinh^2 \beta} + \frac{\omega^2}{\cosh^2 \beta} = 1$$

where β is a design factor determined by N and ε as

$$\beta = \frac{1}{N} \sinh^{-1} \frac{1}{\varepsilon}$$

The geometric construction is shown in Figure 10.28.

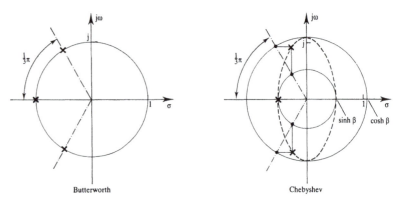

Figure 10.28 Chebyshev pole locations.

Tables

For everyday use Chebyshev parameters have been extensively tabulated in handbooks in terms of the number N of poles and the desired ripple, usually expressed in dB. They tabulate either the polynomial coefficients a_n of $H(s)$ as expressed in the form of Equation 10.24 or factorize the denominator into second-order sections (plus one first-order section for N odd) as in Equation 10.23, or give the s-plane locations of the poles as in Equation 10.22.

Using such tables, a fifth-order Chebyshev filter with 1 dB ripple would be described directly in the forms

$$H_5(s) = \frac{0.1228}{s^5 + 0.9368s^4 + 1.6888s^3 + 0.9744s^2 + 0.5805s + 0.1228}$$

or

$$H_5(s) = \frac{0.1228}{(s + 0.2895)(s^2 + 0.1789s + 0.9883)(s^2 + 0.4684s + 0.4293)}$$

or by the poles

$$p_0 = -0.2895$$

$$p_1, p_1^* = -0.2342 \pm j0.6119$$

$$p_2, p_2^* = -0.0895 \pm j0.9901$$

These values are used throughout Section 10.6 for illustration.

Elliptic or Cauer

An even steeper rate of decay is obtained in the transition band when the magnitude is also forced to oscillate in the stopband. This is achieved by using elliptic functions $R_N(\omega, L)$ in Equation 10.13, to give the Cauer response

$$A_N(\omega) = \frac{1}{\sqrt{1 + \varepsilon^2 R_N^2(\omega, L)}}$$

where ε has the same meaning as in Chebyshev responses and L is a factor relating the passband and stopband tolerance bands.

The treatment of elliptic functions falls outside the scope of this book, suffice it to say that they are rational polynomials and that by the process described in Figure 10.22 the denominator polynomial of $R_N(\omega, L)$ makes its way into the numerator of $A_N(\omega)$, thereby creating system zeros in the stopband.

We illustrate in Figure 10.29 with a fifth-order elliptic function. Its main difference with the corresponding Chebyshev polynomial is a very fast growth to infinity outside bandedge, followed by an oscillation between $\pm\infty$ and $\pm L$. By the process of Figure 10.22 those frequencies for which $R_N(\omega, L)$ reaches infinity become zeros of $A_N(\omega)$, while the frequencies at which the limit $\pm L$ is touched also represent points where $A_N(\omega)$ touches the stopband envelope.

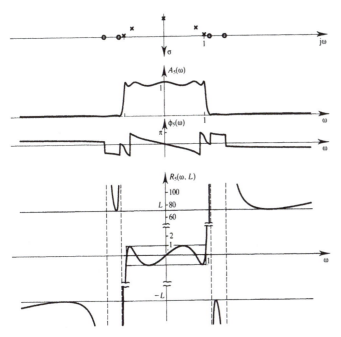

Figure 10.29   Elliptic function and response.

Conclusions

Both the Butterworth and Chebyshev characteristics are sometimes called 'all-pole' filters because all their zeros are at infinity. Each pole pair can be associated with a pair of such zeros and realized as a second-order lowpass section of the 'all-pole' form shown in the top left of Figure 10.18. To realize elliptic filters each pole pair is associated with a pair of zeros on the imaginary axis and realized as a second-order section of the lowpass notch type shown at the top right of Figure 10.18.

The Butterworth response is optimal in terms of flatness at the origin. The Chebyshev response spreads the magnitude deviations over the full width of the passband and is optimal in the sense that, for a given order N, it gives the fastest rate of decay at bandedge attainable with 'all-pole' lowpass sections. The elliptic characteristic improves the rate of decay by placing zeros in the stopband.

But every improvement in magnitude selectivity is accompanied by a degradation of phase linearity, particularly at bandedge. The elliptic response is the worst, followed by Chebyshev and Butterworth. When phase linearity is important the Bessel response is best. It optimizes the flatness of the group delay at the origin, where the Butterworth response optimizes amplitude flatness. The Bessel system function is of the form

$$H_N(s) = \frac{B_N(0)}{B_N(s)}$$

where $B_N(s)$ is the Nth-order Bessel polynomial and the numerator $B_N(0)$ is a constant that normalizes the system function. The polynomials can be generated recursively by the formula

$$B_N(s) = (2N - 1)B_{N-1}(s) + s^2 B_{N-2}(s)$$

starting with the initial conditions $B_0(s) = 1$ and $B_1(s) = s + 1$. Phase linearity improves with increasing N, but the magnitude response is poor.

10.6 Frequency mapping

The classical approximation methods of the preceding section are formulated to yield lowpass filters. We now present a method for generating other types of filter characteristics from those lowpass approximations.

We identify all functions and variables associated with the lowpass prototype with the circumflex (^) and map all $\hat{s}$-plane features onto the target s-plane by using a truncated subset of the general mapping law,

$$\hat{s} = f_x(s) = k \frac{s(s^2 + \omega_2^2)(s^2 + \omega_4^2) \ldots (s^2 + \omega_n^2)}{(s^2 + \omega_1^2)(s^2 + \omega_3^2) \ldots (s^2 + \omega_{n-1}^2)} \quad (10.25)$$

$$0 \leqslant \omega_1 < \omega_2 < \omega_3 < \omega_4 < \ldots < \omega_{n-1} < \omega_n$$

where the subscript x identifies the type of mapping. We start with the trivial lowpass-to-lowpass case to identify general mapping properties and requirements and later extend these to more elaborate types.

These mapping laws relate all aspects of the two systems, from the location of poles and zeros to the shape of the associated functions, such as $H(s)$, as they are transferred from one s-plane to the other. Most of the section deals with analysis, where the mapping parameters k, ω_1, etc. are known. The synthesis problem, determining those parameters, is addressed in Section 10.6.5.

Non-linear complex number operations of the form $1/s$ and s^2, although algebraically simple, are not easy to imagine. Graphical representations are therefore used extensively to introduce underlying concepts and to unify otherwise disjointed results.

Note that in Section 7.2.1 the change of frequency variable $z = e^{sT}$ involved a similar s-plane mapping concept, which will surface again in Section 11.4, together with other frequency mapping laws.

10.6.1 Lowpass to lowpass

The lowpass-to-lowpass mapping law takes the simple form

$$\hat{s} = f_L(s) = ks \quad k > 0 \quad (10.26)$$

which represents a one-to-one mapping of points of the s-plane onto points of the $\hat{s}$-plane, and vice versa. Writing both variables generically as sums of real and imaginary parts

$$\hat{\sigma} + j\hat{\omega} = k\sigma + jk\omega$$

shows that these parts too are scaled by k, as $\hat{\sigma} = k\sigma$ and $\hat{\omega} = k\omega$. The inverse mapping law is simply

$$s = f_L^{-1}(s) = \frac{\hat{s}}{k}$$

A grid of equally spaced straight lines of one plane maps onto a similar grid of the other plane, and a polar grid maps onto a similar polar grid, as shown in Figure 10.30.

These grids serve to relate features of the two planes, as illustrated by the five-pole cluster of the fifth-order Chebyshev filter characteristics given in the last section under the heading 'Tables'. The correspondence in the two planes is self-evident. In practice, this process

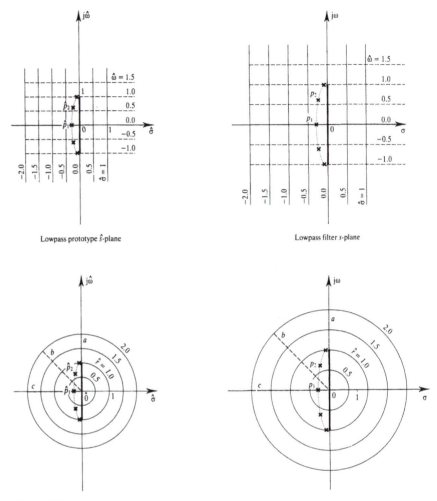

Figure 10.30 Lowpass-to-lowpass mapping: $\hat{s} = ks$.

scales the normalized frequency of a prototype filter to the actual frequency specification.

Although rather trivial, this law embodies the essential requirements of other mapping laws. A detailed examination provides a framework for the general mapping problem. In this context it is significant that for $\hat{\sigma} = 0$ we have $\sigma = 0$, so that the imaginary axis $\hat{s} = j\hat{\omega}$ maps onto the imaginary axis $s = j\omega$. Also significant is that poles from the stable half-plane $\hat{\sigma} < 0$ map onto the stable half-plane $\sigma < 0$.

System function

We wish to show that by mapping the frequency variable according to the law 10.26 all values of the prototype system function $\hat{H}(\hat{s})$ are transferred unscaled to the mapped locations on the s-plane, thus generating the target system function $H(s)$ as

$$H(s) = \hat{H}(\hat{s})|_{\hat{s}=ks} \tag{10.27}$$

Consider a second-order lowpass section with poles at $\hat{s} = \hat{p}_1$ and $\hat{s} = \hat{p}_2$,

$$\hat{H}(\hat{s}) = \frac{\hat{p}_1\hat{p}_2}{(\hat{s} - \hat{p}_1)(\hat{s} - \hat{p}_2)} \tag{10.28}$$

which is of the form of Equation 4.46, with $c_2 = c_1 = 0$ and $c_0 = \hat{p}_1\hat{p}_2$, so that it is normalized at the origin to $\hat{H}(0) = 1$. Evaluating for $\hat{s} = ks$ yields

$$\hat{H}(ks) = \frac{\hat{p}_1\hat{p}_2}{(ks - \hat{p}_1)(ks - \hat{p}_2)}$$

which is a function in the variable ks. To write it more explicitly in terms of s we divide numerator and denominator by k^2 giving

$$H(s) = \frac{(\hat{p}_1/k)(\hat{p}_2/k)}{(s - \hat{p}_1/k)(s - \hat{p}_2/k)} = \frac{p_1 p_2}{(s - p_1)(s - p_2)} \tag{10.29}$$

Although the functions $H(s)$ and $\hat{H}(ks)$ are different, their values at corresponding points $\hat{s} = ks$ are the same (e.g. $H(0) = 1$), which confirms the relationship 10.27. It also confirms that a pair of poles $\hat{p}_1$ and $\hat{p}_2$ of the $\hat{s}$-plane maps onto a similar pair $p_1 = \hat{p}_1/k$ and $p_2 = \hat{p}_2/k$ of the s-plane.

This relationship also applies with the notation of Equation 4.48 of second-order systems, in terms of pole-frequency and pole-Q notation. Thus expressing Equation 10.28 as

$$\hat{H}(\hat{s}) = \frac{\hat{\omega}_r^2}{\hat{s}^2 + (\hat{\omega}_r/\hat{Q})\hat{s} + \hat{\omega}_r^2} \tag{10.30}$$

evaluating for $\hat{s} = ks$ and dividing numerator and denominator by k^2 yields

$$H(s) = \frac{(\hat{\omega}_r/k)^2}{s^2 + [(\hat{\omega}_r/k)/\hat{Q}]s + (\hat{\omega}_r/k)^2} = \frac{\omega_r^2}{s^2 + (\omega_r/Q)s + \omega_r^2} \tag{10.31}$$

The pole-frequencies are thus related by the same mapping law, $\omega_r = \hat{\omega}_r/k$, and the pole-$Q$ remains the same, $Q = \hat{Q}$. In the s-planes this is represented by proportional circles and lines of the same gradient, as shown in Figure 10.31.

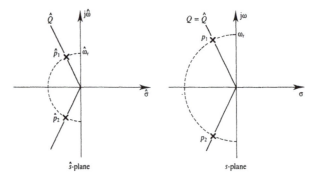

Figure 10.31   Mapped poles of second-order section.

The relationship 10.27 also applies to higher-order systems, and to the partitioning of such systems into cascaded first-order and second-order sections.

Frequency response

Since specification and design are usually formulated in terms of the system's frequency response $H(j\omega)$, the mapping properties of the imaginary axis $s = j\omega$ are of particular interest.

We found earlier that the imaginary axis of one s-plane maps onto the imaginary axis of the other, and that points on these axes are related as

$$\hat{\omega} = k\omega \tag{10.32}$$

This is shown in the construction of Figure 10.32, where the target lowpass response $H_L(j\omega)$ on the right is derived from the prototype response $\hat{H}(j\hat{\omega})$ on the left according to the subset of Equation 10.27,

$$H_L(j\omega) = \hat{H}(j\hat{\omega})|_{\hat{\omega}=k\omega}$$

The upper half of the figure relates the corresponding amplitudes $\hat{A}(\hat{\omega})$ and $A_L(\omega)$. The vertical planes containing these are linked by an auxiliary vertical plane whose orientation in relation to the frequency axes ω and $\hat{\omega}$ represents the gradient k of the linear relationship 10.32.

The points $\hat{O}$, $\hat{R}$ and $\hat{Q}$ of the $\hat{\omega}$-axis, which represent the prototype's origin and band-edges, map onto corresponding points of the target's axis. Two consecutive projections involving the auxiliary plane transfer the associated amplitude values $\hat{A}(\hat{\omega})$, without change, to the new locations, thus generating $A_L(\omega)$.

The lower half of Figure 10.32 repeats the construction for the phase response $\phi(\omega)$. Note that the symmetries are faithfully preserved.

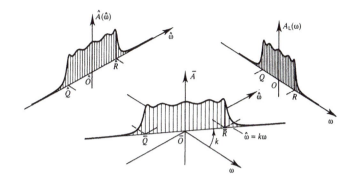

Lowpass prototype $\hat{H}(\hat{\omega})$ Lowpass filter $H_L(\omega)$

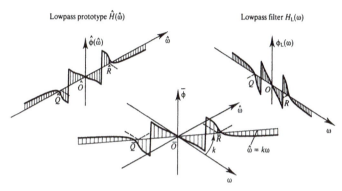

Figure 10.32 Frequency response – lowpass-to-lowpass mapping: $\hat{\omega} = k\omega$.

The same construction can also be used to transfer real and imaginary parts of $\hat{H}(j\hat{\omega})$.

10.6.2 Lowpass to highpass

The next subset of Equation 10.25 brings in the denominator term $(s^2 + \omega_1^2)$. Setting $\omega_1 = 0$ yields the reciprocal mapping law

$$\hat{s} = f_H(s) = \frac{k}{s} \qquad k > 0 \tag{10.33}$$

which turns a lowpass prototype into a highpass filter.

Expressing both variables of Equation 10.33 in exponential form, in terms of magnitude S and angle θ of their complex s-plane locations, yields

$$\hat{S}e^{j\hat{\theta}} = \frac{k}{S}\,e^{-j\theta}$$

which implies reciprocal magnitudes and phase inversion,

$$\hat{S} = \frac{k}{S} \quad \text{and} \quad \hat{\theta} = -\theta$$

Circles centred on the s-plane origin map onto similar concentric circles of the $\hat{s}$-plane. Their radii are inversely related, and angular positions are symmetrical about the real axis, as shown in the lower half of Figure 10.33.

To map the s-plane imaginary axis onto the $\hat{s}$-plane we express $\hat{s}$ generically by its real and imaginary parts and evaluate Equation 10.33 for $s = j\omega$,

$$\hat{\sigma} + j\hat{\omega} = \frac{k}{j\omega} = -j\frac{k}{\omega}$$

Equating real and imaginary parts gives

$$\hat{\sigma} = 0 \quad \text{and} \quad \hat{\omega} = -\frac{k}{\omega}$$

Thus, the s-plane imaginary axis $s = j\omega$ maps onto the imaginary axis $\hat{s} = j\hat{\omega}$ of the prototype plane, in a one-to-one relationship. Points on one axis map onto the opposite side of the other, at the inverse of the distance, and points at the origin and at infinity are interchanged.

We would similarly find that points located on the real axis of one plane map onto points of the other real axis, as $\hat{\sigma} = k/\sigma$, that is, on the same side. Again, points at the origin and at infinity are interchanged.

A Cartesian grid of one plane maps onto a grid of orthogonally intersecting circles of the other, as shown in the top half of Figure 10.33. Each circle passes through the origin, which represents points at plus and minus infinity of all lines, that is, the 'point at infinity' of Section 4.2.6.

This relationship is easily shown. Writing Equation 10.33 as the product

$$(\hat{\sigma} + j\hat{\omega})(\sigma + j\omega) = (\hat{\sigma}\sigma - \hat{\omega}\omega) + j(\hat{\sigma}\omega + \hat{\omega}\sigma) = k$$

and equating real and imaginary parts yields

$$\hat{\sigma}\sigma - \hat{\omega}\omega = k \quad \text{and} \quad \hat{\sigma}\omega + \hat{\omega}\sigma = 0 \qquad (10.34)$$

To represent a line of constant $\hat{\omega}$ we eliminate $\hat{\sigma}$ from these, to give

$$\omega^2 + \sigma^2 + \frac{k\omega}{\hat{\omega}} = 0$$

and adding and subtracting the term $k^2/4\hat{\omega}^2$ yields

$$\sigma^2 + \left(\omega + \frac{k}{2\hat{\omega}}\right)^2 = \left(\frac{k}{2\hat{\omega}}\right)^2$$

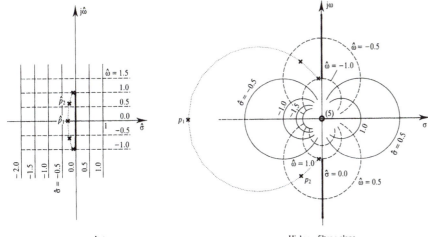

Lowpass prototype $\hat{s}$-plane   Highpass filter s-plane

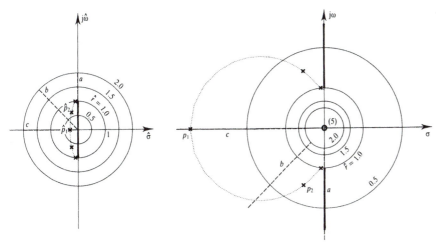

Figure 10.33   Lowpass-to-highpass mapping: $\hat{s} = k/s$.

which is the equation of a circle of the form

$$x^2 + (y - r)^2 = r^2$$

that maps a line of constant $\hat{\omega}$ as a circle of radius $r = k/2\hat{\omega}$, with vertical offset $\omega = -r$.

Similarly, eliminating $\hat{\omega}$ from Equations 10.34 yields

$$\left(\sigma - \frac{k}{2\hat{\sigma}}\right)^2 + \omega^2 = \left(\frac{k}{2\hat{\sigma}}\right)^2$$

which maps a line of constant $\hat{\sigma}$ as a circle of radius $r = k/2\hat{\sigma}$, with horizontal offset $\sigma = r$. Note that lines of negative $\hat{\sigma}$ map onto the side of negative σ. Hence stable poles map as stable poles.

System function

We now show that the reciprocal mapping law 10.33 converts a lowpass prototype into a highpass characteristic. Using the earlier second-order lowpass section for illustration, we evaluate Equation 10.28 for $\hat{s} = k/s$, to give

$$\hat{H}\left(\frac{k}{s}\right) = \frac{\hat{p}_1\hat{p}_2}{(k/s - \hat{p}_1)(k/s - \hat{p}_2)}$$

which, multiplying numerator and denominator by $s^2/\hat{p}_1\hat{p}_2$, is re-arranged as

$$H_H(s) = \frac{s^2}{(s - k/\hat{p}_1)(s - k/\hat{p}_2)} = \frac{c_2 s^2}{(s - p_1)(s - p_2)} \tag{10.35}$$

This has the form of a highpass section, with poles $p_i = k/\hat{p}_i$ and amplitude scaler $c_2 = 1$. Note that this mapping brings to the origin the 'zeros at infinity' of the lowpass section.

If we were to express Equation 10.35 with ω_r and Q notation as

$$H_H(s) = \frac{c_2 s^2}{s^2 + (\omega_r/Q)s + \omega_r^2}$$

we would find that the pole-frequency ω_r relates to that of the lowpass prototype of Equation 10.30 as $\omega_r = k/\hat{\omega}_r$, so that $Q = \hat{Q}$ and $c_2 = 1$.

Based on this typical result we conclude that the reciprocal mapping law 10.33 converts the general lowpass prototype $\hat{H}(\hat{s})$ to a related highpass system $H_H(s)$ by the expression

$$H_H(s) = \hat{H}(\hat{s})|_{\hat{s}=k/s}$$

Figure 10.33 illustrates this relationship in terms of the poles and zeros of a fifth-order Chebyshev filter. Note that the ellipse bearing the poles is severely distorted, poles map to stable poles on the opposite side of the real axis, and the five 'zeros at infinity' map onto the s-plane origin.

Frequency response

The imaginary axes map onto each other, and points located on these axes are related by the function $\hat{\omega} = -k/\omega$. The frequency response $\hat{H}(j\hat{\omega})$ is therefore transferred as

$$H_H(j\omega) = \hat{H}(j\hat{\omega})|_{\hat{\omega}=-k/\omega}$$

The function relating $\hat{\omega}$ to ω is a hyperbola in the $\omega\hat{\omega}$-plane of Figure 10.34 and it generates two curved surfaces. Projecting the amplitude response $\hat{A}(\hat{\omega})$ onto these surfaces tears it into half. A second projection

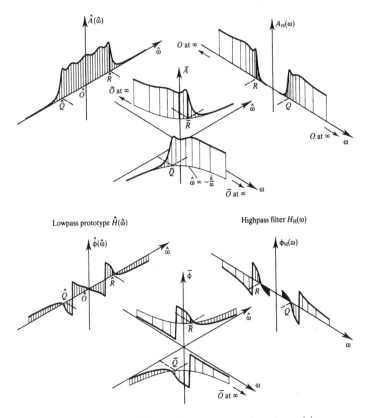

Lowpass prototype $\hat{H}(\hat{\omega})$ Highpass filter $H_H(\omega)$

Figure 10.34 Lowpass-to-highpass frequency mapping: $\hat{\omega} = -k/\omega$.

reassembles the two halves as a highpass response $A_H(\omega)$ on the target frequency axis.

The nonlinearity of the process is stressed by hatching the surface $\hat{A}(\hat{\omega})$ uniformly and mapping the hatching lines. Note that the point at the origin maps on plus and minus infinity and vice versa, thus confirming that 'zeros at infinity' are brought to the origin.

The lower half of Figure 10.34 repeats the construction for phase. As the function $\hat{\omega} = -k/\omega$ is again odd, with positive gradient, the symmetries of the frequency response are preserved, such that phase with negative gradient maps onto phase with negative gradient.

10.6.3   Lowpass to bandpass

Truncating Equation 10.25 at the factor in ω_2 and setting $\omega_1 = 0$ yields the next mapping law,

$$\hat{s} = f_B(s) = \frac{k(s^2 + \omega_2^2)}{s} \tag{10.36}$$

which converts a lowpass prototype to a bandpass filter.

The quadratic factor complicates matters, causing two points of the s-plane to be mapped onto one point of the prototype plane. The points $s = \pm j\omega_2$ are characteristic in that, by annulling the quadratic term, they both map onto the $\hat{s}$-plane origin.

To derive the inverse mapping law $f_B^{-1}(\hat{s})$, we write Equation 10.36 as

$$s^2 - \frac{\hat{s}}{k}s + \omega_2^2 = 0 \tag{10.37}$$

and find its two roots. The complex square roots of a complex number $z = Ze^{j\theta}$ are $z_1 = \sqrt{Z}\,e^{j\theta/2}$ and $z_2 = -z_1$. Using the principal value z_1, we write

$$s_1, s_2 = \frac{\hat{s}}{2k} \pm \sqrt{\left(\frac{\hat{s}}{2k}\right)^2 - \omega_2^2} \tag{10.38}$$

Each point $\hat{s}$ of the prototype plane maps onto two points of the s-plane. The latter are inversely related to each other, as seen from their product

$$s_1 s_2 = \left(\frac{\hat{s}}{2k}\right)^2 - \left(\frac{\hat{s}}{2k}\right)^2 + \omega_2^2 = \omega_2^2$$

where ω_2 is a real number, so that $s_2 = \omega_2^2/s_1$, or, in terms of magnitude and angle,

$$S_2 = \frac{\omega_2^2}{S_1} \quad \text{and} \quad \theta_2 = -\theta_1 \tag{10.39}$$

This expresses a form of inverse symmetry, or 'reflection' about the circle of radius ω_2 for magnitude, and about the real axis for angle, as illustrated in Figure 10.35 for a typical point $\hat{s}$.

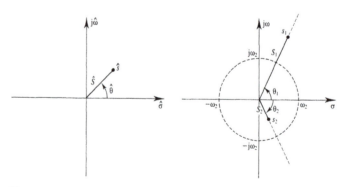

Figure 10.35 Mapping a typical point $\hat{s}$.

The point at infinity $\hat{s} = \infty$ maps onto both the point at infinity $s_1 = \infty$ and the origin $s_2 = 0$, as the two terms of Equation 10.38 are infinite quantities of the same order. And this is consistent with Equation 10.39.

Imaginary axis

The imaginary axes map fully onto each other. This is clear from evaluating Equation 10.36 for $s = j\omega$

$$\hat{\sigma} + j\hat{\omega} = \frac{k(-\omega^2 + \omega_2^2)}{j\omega} = j\frac{k(\omega^2 - \omega_2^2)}{\omega}$$

Equating real and imaginary parts we have $\hat{\sigma} = 0$ and

$$\hat{\omega} = \frac{k(\omega^2 - \omega_2^2)}{\omega} \tag{10.40}$$

For the inverse relationship this is rearranged as

$$\omega^2 - \frac{\hat{\omega}}{2k}\omega - \omega_2^2 = 0$$

which yields the real roots

$$\omega_1, \omega_2 = \frac{\hat{\omega}}{2k} \pm \sqrt{\left(\frac{\hat{\omega}}{2k}\right)^2 + \omega_2^2} \tag{10.41}$$

The correspondence between points of the two axes is represented in Figure 10.36 for later use with the frequency response.

Real axis

Evaluating Equation 10.38 for points located on the real prototype axis $\hat{s} = \hat{\sigma}$, we obtain two points of the s-plane,

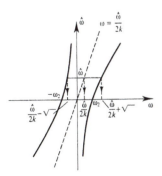

Figure 10.36   Link between points of the imaginary axes.

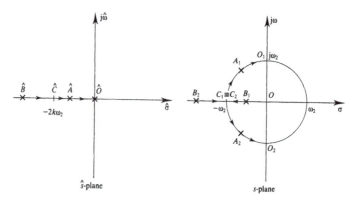

Figure 10.37 Mapping four typical points of the $\hat{\sigma}$-axis.

$$\sigma + j\omega = \frac{\hat{\sigma}}{2k} \pm \sqrt{\left(\frac{\hat{\sigma}}{2k}\right)^2 - \omega_2^2} \qquad (10.42)$$

The argument of the square root is real, but can take positive or negative values. When $|\hat{\sigma}| = 2k\omega_2$, such as point $\hat{C}$ of Figure 10.37, the two s-plane points are real and coincide on $|\sigma| = \omega_2$. For $|\hat{\sigma}| > 2k\omega_2$, such as point $\hat{B}$, the two solutions fall on the real s-plane axis. When $|\hat{\sigma}| < 2k\omega_2$, such as point $\hat{A}$, the square root is pure imaginary, and Equation 10.42 becomes

$$\sigma + j\omega = \frac{\hat{\sigma}}{2k} \pm j\sqrt{\omega_2^2 - \left(\frac{\hat{\sigma}}{2k}\right)^2}$$

and equating real and imaginary parts yields

$$\sigma = \frac{\hat{\sigma}}{2k} \quad \text{and} \quad \omega = \sqrt{\omega_2^2 - \left(\frac{\hat{\sigma}}{2k}\right)^2}$$

These are the parametric equations of an s-plane circle of radius ω_2 and centre at the origin, otherwise expressed as

$$\sigma^2 + \omega^2 = \left(\frac{\hat{\sigma}}{2k}\right)^2 + \omega_2^2 - \left(\frac{\hat{\sigma}}{2k}\right)^2 = \omega_2^2$$

This circle is characteristic of the mapping law 10.36 and represents a reference line for the inverse symmetry mentioned earlier.

Grids

The regular $\hat{s}$-plane grids of Figure 10.38 were mapped by applying Equation 10.38 to their individual points. An asymmetric number of grid lines is shown and these are cross-referenced to the s-plane. Despite

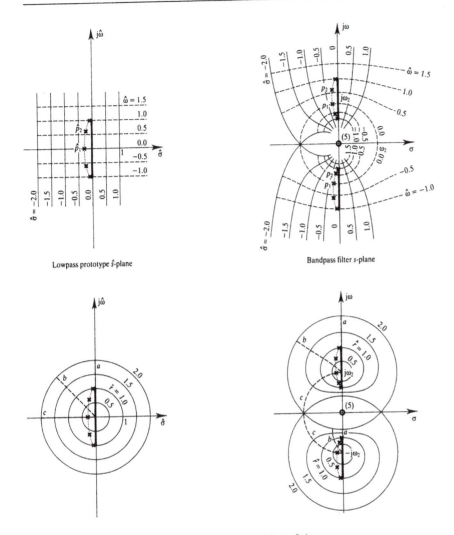

Figure 10.38 Lowpass-to-bandpass mapping $\hat{s} = k(s^2 + \omega_2^2)/s$.

appearances, each grid maps onto two non-overlapping grids separated by the real axis, and this should be clear from Figure 10.35.

The five Chebyshev poles illustrated in Figure 10.38 map asymmetrically too. This is not obvious from the overall results of causal systems, as all features of the prototype plane are already symmetrical.

System function

We start with a **first-order lowpass prototype** of the form

$$\hat{H}(\hat{s}) = \frac{-\hat{p}_1}{\hat{s} - \hat{p}_1} \tag{10.43}$$

Evaluating for $\hat{s} = k(s^2 + \omega_2^2)/s$ gives

$$\hat{H}\left(k\frac{s^2 + \omega_2^2}{s}\right) = \frac{-\hat{p}_1}{k(s^2 + \omega_2^2)/s - \hat{p}_1}$$

which, rearranged in terms of s yields a typical **second-order bandpass filter**

$$H_B(s) = \frac{-(\hat{p}_1/k)s}{s^2 - (\hat{p}_1/k)s + \omega_2^2} = \frac{c_1 s}{(s - p_{11})(s - p_{12})} \tag{10.44}$$

The first denominator is of the form of Equation 10.37. According to Equation 10.38 its roots p_{11} and p_{12} are

$$p_{11}, p_{12} = \frac{\hat{p}_1}{2k} \pm \sqrt{\left(\frac{\hat{p}_1}{2k}\right)^2 - \omega_2^2} \tag{10.45}$$

The numerator can be similarly interpreted as mapping the prototype 'zero at infinity' as two s-plane zeros, one 'at infinity', the other at the origin. The resulting scaling coefficient $c_1 = -\hat{p}_1/k$ shows that the value $\hat{H}(0) = 1$ is mapped unscaled, i.e. $H_B(j\omega_2) = H_B(-j\omega_2) = \hat{H}(0) = 1$.

In real systems the single pole $\hat{p}_1$ is on the real axis $\hat{p}_1 = \hat{\sigma}_1$ so that it is mapped according to Equation 10.42. The real point $\hat{A}$ of Figure 10.37 represents a typical pole location, which maps onto symmetric points A_1 and A_2 of the circle of radius ω_2. In terms of ω_r and Q notation, this means that $\omega_r = \omega_2$ and $\omega_r/Q = -\hat{\sigma}_1/k$.

But Equations 10.44 and 10.45 are also valid when $\hat{p}_1$ is complex, for instance, when it belongs to a conjugate pair. Thus, interpreting the **second-order lowpass prototype** of Equation 10.28 as the product of two first-order sections,

$$\hat{H}(\hat{s}) = \frac{-\hat{p}_1}{\hat{s} - \hat{p}_1} \frac{-\hat{p}_2}{\hat{s} - \hat{p}_2} \tag{10.46}$$

each of these converts to a second-order section, with the result

$$H_B(s) = \frac{c_{11} s}{(s - p_{11})(s - p_{12})} \frac{c_{12} s}{(s - p_{21})(s - p_{22})}$$

which is a **fourth-order bandpass filter**. The individual poles $\hat{p}_1$ and $\hat{p}_2$ map asymmetrically, as shown in Figure 10.39. But given their conjugate symmetry $\hat{p}_2 = \hat{p}_1^*$, and the inverse symmetry (Equation 10.39) of the mapped poles, these can be recombined into conjugate pairs yielding the system function

$$H_B(s) = \frac{c_{11} s}{(s - p_{11})(s - p_{22})} \frac{c_{12} s}{(s - p_{21})(s - p_{12})}$$

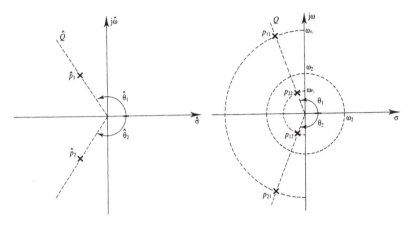

Figure 10.39 Conjugate and inverse symmetries of mapped poles.

which represents **two-cascaded second-order bandpass sections** of standard realizable form.

It is clear from Equation 10.39 and from Figure 10.39 that the pole-frequencies of the two sections are inversely related or 'reflected' about the circle of radius ω_2, that is, $\omega_{r_2} = \omega_2/\omega_{r_1}$, and that their pole-$Q$s are the same. Note once again that stable poles map to stable poles.

We conclude that the mapping law 10.36 converts an arbitrary lowpass prototype $\hat{H}(\hat{s})$ into a bandpass filter $H_B(s)$ by the relationship

$$H_B(s) = \hat{H}(\hat{s})\big|_{\hat{s}=k(s^2+\omega_2^2)/s}$$

whose poles and zeros are mapped versions (Equation 10.38) of the prototype poles and zeros.

Frequency response

The function 10.40 relating points of the two imaginary axes was represented in Figure 10.36. The latter is transferred to the $\omega\hat{\omega}$-plane of Figure 10.40, where its two lobes generate two vertical surfaces. The usual double projection involving these surfaces distorts and duplicates the lowpass characteristic, thus converting the prototype lowpass frequency response $\hat{H}(j\hat{\omega})$ into the bandpass filter response $H_B(j\omega)$.

The resulting distortions of magnitude and phase are illustrated in Figure 10.40 with the usual five-pole Chebyshev response. Each lobe is asymmetrical about the two points $\omega = \pm\omega_2$ on which the origin of the $\hat{\omega}$-axis is mapped. But the symmetries of $\hat{H}(j\hat{\omega})$ give rise to an overall even symmetry for the magnitude $A_B(\omega)$ and overall odd symmetry for the phase $\phi_B(\omega)$.

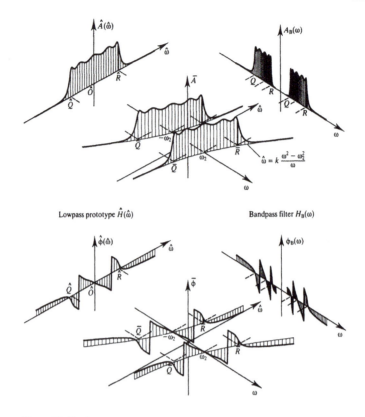

Figure 10.40 Lowpass-to-bandpass frequency mapping: $\hat{\omega} = k(\omega^2 - \omega_2^2)/\omega$.

10.6.4 Lowpass to bandstop

Truncating Equation 10.25 at the first denominator factor yields a law that is virtually the inverse of the bandpass case, namely

$$\hat{s} = f_S(s) = \frac{ks}{s^2 + \omega_1^2} \tag{10.47}$$

Now the points $s = \pm j\omega_1$ are characteristic in that they map to the point at infinity of the $\hat{s}$-plane.

Proceeding as in Section 10.6.3, the inverse mapping law $f_S^{-1}(\hat{s})$ becomes

$$s_1, s_2 = \frac{k}{2\hat{s}} \pm \sqrt{\left(\frac{k}{2\hat{s}}\right)^2 - \omega_1^2} \tag{10.48}$$

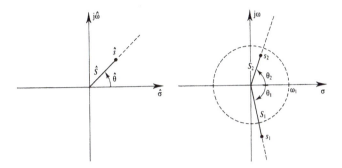

Figure 10.41   Mapping a typical point $\hat{s}$.

which is formally similar to Equation 10.38, but the terms in s are reciprocal functions. This reflects on the mapped roots, illustrated in Figure 10.41, which are reciprocal in character to the roots of Figure 10.35.

The roots are again inversely related to each other, $s_2 = \omega_1^2/s_1$, also expressed as

$$ S_2 = \frac{\omega_1^2}{S_1} \quad \text{and} \quad \theta_2 = -\theta_1 \tag{10.49} $$

where the symmetry is now about the circle of radius ω_1 for magnitude, and about the real axis for angle, as seen in Figure 10.41. From Equation 10.47 we know that the point $\hat{s} = 0$ maps onto $s_1 = 0$, and by Equations 10.49 it also maps onto $s_2 = \infty$. As seen earlier, the point $\hat{s} = \infty$ maps onto $s = \pm j\omega_1$.

Imaginary axis

Interpreting Equation 10.47 for $s = j\omega$ shows that the imaginary axes map onto each other according to

$$ \hat{\omega} = \frac{k\omega}{-\omega^2 + \omega_1^2} $$

which, written in quadratic form, yields the roots

$$ \omega_1, \omega_2 = -\frac{k}{2\hat{\omega}} \pm \sqrt{\left(\frac{k}{2\hat{\omega}}\right)^2 + \omega_1^2} \tag{10.50} $$

This relationship is interpreted in Figure 10.42.

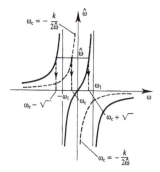

Figure 10.42 Relationship between points of imaginary axes.

Real axis

Points of the real axis $s = \sigma$ map onto the real prototype axis, according to the expression

$$\hat{\sigma} = \frac{k\sigma}{\sigma^2 + \omega_1^2}$$

As in the bandpass case, the inverse is not necessarily true. Interpreting Equation 10.48 for points of the real axis $\hat{s} = \hat{\sigma}$, we have

$$\sigma + j\omega = \frac{k}{2\hat{\sigma}} \pm \sqrt{\left(\frac{k}{2\hat{\sigma}}\right)^2 - \omega_1^2} \qquad (10.51)$$

The root argument can take positive, zero or negative values. The latter occurs for $|\hat{\sigma}| > k/2\omega_1$, when the image points form a conjugate pair on the circle of radius ω_1. The three cases are illustrated in Figure 10.43 for points of negative $\hat{\sigma}$.

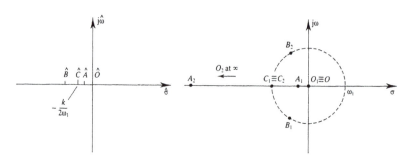

Figure 10.43 Mapping four typical points of the $\hat{\sigma}$-axis.

Grids

The regular $\hat{s}$-plane grids of Figure 10.44 were mapped onto the s-plane by the relationship 10.48. The poles of the Chebyshev response are duplicated and rearranged as two groups, loosely resembling both the poles of the lowpass filter of Figure 10.30 as well as those of the highpass filter of Figure 10.33. The 'zeros at infinity' are duplicated and mapped onto $s = \pm j\omega_1$.

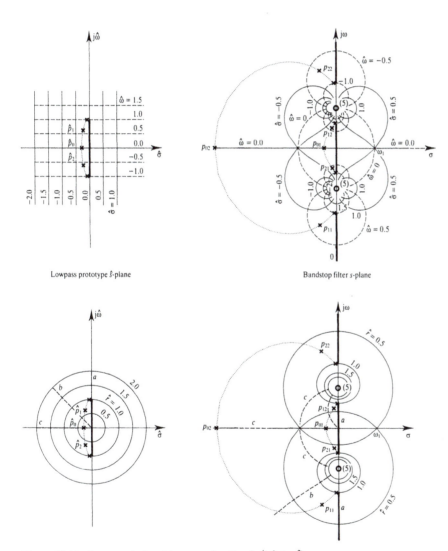

Figure 10.44 Lowpass-to-bandstop mapping $\hat{s} = ks/(s^2 + \omega_1^2)$.

System function

The first-order lowpass prototype of Equation 10.43 evaluated for $\hat{s} = ks/(s^2 + \omega_1^2)$ gives

$$\hat{H}\left(\frac{ks}{s^2 + \omega_1^2}\right) = \frac{-\hat{p}_1}{ks/(s^2 + \omega_1^2) - \hat{p}_1}$$

which represents a **second-order notch filter** of standard form

$$H_S(s) = \frac{s^2 + \omega_1^2}{s^2 - (k/\hat{p}_1)s + \omega_1^2} = \frac{s^2 + c_2}{(s - p_{11})(s - p_{12})} \qquad (10.52)$$

The denominator roots are given by Equation 10.48 as

$$p_{11}, p_{12} = \frac{k}{2\hat{p}_1} \pm \sqrt{\left(\frac{k}{2\hat{p}_1}\right)^2 - \omega_1^2} \qquad (10.53)$$

and the 'zero at infinity' maps to two s-plane zeros at $s = \pm j\omega_1$.

In real systems the real pole $\hat{p}_1 = \hat{\sigma}_1$ maps according to Equation 10.51, typically on the real s-plane axis, see point $\hat{A}$ of Figure 10.43, but can fall on the circle of radius ω_1, see point $\hat{B}$ of Figure 10.43.

A **second-order lowpass section**, interpreted as the product of two complex first-order terms (Equation 10.46), each of which converts to a second-order notch section, gives rise to a **fourth-order bandstop filter**

$$H_S(s) = \frac{s^2 + c_{21}}{(s - p_{11})(s - p_{12})} \frac{s^2 + c_{22}}{(s - p_{21})(s - p_{22})}$$

As in the bandpass case, the conjugate symmetry $\hat{p}_2 = \hat{p}_1^*$, with the inverse symmetry in Equation 10.49 of the mapped poles, yields the system function

$$H_S(s) = \frac{s^2 + c_{21}}{(s - p_{11})(s - p_{22})} \frac{s^2 + c_{22}}{(s - p_{21})(s - p_{12})} \qquad (10.54)$$

which represents **two realizable second-order notch sections**. Again, the pole-frequencies of these two sections are inversely related as $\omega_{r_2} = \omega_1^2/\omega_{r_1}$, their pole-$Q$s are the same, and $c_{21} = c_{22} = \omega_1^2$.

Frequency response

The effects of this mapping on the frequency response are interpreted in Figure 10.45, which has the $\omega\hat{\omega}$-plane of Figure 10.42 at its core. The continuous lobe at the centre replicates a distorted lowpass section, while the two outer half-lobes combine to give a distorted highpass section.

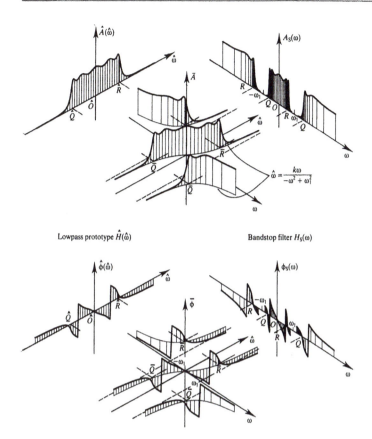

Lowpass prototype $\hat{H}(\hat{\omega})$

Bandstop filter $H_S(\omega)$

Figure 10.45 Frequency response of bandstop filter.

10.6.5   Recapitulation

We have now examined the effects of applying the simplest subsets of the mapping law 10.25 repeated here,

$$\hat{s} = f_x(s) = k\frac{s(s^2 + \omega_2^2)(s^2 + \omega_4^2) \ldots (s^2 + \omega_n^2)}{(s^2 + \omega_1^2)(s^2 + \omega_3^2) \ldots (s^2 + \omega_{n-1}^2)} \tag{10.55}$$

$$0 \leqslant \omega_1 < \omega_2 < \omega_3 < \omega_4 < \ldots \omega_{n-1} < \omega_n$$

In every case the imaginary axes map fully onto each other, their points being related by a similar expression

$$\hat{\omega} = \frac{1}{j}f_x(j\omega) = k\frac{\omega(-\omega^2 + \omega_2^2)(-\omega^2 + \omega_4^2) \ldots (-\omega^2 + \omega_n^2)}{(-\omega^2 + \omega_1^2)(-\omega^2 + \omega_3^2) \ldots (-\omega^2 + \omega_{n-1}^2)}$$

$$\tag{10.56}$$

The graphical interpretations of five of these expressions are collected in the later Figure 10.47. Note that the symmetrical roots of the numerator determine the zeros of these functions, and those of the denominator determine the poles. The single factor ω makes the symmetry odd, and the sign of k is such that the function gradient is positive. These conditions are necessary to preserve all the symmetries of $\hat{H}(\hat{s})$.

The principle for inverting the laws 10.55 is simple. Multiply both sides by the denominator, expand the products and gather terms in powers of s. This yields a polynomial, whose roots provide the required inversion. This is how we found the algebraic solutions for the four cases studied earlier. For more elaborate mapping laws the algebra is more involved and numerical techniques are more expedient.

If the applicable parameters $k, \omega_1, \ldots, \omega_n$ are given, then the law 10.55 with its subset 10.56 and the corresponding inversions provide the necessary tools to deal with all aspects of the conversion. The system function can be converted by evaluating the prototype as

$$H_x(s) = \hat{H}(\hat{s})\big|_{\hat{s}=f_x(s)}$$

Usually it is simpler, and conceptually clearer, to map the prototype poles and zeros, including those 'at infinity', by the inversion law

$$p_i = f_x^{-1}(\hat{p}_i) \qquad z_i = f_x^{-1}(\hat{z}_i)$$

and reconstitute $H_x(s)$ by pairing the resulting poles and zeros into realizable sections.

The frequency response $H_x(j\omega)$ can be similarly obtained, either by evaluating $\hat{H}(j\hat{\omega})$ according to Equation 10.56, or by interpreting $H_x(s)$ for $s = j\omega$. Graphical constructions involving the $\omega\hat{\omega}$-planes of Figure 10.47 (shown later) provide an intuitive guide.

In the conversion process the basic character of the frequency response is retained. For instance, for every Butterworth or Chebyshev lowpass prototype there is a set of related highpass, bandpass and other filter forms, which are identified with the same label.

Although we used lowpass prototypes for illustration, the above mapping laws are equally valid for converting other systems. For example, applying the mapping law $\hat{s} = k/s$ to a highpass filter would lead to a lowpass filter.

In certain system realizations, such as passive RC or RL networks, these mapping laws form the basis of synthesis techniques. The lowpass characteristic is implemented by a prototype network, whose components are then transformed by impedance transformations based on the same mapping laws, thus generating the corresponding highpass, bandpass and other networks.

Such realizations, or others involving active circuits, are outside the scope of this book. For a treatment sympathetic to poles and zeros the reader is referred to Van Valkenburg (1982).

10.6.6   Design process and parameters

So far we assumed that the mapping parameters k, $\omega_1, \ldots, \omega_n$ were known. We now show that finding the m unknown parameters reduces to formulating and solving a system of m linear equations.

Design process

A filter is specified by two kinds of data. We would typically specify an amplitude response tolerance envelope, such as the bandpass envelope $E(\omega)$ of the lower left corner of Figure 10.46, with the requirement that the frequency response is of a certain type, for instance Chebyshev.

The design process involves three steps, indicated by the solid arrows of Figure 10.46. The specified filter envelope $E(\omega)$ is first converted to an auxiliary lowpass prototype envelope $\hat{E}(\hat{\omega})$. The required type of lowpass filter is then fitted into this envelope, and the

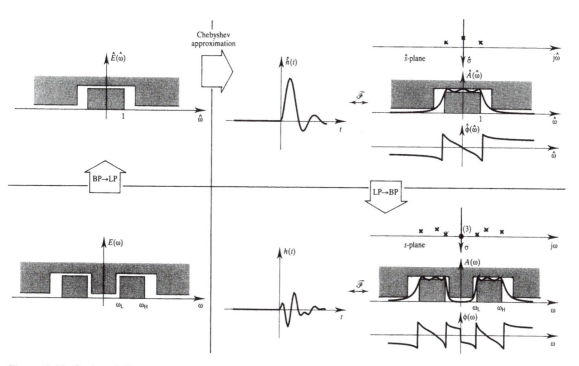

Figure 10.46   Design of Chebyshev bandpass filter.

resulting prototype $\hat{H}(j\hat{\omega})$ is finally back-converted to the original frequency variable.

The second step, prototype approximation, was the subject of Section 10.5. The type of the characteristic to be fitted forms part of the specification so that the problem reduces to finding the minimum number of prototype poles and zeros and their $\hat{s}$-plane locations. This approximation is independent of any subsequent frequency mapping.

The final step consists of frequency mapping the prototype $\hat{s}$-plane back to the target s-plane, according to the law $s = f_x^{-1}(\hat{s})$, as described earlier in this section.

This leaves the first step, converting $E(\omega)$ to $\hat{E}(\hat{\omega})$, which involves mapping the key frequencies of the specification according to the mapping law $\hat{s} = f_x(s)$. The two mapping processes are therefore complementary, one anticipating the subsequent frequency distortions of the other.

Design parameters

Finding the mapping parameters reduces to matching key points of the two imaginary frequency axes through the appropriate mapping law 10.56.

The number of points that can be independently matched equals the number m of unknown parameters. Each matched pair gives rise to a linear equation in m unknowns, and the solutions of the resulting system of m linear equations provide the required parameters. When m is small, the system can be solved algebraically. For larger m, this becomes unmanageable and calls for numerical methods.

We now build up typical systems of linear equations, for increasing m. The mapping laws are arranged in Figure 10.47, using the $\omega\hat{\omega}$-planes derived earlier. An ideal prototype is shown for reference on the left side of each diagram, with the derived ideal filter at the top. The normalized lower and upper band edges $\hat{Q}$ and $\hat{R}$ of the prototype are the key points to be matched. Their mapped locations, as well as those of the prototype origin $\hat{O}$, are indicated in each case.

Lowpass to lowpass

The mapping law in this case is the trivial relationship

$$\hat{\omega} = k\omega$$

with one unknown, k, so that we can choose only one point for matching. For example, the bandedge $\omega = \omega_c$, represented by R, maps onto the prototype at $\hat{R}$, whose position is $\hat{\omega} = 1$, so that

$$1 = k\omega_c \quad \text{hence} \quad k = \frac{1}{\omega_c}$$

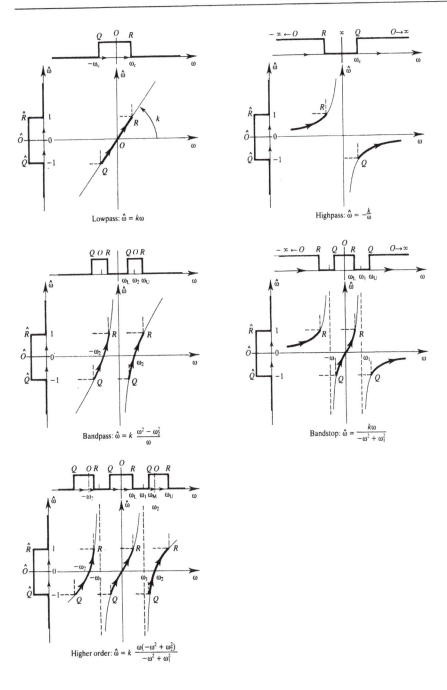

Lowpass: $\hat{\omega} = k\omega$

Highpass: $\hat{\omega} = -\dfrac{k}{\omega}$

Bandpass: $\hat{\omega} = k\,\dfrac{\omega^2 - \omega_2^2}{\omega}$

Bandstop: $\hat{\omega} = \dfrac{k\omega}{-\omega^2 + \omega_1^2}$

Higher order: $\hat{\omega} = k\,\dfrac{\omega(-\omega^2 + \omega_2^2)}{-\omega^2 + \omega_1^2}$

Figure 10.47 Determining the design parameters.

Lowpass to highpass

The reciprocal law

$$\hat{\omega} = -\frac{k}{\omega}$$

also has one unknown. The bandedge Q, of frequency $\omega = \omega_c$, maps onto the lower prototype bandedge $\hat{Q}$, located at $\hat{\omega} = -1$, hence

$$-1 = -\frac{k}{\omega_c} \quad \text{and} \quad k = \omega_c$$

Lowpass to bandpass

The mapping law 10.40

$$\hat{\omega} = k\frac{\omega^2 - \omega_2^2}{\omega}$$

has two unknowns, k and ω_2, and two points can be selected. For instance, Q and P of the positive half-axis map onto $\hat{Q}$ and $\hat{P}$ of the prototype. This is expressed by the system of linear equations

$$-1 = k\frac{\omega_L^2 - \omega_2^2}{\omega_L}$$

$$+1 = k\frac{\omega_U^2 - \omega_2^2}{\omega_U}$$

Making a change of variable, $1/k = x_1$ and $\omega_2^2 = x_2$, the above system takes the more familiar form

$$-\omega_L x_1 + x_2 = \omega_L^2$$

$$\omega_U x_1 + x_2 = \omega_U^2$$

Solving for x_1 and x_2 and changing back to the original variables yields the desired parameters

$$k = \frac{1}{\omega_U - \omega_L} \quad \text{and} \quad \omega_2^2 = \omega_L \omega_U \qquad (10.57)$$

Lowpass to bandstop

The mapping law is

$$\hat{\omega} = \frac{k\omega}{-\omega^2 + \omega_1^2}$$

which also has two unknown parameters, k and ω_1. Matching R and Q of the positive half of the ω-axis, we write

$$+1 = \frac{k\omega_L}{-\omega_L^2 + \omega_1^2}$$

$$-1 = \frac{k\omega_U}{-\omega_U^2 + \omega_1^2}$$

The change of variables $k = x_1$ and $\omega_1^2 = x_2$ gives

$$\omega_L x_1 - x_2 = -\omega_L^2$$

$$\omega_U x_1 + x_2 = \omega_U^2$$

which yields

$$k = \omega_U - \omega_L \quad \text{and} \quad \omega_1^2 = \omega_L \omega_U$$

Lowpass to higher forms

The algebra of the preceding cases was simple enough to seek the general algebraic solution of the system. For higher-order mapping the algebraic manipulations are more involved and call for numerical methods. We use this example to demonstrate how the procedure of matching arbitrary points leads to systems of equations that stay linear for higher-order mapping.

The law

$$\hat{\omega} = k\omega \frac{-\omega^2 + \omega_2^2}{-\omega^2 + \omega_1^2}$$

has three unknowns, k, ω_1 and ω_2. The point R of frequency $\omega = \omega_L$ maps onto $\hat{R}$ by the expression

$$+1 = k\omega_L \frac{-\omega_L^2 + \omega_2^2}{-\omega_L^2 + \omega_1^2}$$

which can be rearranged as

$$\omega_L^2 \frac{1}{k} + \omega_L \omega_2^2 - \frac{\omega_1^2}{k} = \omega_L^3$$

With the change of variables, $1/k = x_1$, $\omega_2^2 = x_2$ and $\omega_1^2/k = x_3$, we can write this equation, and similarly the other two equations of the system, in the form

$$\omega_L^2 x_1 + \omega_L x_2 - x_3 = \omega_L^3$$

$$-\omega_M^2 x_1 + \omega_M x_2 + x_3 = \omega_M^3$$

$$\omega_U^2 x_1 + \omega_U x_2 - x_3 = \omega_U^3$$

Again, this is a system of linear equations, which can be solved by standard numerical methods for x_1, x_2 and x_3. These can then be changed back to the required numerical values of k, ω_1 and ω_2.

This example illustrates the method for writing the equations, and obtaining the design parameters of more elaborate mapping laws. Attempting to match more points than the number of unknown parameters would be futile, as it would only lead to a system of equations that is indeterminate.

Example 10.2

We convert the 3rd-order lowpass Chebyshev filter of Example 10.1 into a bandpass filter with lower and upper bandedges of 1 kHz and 2 kHz. Mapping the prototype bandedges ± 1 rad/s to $\omega_L = 6283$ rad/s and $\omega_U = 12\,566$ rad/s yields the design parameters k and ω_2 as in Equations 10.57,

$$k = \frac{1}{\omega_U - \omega_L} = \frac{1 \text{ rad/s}}{6283 \text{ rad/s}} = 0.1592 \times 10^{-3}$$

and

$$\omega_2^2 = \omega_L \omega_U = 78.95 \times 10^6 \ (\text{rad/s})^2 \qquad \text{hence} \qquad \omega_2 = 8886 \text{ rad/s}$$

These values can be verified by mapping the bandedge frequencies, which are located on the imaginary axis, according to Equation 10.41.

The three poles of the prototype map by Equation 10.45 onto six poles as

$$\hat{p}_0 = -0.4942 \qquad \rightarrow \qquad \begin{cases} p_{01} = -1552.5 - j8749. \\ \\ p_{02} = -1552.5 + j8749. \end{cases}$$

$$\hat{p}_1 = -0.2471 + j0.9660 \qquad \rightarrow \qquad \begin{cases} p_{11} = -524.6 - j6326. \\ \\ p_{12} = -1027.9 + j12395. \end{cases}$$

$$\hat{p}_2 = -0.2471 - j0.9660 \qquad \rightarrow \qquad \begin{cases} p_{21} = -1027.9 - j12395. \\ \\ p_{22} = -524.6 + j6326. \end{cases}$$

while the three 'zeros at infinity' map onto three zeros at the origin plus three 'at infinity'.

Interpreting the prototype lowpass filter $\hat{H}(\hat{s})$ as a product of a first-order section of the form of Equation 10.43 and a second-order section of the form of Equation 10.46, the bandpass system function $H_B(s)$ can be compiled from the calculated poles and zeros as

$$H_B(s) = \frac{(0.4913/k^3)s^3}{(s - p_{01})(s - p_{02}) \times (s - p_{11})(s - p_{22}) \times (s - p_{21})(s - p_{12})}$$

$$(10.58)$$

whose poles have been grouped into three complex conjugate pairs.

Exercises

10.1 Graphically add all the characterizations of the linear-phase ideal lowpass and highpass filters of Figure 10.5. Verify that the resulting impulse and frequency responses form a Fourier pair. Similarly, add the bandpass and bandstop responses and check for consistency.

10.2 Adapt the graphics of Figure 10.7 to derive a finite length impulse response associated with a linear-phase bandpass characteristic. Could such a response be realized by means of the lumped parameter systems of Chapter 4?

10.3 Relate the characterizations of the band-limited differentiator and integrator of Figure 10.14 to those of the ideal filter of the same bandwidth in terms of appropriate Fourier transform properties.

10.4 Adapt the graphical derivations of Figure 10.13 for the case of the finite impulse response, linear phase, band-limited Hilbert transformer.

10.5 Discuss the inherent limitations of lumped parameter systems, described in Chapter 4 by rational polynomials $H(s)$, with regard to attaining a frequency response with exactly zero magnitude or exactly linear phase over any finite frequency range.

10.6 Discuss the usage and limitations of series RC networks to approximate the ideal differentiator and the ideal integrator.

10.7 Discuss the relationship between a specified polynomial family $F_N(\omega)$ and the resulting lowpass frequency response $H(j\omega)$, as expressed in

$$A(\omega) = |H(j\omega)| = \frac{1}{\sqrt{1 + \varepsilon^2 F_N^2(\omega)}}$$

and explain how these relate to the specification envelope $E(\omega)$.

10.8 Describe the process leading from a specified magnitude response, represented by a real function of real variable $A(\omega)$, to the poles and zeros of a stable system response $H(s)$, which is a complex function of complex variable.

10.9 Derive the system functions $H(s)$ of the 4th and 5th-order lowpass Butterworth filters. Find and plot their poles and sketch their frequency responses $H(j\omega)$.

10.10 Find the system function $H(s)$ and the poles of a 3rd-order Chebyshev lowpass filter with 0.5 dB bandpass ripple.

10.11 Map the poles and the 'zeros at infinity' of the Chebyshev response of Exercise 10.10 according to the following mapping laws,

(a) lowpass to lowpass, with $k = 10^{-3}$
(b) lowpass to highpass, with $k = 10^3$ $(\text{rad/s})^2$
(c) lowpass to bandpass, with $k = 10^{-3}$, $\omega_2 = 2 \times 10^3$ rad/s
(d) lowpass to bandstop, with $k = 10^3$ $(\text{rad/s})^2$, $\omega_1 = 2 \times 10^3$ rad/s

Use the mapped poles and zeros to build up the resulting system functions $H(s)$.

10.12 Show that the lowpass-to-highpass mapping law $\hat{s} = k/s$ converts a second-order lowpass notch filter into a second-order highpass notch filter, and vice versa. Sketch the conversion of the amplitude response $A(\omega)$ on the basis of Figure 10.34.

10.13 Find the applicable design parameters k, ω_1, ω_2, including their units, that convert the normalized lowpass bandwidth ± 1 rad/s to the following bandedge values

(a) lowpass to lowpass, for $\omega_c = 3 \times 10^6$ rad/s
(b) lowpass to highpass, for a bandedge of 10 kHz
(c) lowpass to bandpass, for $\omega_L = 1000$ rad/s, $\omega_U = 1200$ rad/s
(d) lowpass to bandstop, for $\omega_L = 3000$ rad/s, $\omega_U = 4500$ rad/s

CHAPTER 11

Discrete-time Systems and System Simulation

We now extend the approximations of the preceding chapter to discrete-time systems. Continuing advances in digital technology and design techniques are directed towards higher speeds, greater accuracy, higher reliability and lower costs, thus shifting the boundaries between cost-effective analogue and digital implementations.

Besides such directly comparable factors, digital realizations offer parameter stability and these parameters can be easily modified to change the system's characteristics. This flexibility permits new applications in speech, image and radar processing that were not feasible with analogue systems.

A practical limitation of digital implementations is the speed of operation of digital signal processors and of the associated analogue-to-digital converters. Signals with very wide bandwidths still require continuous-time systems.

To stress fundamental similarities between systems of different classes, we treat the discrete-time system as if embedded, suitably interfaced, in a larger continuous-time system. We therefore concentrate on **system simulation methods** that extend the results of Chapter 10. The vast body of knowledge accumulated over half a century of continuous-time system design can thus be transferred to the synthesis of discrete-time systems. This gives a better appreciation of signals and systems as a whole.

In Section 11.2 we present windowing techniques for non-recursive structures, which lead to realizable approximations of the linear-phase finite impulse responses of Sections 10.2 and 10.3.

The remainder of the chapter is devoted to recursive structures. In Section 10.3 we briefly examine so-called time domain methods. Thereafter we concentrate on frequency domain substitution methods.

In essence, these involve mapping laws that transfer s-plane features onto the z-plane, but distorting their locations by a process that is conceptually similar to the frequency mapping laws of Section 10.6.

In Section 11.4 we employ concepts from numerical analysis to introduce and organize the most familiar simulation methods. These are consolidated in Section 11.5 into a common mapping method, which reveals the full time domain and frequency domain implications of system simulation.

The bilinear mapping method, also called bilinear transformation or the bilinear z-transform method, is examined in detail in Section 11.6. We use it to illustrate the usage of mapping techniques and to investigate the role of simulation in the overall scheme of designing a discrete-time system. In Section 11.7 we bring together the various design steps and integrate these with the earlier developments of Chapter 10.

11.1   System approximation

In Section 10.4 we interpreted the system approximation problem in terms of finding a realizable, lumped parameter, continuous-time system that fits the specification envelope. We now widen that interpretation by including approximations realized with discrete-time systems or sampled-time systems.

Consider a continuous-time system specification, which for a given input signal $x(t)$ produces a given output signal $y(t)$, within a specified tolerance envelope $\varepsilon(t)$, as indicated at the top of Figure 11.1.

In Chapter 10 we approximated the specification by a continuous-time, lumped parameter system $h(t)$, whose output $y_C(t)$ satisfied the tolerance envelope $\varepsilon(t)$. But if a sampled-time system $h_T(t)$ can be found, whose output $y_T(t)$ satisfies $\varepsilon(t)$, this too would represent a valid approximation. Similarly, a discrete-time system $h_d[nT]$, which, together with the necessary signal conditioning units, produces an output $y_D(t)$ that falls within $\varepsilon(t)$, would also be valid. These three choices are indicated in Figure 11.1.

Thus, although the specified process is of continuous time, the actual processing could be performed by a continuous-time, or a sampled-time or a discrete-time system. Such attitude keeps track of the continuous-time objectives of the overall system.

This chapter addresses the discrete-time alternative in depth, treating it as a **system simulation** problem in which a real or hypothetical continuous-time system is replaced by an equivalent discrete-time system.

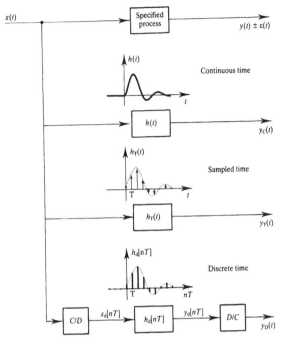

Figure 11.1   Three approximations to a process requirement.

11.1.1   Discrete-time simulation of systems

A basic schematic of the overall simulation problem is indicated in Figure 11.2. Given a continuous-time system $h(t)$, with input $x(t)$ and output $y(t)$, we wish to replace it by an equivalent discrete-time system $h_d[nT]$. Ideally, when the input $x_d[nT]$ is a sampled version of $x(t)$, the output $y_d[nT]$ should be a similar sampled version of $y(t)$. In practice there is always a simulation error $e(t)$, and the reconstructed signal $y_D(t)$ can never be indentical to $y(t)$.

The overall simulation process is enclosed in the box of Figure 11.2. It consists principally of the discrete-time simulation of the system $h(t)$, but also involves the conversion of the input and output signals from continuous to discrete time, and vice versa, and the necessary conditioning.

For brevity we worded the simulation problem as if it were a time domain process. But all the processes invoked have counterparts in the frequency domain, where the simulation is usually formulated. To put

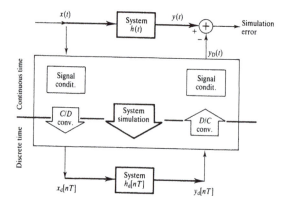

Figure 11.2   Basic schematic of overall simulation.

these aspects into perspective, we view each element of the basic schematic in the equivalent representations of Figure 11.3. The full continuous-time process is enclosed in the upper box, where domains are related by the Fourier and Laplace transforms. Similarly, the full discrete-time process is enclosed in the lower box, where the discrete-time Fourier transform and the z-transform apply.

The class-conversion of the input signal could follow any of the three representations. But we would normally interpret the signal conditioning operation represented by the anti-aliasing filter as a frequency domain process, while the subsequent time-sampling operation would be interpreted as a time domain process, as in Chapter 6.

This leaves the core problem of simulating the system itself. One possible approach is to convert the impulse response $h(t)$ to discrete time, using the same concepts and sampler $s_T[nT]$ as with the input signal. This is intuitively simple and, as we will see later, constitutes the **time-sampling method** of simulation.

Other methods are formulated in the frequency domain. Our immediate objective is to present the most common simulation methods, examining the concepts they embody in both time and frequency. But equally important is our aim of keeping track of the fundamental relationships between continuous- and discrete-time representations, to gain insight into the merits of the different simulation strategies.

A system is simulated for two main purposes, either to replace a continuous-time system by a more effective discrete-time system, or as an analytical design aid, to verify the performance of a continuous-time system design. The simulation methods presented in this chapter apply to both.

11.1.2 Primitive approximations of continuous-time systems

Before introducing formal approximation methods, we establish an intuitive correspondence between some simple discrete-time systems and possible continuous-time processing functions.

Although the frequency response of a discrete-time system extends periodically to infinite frequency, its essential properties are fully described in the fundamental cycle $-\frac{1}{2}\omega_s < \omega < \frac{1}{2}\omega_s$, where $\omega_s = 2\pi/T$ is the sampling frequency. When such a system is used to simulate a continuous-time system, as in Figure 11.3, its fundamental cycle determines the upper bandwidth of admissible continuous-time input signals, thereby establishing a common frequency range for meaningful comparisons.

In this context the comparison of lowpass or bandpass filters is self-evident. In contrast, a discrete-time filter is said to have a highpass characteristic when it does not distort the high frequency components of the fundamental cycle, that is, those in the vicinity of $\pm\frac{1}{2}\omega_s$. Comparisons can then be made with band-limited continuous-time highpass systems. Similarly for the operators of Section 10.3, in their band-limited forms.

To avoid aliasing problems, the overall system simulation must ensure that the input signal is suitably conditioned, in that all frequency components higher than half-sampling frequency are removed before the signal is applied.

We illustrate with some crude approximations attainable with two first-order non-recursive systems.

Example 11.1

The responses of the system shown in the upper half of Figure 11.4 are given in Section 8.2.1 as

$$h_1[nT] = \delta[nT] + \delta[nT - T] \quad \xrightarrow{\;\mathscr{Z}\;} \quad \bar{H}_1(z) = 1 + z^{-1} = \frac{z + 1}{z}$$

Evaluating $\bar{H}_1(z)$ on the unit circle we express the frequency response as

$$\bar{H}_1(e^{j\omega T}) = 1 + e^{-j\omega T} = e^{-j\omega T/2}(e^{j\omega T/2} + e^{-j\omega T/2})$$

$$= 2e^{-j\omega T/2} \cos \tfrac{1}{2}\omega T$$

so that the magnitude and phase of the fundamental cycle are

$$A_1(\omega) = 2|\cos \tfrac{1}{2}\omega T| \quad \text{and} \quad \phi_1(\omega) = -\tfrac{1}{2}\omega T$$

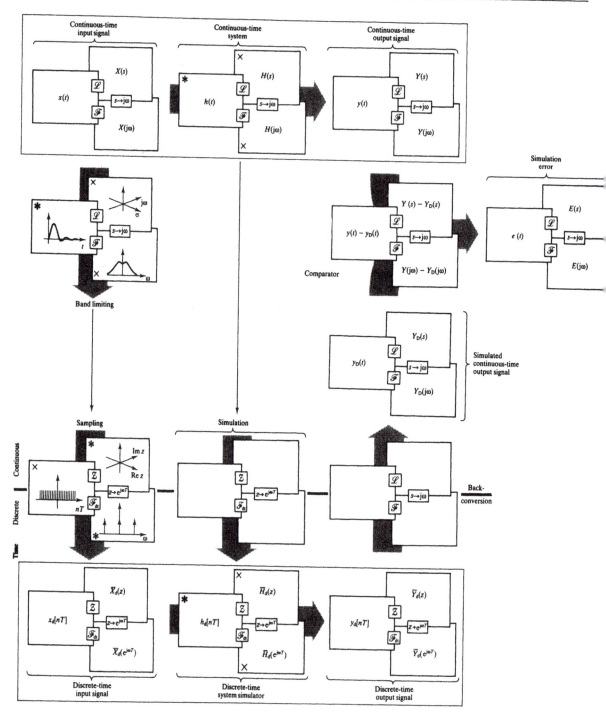

Figure 11.3   Discrete-time simulation of continuous-time system.

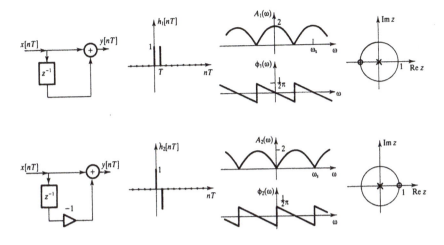

Figure 11.4   Primitive approximation of continuous-time functions.

and these are repeated periodically as shown in Figure 11.4. The null at half sampling frequency attenuates high frequency components, while in the immediate vicinity of the origin the magnitude is approximately constant with zero phase. This is a crude approximation of a lowpass filter.

Example 11.2

The system of the lower half of Figure 11.4 has the responses

$$h_2[nT] = \delta[nT] - \delta[nT - T] \quad \xrightarrow{\mathscr{Z}} \quad \bar{H}_2(z) = 1 - z^{-1} = \frac{z - 1}{z}$$

where $\bar{H}_2(z)$ leads to the frequency response

$$\bar{H}_2(e^{j\omega T}) = 1 - e^{-j\omega T} = e^{-j\omega T/2}(e^{-j\omega T/2} - e^{-j\omega T/2})$$

$$= 2j e^{-j\omega T/2} \sin \tfrac{1}{2}\omega T$$

The amplitude and phase of the fundamental cycle are

$$A_2(\omega) = 2|\sin \tfrac{1}{2}\omega T|$$

and

$$\phi_2(\omega) = \begin{cases} \tfrac{1}{2}\pi - \tfrac{1}{2}\omega T & 0 < \omega < \tfrac{1}{2}\omega_s \\ -\tfrac{1}{2}\pi - \tfrac{1}{2}\omega T & -\tfrac{1}{2}\omega_s < \omega < 0 \end{cases}$$

which are repeated periodically as shown in Figure 11.4.

This system can be given two interpretations. Firstly, by its behaviour at high frequencies, where the amplitude is approximately constant with zero phase in the vicinity of half sampling frequency while attenuating low frequencies, it can be considered a highpass filter. On the other hand, its low frequency behaviour is comparable to that of the continuous-time approximation of the differentiator given in Figure 10.21. For low frequencies it can therefore also be considered a primitive approximation of a differentiator.

The recursive first-order and second-order systems of Sections 8.2 and 8.3 can be interpreted as simple lowpass, highpass or bandpass filters, by analogy to the continuous-time interpretations of Section 10.4.3, as for instance in Figure 10.18. Such analogies will emerge in later sections of this chapter in the context of system simulation, as a consequence of some simple discrete-time approximations of differentiators and integrators.

11.1.3 Simulation methods and numerical optimization

Different system structures lend themselves to different simulation techniques. We will examine the so-called windowing method for finite impulse response systems and concentrate mainly on frequency mapping techniques for infinite impulse response systems. This choice was made on pedagogic grounds, to bring out fundamental relationships between continuous-time and discrete-time systems and their representations.

But it must be understood that in practice much of the design work is performed with the aid of highly effective numerical methods. We will only give a brief outline of such methods.

Numerical optimization methods

The system is usually specified by the frequency response amplitude $A(\omega)$ over a given frequency range. A structure is assumed for realization, including a guess of the order of the system. Some parameters x_i are selected for variation, typically polynomial coefficients or pole-zero locations, and an expression is written in terms of these for the frequency response of the assumed system, in the form $\bar{A}(\omega; x_1, x_2, \ldots, x_k)$.

The parameters x_i are then modified in some controlled fashion, with the objective of minimizing an error function

$$\varepsilon = A(\omega) - \bar{A}(\omega; x_1, x_2, \ldots, x_k)$$

over the frequency interval to be approximated, according to some criterion deemed to be optimal. Methods differ in the choice of optimality criterion, which is typically formulated in terms of a least squares approximation, or of maximum deviation or flatness.

Unfortunately, numerical expediency is gained at the cost of conceptual clarity regarding signal and system fundamentals.

11.2 Non-recursive systems

A non-recursive structure only has forward loops. These lead to a system function $\bar{H}(z)$ that is an 'all-numerator' function in z^{-1} and an impulse response $h[nT]$ of finite length. Furthermore, the coefficients $h[nT]$ of the impulse response are also the coefficients of the system function $\bar{H}(z)$ and are the same as the block diagram multiplier coefficients b_n.

This direct relationship makes this structure ideal for introducing the concept of discrete-time simulation of continuous-time systems. In Section 11.2.2 we present one of the most remarkable features of non-recursive systems, their ability to actually implement in discrete time the ideal linear-phase systems derived earlier in Chapter 10.

11.2.1 Reappraisal of non-recursive structure

We first examine the general Nth-order non-recursive structure of Figure 11.5, with a view to simulation. It represents the input half of the direct form of Figure 8.10, which was identified there with the system function $\bar{H}_b(z)$.

Impulse response

Starting with the system at rest, that is, with the outputs from all N delay elements set to zero,

$$x[nT - mT] = 0 \qquad 1 \leqslant m \leqslant N$$

we clock a discrete-time impulse function $\delta[nT]$ through the system. The signal arriving at the system's input is the sequence

$$\{x[nT]\} = \ldots 0, 0, 1, 0, 0, 0, \ldots$$
$$\uparrow$$

where the unit value $x[0] = 1$ arrives at time $nT = 0$, identified with an

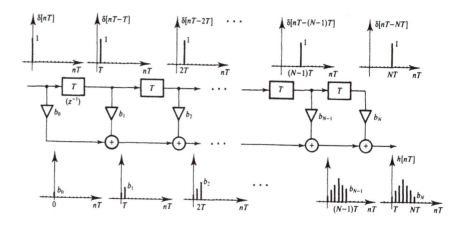

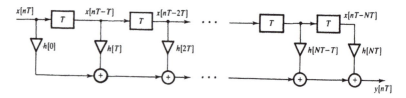

Figure 11.5   Non-recursive system, link to impulse response.

arrow. Observing the outputs of the delay elements T, these are the progressively delayed impulses $\delta[nT - mT]$ indicated above each delay output in the upper half of Figure 11.5.

The delayed impulses become the inputs of the multipliers b_m, whose outputs, before summing, are $b_m \delta[nT - mT]$. The signals emerging from the summing points represent the cumulative output of all preceding multipliers, as indicated in Figure 11.5. The overall system output $y[nT]$ is therefore the sequence

$$\{y[nT]\} = \ldots, 0, 0, b_0, b_1, b_2, \ldots, b_{N-1}, b_N, 0, 0, 0, \ldots$$
$$\uparrow$$

whose first non-zero value b_0 occurs at time $nT = 0$, when the unit value of the impulse sequence $\{\delta[nT]\}$ arrives at the system's input.

Since the input signal was the discrete-time impulse $x[nT] = \delta[nT]$, the output signal is, by definition, the system's impulse response, that is,

$$x[nT] = \delta[nT] \xrightarrow{h[nT]} y[nT] = h[nT]$$

Interpreting the output sequence $\{y[nT]\}$ in terms of the scaled and delayed impulses yields the finite sum

$$h[nT] = b_0\delta[nT] + b_1\delta[nT - T] + b_2\delta[nT - 2T]$$
$$+ \ldots + b_N\delta[nT - NT]$$
$$= \sum_{m=0}^{N} b_m\delta[nT - mT] \tag{11.1}$$

System function

The z-transform of the impulse response $h[nT]$ is the system function

$$\bar{H}(z) = b_0 + b_1 z^{-1} + b_2 z^{-2} + \ldots + b_{N-1} z^{-(N-1)} + b_N z^{-N}$$
$$= \sum_{m=0}^{N} b_m z^{-m} \tag{11.2}$$

A term by term comparison with the generic form of Equation 7.14 of the z-transform of an arbitrary function $h[mT]$

$$\bar{H}(z) = \sum_{m=-\infty}^{\infty} h[mT]z^{-m}$$
$$= \ldots + h[0] + h[T]z^{-1} + h[2T]z^{-2}$$
$$+ \ldots + h[mT]z^{-m} + \ldots$$

yields the correspondence

$$h[mT] = \begin{cases} b_m & 0 \leqslant m \leqslant N \\ 0 & \text{elsewhere} \end{cases}$$

This means that the multiplier coefficients b_m of the upper half of Figure 11.5 represent the coefficients of the system function $\bar{H}(z)$ in Equation 11.2 as well as the values of the impulse response $h[mT]$ in Equation 11.1. This identity is reflected in the lower half of Figure 11.5.

System poles and zeros

Multiplying and dividing Equation 11.2 by z^N gives an explicit function in z,

$$\bar{H}(z) = \frac{b_0 z^N + b_1 z^{N-1} + \ldots + b_{N-1}z + b_N}{z^N} = \frac{1}{z^N}\sum_{m=0}^{N} b_m z^{N-m}$$
$$\tag{11.3}$$

which identifies N poles π_i at the z-plane origin and N zeros ζ_i determined by the roots of the numerator polynomial.

Frequency response

Evaluating the system response of Equation 11.2 on the z-plane's unit circle $z = e^{j\omega T}$ extends the identity of multiplier coefficients and impulse response values to the system's frequency response

$$\bar{H}(e^{j\omega T}) = \sum_{m=0}^{N} b_m e^{-j\omega mT}$$

$$= b_0 + b_1 e^{-j\omega T} + b_2 e^{-j\omega 2T} + \ldots + b_{N-1} e^{-j\omega(N-1)T}$$

$$+ b_N e^{-j\omega NT} \tag{11.4}$$

Impulse response method

Interpreting the operations of the lower half of Figure 11.5 for an arbitrary input $x[nT]$ gives the output as the sum

$$y[nT] = h[0]x[nT] + h[T]x[nT - T] + h[2T]x[nT - 2T]$$

$$+ \ldots + h[NT - T]x[nT - (N - 1)T]$$

$$+ h[NT]x[nT - NT]$$

$$= \sum_{m=0}^{N} h[mT]x[nT - mT]$$

This represents the discrete-time convolution

$$y[nT] = x[nT] * h[nT]$$

which simply confirms the validity of the impulse response method of expression 9.4 for this system structure.

Implications for system simulation

These responses have many desirable characteristics in regard to system simulation. Firstly, the impulse response $h[nT]$ is zero for negative time $nT < 0$, so that the system is always causal. Secondly, the length of the impulse response is limited by the number of delay elements to $N + 1$ values. The output is bounded in amplitude and dies out N delays after the last non-zero value of the input signal is received, making the system inherently stable. Thirdly, the system is eminently realizable. It can be implemented directly in hardware or software, as indicated by the block diagram of Figure 11.5.

These properties suggest a simple means for deriving a non-recursive discrete-time approximation from a known continuous-time approximation of the types seen in Chapter 10. It consists, in essence, of sampling the impulse response $h(t)$ of the given system, and assigning the consecutive sample values $h[nT]$ to the coefficients b_n of the block diagram.

The number of delay elements N of the discrete system, although allowed to be large, must be finite. The impulse response $h(t)$ must therefore be suitably windowed either before or after sampling.

Such a process could be applied to arbitrary causal systems. But one of the main attractions of non-recursive systems is that they make it possible to obtain a desired amplitude response **and** linear phase, as we presently show.

11.2.2   Non-recursive systems with linear phase

In Sections 10.2.3 and 10.3.2 we got within sight of realizing linear-phase finite impulse response filters and differentiators. We only need to sample the resulting impulse responses and assign the discrete-time samples $h_d[nT]$ to the multiplier coefficients b_n of Figure 11.5 to realize those systems.

Even-symmetric filters

We repeat the construction of Figure 10.7 in Figure 11.6, with small changes to emphasize discrete-time properties. Thus, to simulate the ideal lowpass filter, we truncate its impulse response $h_L(t)$ with the rectangular window $w(t)$, of width 2τ, which introduces ripples at bandedge transitions of the frequency domain.

We next take discrete-time samples of the windowed impulse response $h_w(t)$. For brevity we indicate this as a multiplication with the discrete-time sampler $s_T[nT]$. This yields a symmetric and acausal impulse response $h_s[nT]$, comprising $N + 1$ samples, where N is related to the window width as $NT = 2\tau$. The corresponding frequency convolution with $\tilde{S}_T(j\omega)$ replicates $H_w(j\omega)$ at multiples of the sampling frequency interval $\omega_s = 2\pi/T$, causing some aliasing. Up to here all functions are real and even, with zero phase.

To make $h_s[nT]$ causal we shift it to $\frac{1}{2}NT = \tau$, by multiplying its frequency representation by $e^{-j\omega NT/2}$, thus transferring the linear phase of the exponential to $\tilde{H}_d(j\omega)$, as expressed in the Fourier pair

$$h_d[nT] = h_s[nT - \tfrac{1}{2}NT] \overset{\mathcal{F}_{dt}}{\longleftrightarrow} \tilde{H}_d(j\omega) = e^{-j\omega NT/2}\,\tilde{H}_s(j\omega)$$

with the result shown at the bottom of Figure 11.6.

For continuity with Figure 10.7 we also show the complex representation of $\tilde{H}_d(j\omega)$, which suggests the process of winding back its phase. The relationship between domains is such that one full turn of

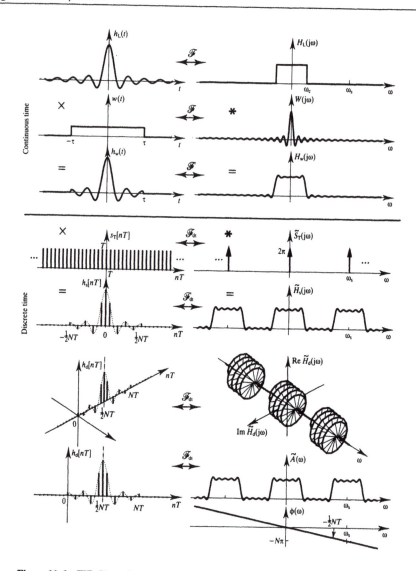

Figure 11.6 FIR filter, linear phase.

the frequency component located at the sampling frequency ω_s advances the time function by one sampling interval T. It thus takes $\frac{1}{2}N$ full revolutions, or a phase value $\phi(\omega_s) = -N\pi$ to make $h_d[nT]$ causal.

To implement the discrete-time filter it is sufficient to allocate the values $h_d[nT]$ to the multipliers b_n of the Nth-order non-recursive structure of Figure 11.5.

Similar results may be obtained for other ideal filters of Section 10.2, to yield characteristic impulse responses

$$h[nT] = 0 \begin{cases} n < 0 \\ n > N \end{cases}$$

that are even-symmetric about the midpoint $t = \tau = \frac{1}{2}NT$, expressed as

$$h[nT] = h[NT - nT]$$

These results are formally similar to those of the continuous-time counterparts of Section 10.2.3, and could be interpreted in terms of sampling those continuous-time results. The reasons for finite length, symmetry about the mid-point and phase linearity of both cases are the same.

As in continuous-time, the width and shape of the time window determine the distortions from the ideal band-limited response. A wider time window would involve a larger number N of samples, hence a better approximation of the ideal frequency response. A smoother window would reduce the ripple content, but at the same time give less clearly defined transitions.

Odd-symmetric operators

Processing the odd-symmetric band-limited impulse responses of Section 10.3 in a similar fashion leads to finite impulse responses $h_d[nT]$ that are odd-symmetric about the mid-sample $h_d[\frac{1}{2}NT]$.

We use the differentiator $d(t)$ of Section 10.3.2 for illustration. The band-limited ideal differentiator $d_b(t)$ is derived at the top of Figure 11.7 from the ideal lowpass filter $h_L(t)$ to provide a familiar reference point for the location of the sampling interval T in relation to the bandwidth ω_c and the sampling frequency $\omega_s = 2\omega_c = 2\pi/T$.

Thus, sampling $d_b(t)$ at intervals T replicates $D_b(j\omega)$ periodically at integers of the sampling frequency ω_s to give $\tilde{D}_T(j\omega)$, as shown in the figure. A subsequent windowing of $d_T[nT]$ smears the sharp discontinuities of the frequency domain, introducing the familiar ripples when the window is rectangular. Adding linear phase, such that $\phi(\omega_s) = -N\pi$ shifts $d_s[nT]$ to yield the causal response $d_d[nT]$ shown in the bottom row of Figure 11.7.

The resulting response is a discrete-time version of that obtained in the bottom right corner of Figure 10.13. The same process, applied to the other odd-symmetric band-limited operators, such as the Hilbert transformer of Figure 10.14, yields similar discrete-time versions of the corresponding linear phase operators. They are all odd-symmetric about the mid-sample at $t = \tau = \frac{1}{2}NT$, expressed as

$$h[nT] = -h[NT - nT]$$

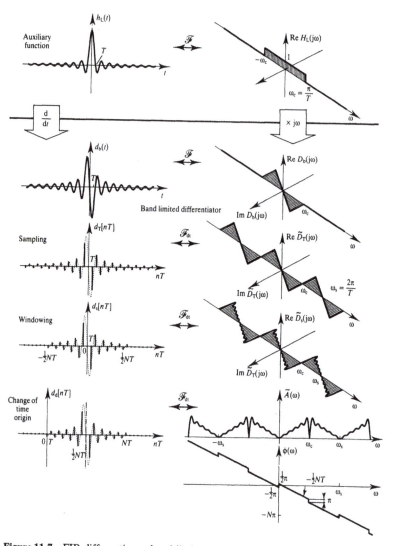

Figure 11.7 FIR differentiator, band limited, linear phase.

11.2.3 *z*-plane symmetries

A real-valued impulse response $h_d[nT]$ with mid-point symmetry imposes rigid symmetry constraints on the system function $\bar{H}_d(z)$ of a linear-phase system, and thereby on the *z*-plane locations of its poles and zeros. Time-shifting the impulse response by m samples simply adds or cancels m poles at the origin, without affecting the locations of the zeros. We can therefore examine the latter in the context of the applicable symmetric function, such as $h_s[nT]$ of Figure 11.6 or $d_s[nT]$ of Figure 11.7.

Given an arbitrary z-transform pair

$$f[nT] \quad \overset{\mathscr{Z}}{\longleftrightarrow} \quad \bar{F}(z)$$

as a result of the time-scaling property of expression 7.37 the time-reversed function $f[-nT]$ transforms as

$$f[-nT] \quad \overset{\mathscr{Z}}{\longleftrightarrow} \quad \bar{F}(z^{-1})$$

Applied to an **even-symmetric** impulse response this yields the pair

$$h_s[nT] = h_s[-nT] \quad \overset{\mathscr{Z}}{\longleftrightarrow} \quad \bar{H}_s(z) = \bar{H}_s(z^{-1})$$

A zero ζ of $\bar{H}_s(z)$ must therefore also be a zero of $\bar{H}_s(z^{-1})$, which implies that both functions must have a zero at $\bar{\zeta} = 1/\zeta$, as shown in the top row of Figure 11.8. This represents an inverse symmetry, or 'reflection about the unit circle', similar to that seen in Section 10.6 for inverse mapping.

The z-plane zeros of a linear-phase system thus exhibit a double symmetry. They need the usual conjugate symmetry about the real axis,

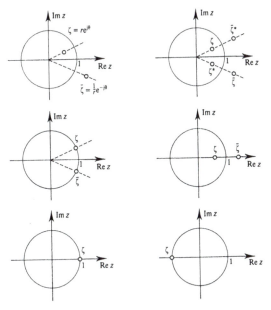

Figure 11.8 Locations of linear-phase system zeros.

to make the impulse response real-valued, and a reflection about the unit circle, for linear phase. An arbitrarily located zero must therefore belong to a set of four, as shown in the top row of Figure 11.8. If a zero is located on either the unit circle or on the real axis, it is self-symmetric with respect to one of these and only needs to belong to a pair, as shown in the middle row of the figure. Finally, a single zero can only be located at an intersection of the real axis and the unit circle, that is at the points $z = 1$ or $z = -1$, as shown in the bottom row of Figure 11.8.

Similarly, for an **odd-symmetric** impulse response, such as $d_s[nT]$ of Figure 11.7, we have

$$d_s[nT] = -d_s[-nT] \quad \xleftrightarrow{\mathscr{Z}} \quad \bar{D}_s(z) = -\bar{D}_s(z^{-1})$$

But any two functions $\bar{F}(z)$ and $-\bar{F}(z)$ only differ in an amplitude scaling factor, so that their pole and zero locations are identical. This leads to the same situation regarding z-plane symmetries as in the case of even-symmetric functions and to identical conclusions.

11.2.4 Odd-N delays

In the preceding developments the number N of delay elements was even. This gave impulse responses with an odd number $(N + 1)$ of samples that allowed symmetry about the mid-sample. The same process is applicable when N is odd, but, since the symmetry point falls half way between two samples, its interpretation requires a slight adjustment involving the concept of half-sample shift. This device was introduced in the context of the fast Fourier transform for shifting the time domain envelope to make components of odd index fit the $\frac{1}{2}$N-point transform (see appendix to Chapter 6).

Using the same concept we shift the time-sampler $s_T[nT]$ of Figure 11.6 by half a sampling interval to give the shifted sampler $s_T[nT + \frac{1}{2}T]$ of Figure 11.9. Its frequency representation $e^{j\omega T/2}\tilde{S}_T(j\omega)$ is essentially the impulse train $\tilde{S}_T(j\omega)$ but with alternating impulses reversed as shown.

Applied to a symmetric impulse response $h_L(t)$ this yields an auxiliary function $h_s[nT + \frac{1}{2}T]$ whose samples are symmetric about the mid-point between adjacent samples. When combined with a symmetric window this gives the desired symmetry for an even number of samples $N + 1$.

The time domain multiplication required for sampling implies a convolution in frequency that produces alternating positive and negative replicas of the real even function $H_L(j\omega)$, see Figure 11.9, whose sum $\tilde{H}(j\omega)$ is a real even function with period $2\omega_s$. The overall effect is as if

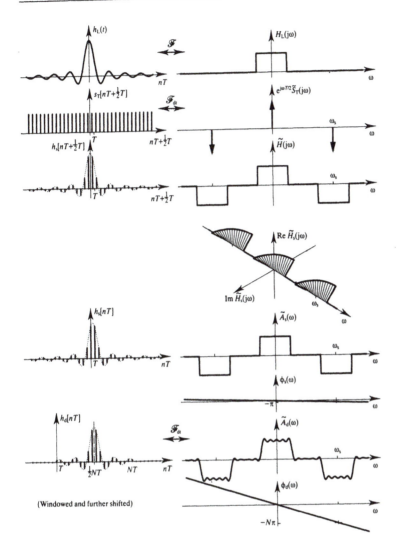

Figure 11.9 Odd-N delays.

the function $h_L(t)$ had been sampled at twice the sampling frequency and all even samples were set to zero.

To return to the usual time variable nT we shift the function $h_s[nT + \frac{1}{2}T]$ back by half a sampling interval, giving $h_s[nT]$. The corresponding frequency domain multiplication by $e^{-j\omega T/2}$ counteracts the earlier effect of shifting the sampler and yields a complex function $\tilde{H}_s(j\omega)$ with the usual period ω_s, as shown in the three-dimensional diagram of Figure 11.9.

To stress phase linearity it is expedient to interpret the amplitude $\tilde{A}_s(\omega)$ of $\tilde{H}_s(j\omega)$ in the sense of Section 10.2.3, that is, as a real function with positive and negative values. This makes the shape of $\tilde{A}_s(\omega)$ identical to that of the real function $\tilde{H}(j\omega)$, with period $2\omega_s$. The corresponding phase $\phi_s(\omega)$ is then represented by an unbroken straight line, whose value at $\omega = \omega_s$ is $-\pi$, which is consistent with the three-dimensional diagram.

From here the interpretation of windowing and further time-shifting is the same as for systems with even N. Windowing introduces ripples at the discontinuities of $\tilde{A}_s(\omega)$ while the additional $\frac{1}{2}(N-1)$ turns required to reach the lower edge of the window gives the linear phase characteristic shown at the bottom of Figure 11.9.

11.3 Recursive systems, time domain methods

For the remainder of this chapter we look at approximations involving recursive systems. With few exceptions, their impulse responses are of infinite length.

We concentrate on methods that yield a discrete-time approximation $\bar{H}_d(z)$ by simulating a continuous-time reference system $H(s)$, known to fit the design specification, which was itself approximated by the methods of Chapter 10. Both these system functions are rational polynomials with real coefficients that represent realizable, causal and stable systems. We seek methods for converting one to the other.

In this section we present 'time domain methods', a description that refers to formulation rather than implementation. They simulate a time domain representation of the continuous-time reference system, usually the impulse response $h(t)$, sometimes the step response $r(t)$, by discrete-time samples.

11.3.1 Time-sampling method

The impulse response $h(t)$ of a continuous-time system is basically a signal, and as such can be time-sampled as indicated in the upper half of Figure 11.10.

This suggests a direct method for discrete-time simulation, namely, sample $h(t)$ with the same discrete-time sampler $s_T[nT]$ used to derive $x_d[nT]$ from $x(t)$, see the lower half of Figure 11.10, and implement a discrete-time system with the resulting function $h_d[nT]$ as its impulse response.

We formulate the simulation in terms of the corresponding impulse response methods of Sections 5.2.3 and 9.2.2. Interpreting the

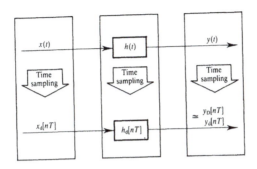

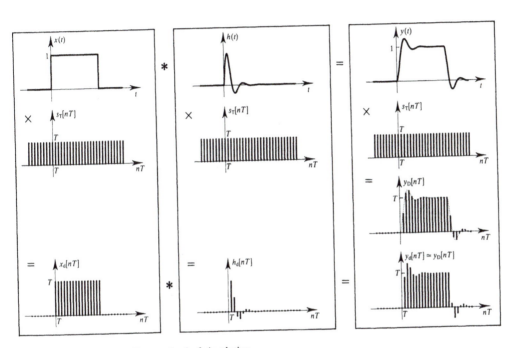

Figure 11.10   Time-sampling method of simulation.

continuous-time output $y(t)$ as the limit of Equation 5.6 evaluated for $t = nT$, with m as the dummy summation variable, we multiply both sides by T and compare the result with the output $y_d[nT]$ expressed in Equation 9.3 as a discrete-time convolution

$$Ty(nT) = \lim_{T \to 0} \sum_{m=-\infty}^{\infty} Tx(mT) \; Th(nT - mT)$$

$$\downarrow \qquad\qquad \downarrow \qquad\qquad \downarrow$$

$$y_d[nT] \qquad = \sum_{m=-\infty}^{\infty} x_d[mT] \; h_d[nT - mT]$$

This implies the sampling conversions $Tx(nT) = x_d[nT]$ and $Th(nT) = h_d[nT]$ on the right side.

A perfect simulation would make the output $y_d[nT]$ of the simulated system identical to samples $y_D[nT] = Ty(nT)$ of the continuous output $y(t)$, as indicated in the lower half of Figure 11.10. Clearly, such identity could only be achieved in the limit, when $T \to 0$. But for a sufficiently small sampling interval T and a suitably band-limited input function $x(t)$ the approximation error can be made negligible, that is, $y_d[nT] - y_D[nT] < e[nT]$.

11.3.2 Frequency implications

Although we formulated this method in the time domain, in practice the continuous-time reference system is specified by the system function $H(s)$. Expanding this function in partial fractions provides the frequency domain background for interpreting the time-sampling method and leads directly to the simulated function $\bar{H}_d(z)$.

The simulation strategy is summarized in Figure 11.11. We assume that $H(s)$ is a proper fraction with simple poles (the method can be extended to include multiple poles) of the form of Equation 4.69,

$$H(s) = \frac{b_{N-1}s^{N-1} + b_{N-2}s^{N-2} + \ldots + b_1 s + b_0}{(s - p_1)(s - p_2) \ldots (s - p_N)} \tag{11.5}$$

and analyse it by partitioning into a parallel connection of first-order sections $T_i(s)$, as expressed by the partial fraction expansion of Equation 4.70, summarized as

$$H(s) = \sum_{i=1}^{N} \frac{c_i}{(s - p_i)} \tag{11.6}$$

The terms of this sum have the form of the Laplace pairs 3.11, so that the impulse response $h(t)$ becomes

$$h(t) = \sum_{i=1}^{N} c_i e^{p_i t} u(t) \tag{11.7}$$

Sampling this response with the sampler $s_T[nT]$ implies transferring the numerical values of the samples $h(nT)$ to corresponding time values of the discrete-time impulse response by the conversion $h_d[nT] = Th(nT)$, yielding

$$h_d[nT] = T \sum_{i=1}^{N} c_i e^{p_i nT} u[nT] \tag{11.8}$$

Interpreting the exponential as $e^{p_i nT} = a^n$, the terms of the impulse

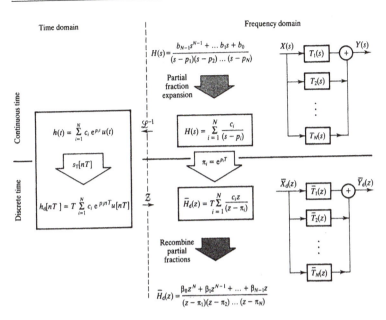

Figure 11.11 Frequency implications of time-sampling method.

response of Equation 11.8 have the form of the z-transform pair in expression 7.27, so that the simulated system function $\overline{H}_d(z)$ becomes

$$\overline{H}_d(z) = T\sum_{i=1}^{N}\frac{c_iz}{z - e^{p_iT}} = T\sum_{i=1}^{N}\frac{c_iz}{(z - \pi_i)} \qquad (11.9)$$

This sum also represents a parallel connection of first-order systems $\overline{T}_i(z)$, as shown in Figure 11.11, and it too can be interpreted as the result of a partial fraction expansion of a function of the form

$$\overline{H}_d(z) = \frac{\beta_0z^N + \beta_1z^{N-1} + \ldots + \beta_{N-1}z}{(z - \pi_1)(z - \pi_2) \ldots (z - \pi_N)} \qquad (11.10)$$

The intermediate results of Equations 11.6 and 11.9 show a direct link between the poles of Equations 11.5 and 11.10. Each reference pole p_i is simply mapped to a corresponding simulated pole π_i as $\pi_i = e^{p_iT}$, which is similar to the standard $\tilde{s}$-plane to z-plane mapping law $z = e^{sT}$, but not fully reversible.

The link between zeros is more involved. Their z-plane positions ζ_i depend on the factors c_i, hence on the zeros z_i of $H(s)$, as well as on all the pole positions π_i, hence on the poles p_i. They cannot be mapped directly.

In the schematic of Figure 11.11 the two impulse responses define the method and provide the conceptual link between $H(s)$ and $\overline{H}_d(z)$. When it comes to implementation, the time domain can be bypassed altogether, by mapping the poles directly, as indicated. The system zeros can then be found from the numerator of Equation 11.10.

The impulse response of Equation 11.8 can also be expressed more succinctly in terms of the mapped poles as

$$h_d[nT] = T\sum_{i=1}^{N} c_i \pi_i^n u[nT]$$

Example 11.3

We use the time-sampling method to simulate a first-order lowpass filter expressed in terms of its real pole p_1 as the Laplace transform pair

$$h(t) = -p_1 e^{p_1 t} u(t) \quad \overset{\mathscr{L}}{\longleftrightarrow} \quad H(s) = \frac{-p_1}{s - p_1}$$

and represented in the upper half of Figure 11.12.

With $N = 1$ and $c_1 = -p_1$ the system response $H(s)$ represents a typical term of the partial fraction expansion of Equation 11.6. The pole p_1 maps onto the simulated pole $\pi_1 = e^{p_1 T}$ and, according to Equation 11.9, yields the discrete-time system function

$$\bar{H}_d(z) = \frac{-p_1 T z}{z - \pi_1}$$

where the simulation introduces a zero ζ_1 at the z-plane origin. Writing

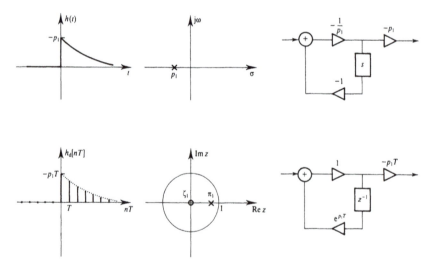

Figure 11.12 Time-sampled simulation of first-order system.

the pole π_1 in terms of p_1, the resulting system can also be expressed as the z-transform pair

$$h_d[nT] = -p_1 T e^{p_1 nT} u[nT] \quad \overset{\mathcal{Z}}{\longleftrightarrow} \quad H(z) = \frac{-p_1 T}{1 - e^{p_1 T} z^{-1}}$$

where the impulse response $h_d[nT]$ is clearly a sampled version of $h(t)$, as seen in the lower half of Figure 11.12.

Second-order section

In general the partial fraction coefficients c_i, which are common to both systems, are complex. In Section 4.4.3 we saw that those associated with a conjugate pair of poles, $p_2 = p_1^*$, also form a conjugate pair, $c_2 = c_1^*$, and that such pairs can be combined into a second-order section $T_i(s)$ with real coefficients. We now simulate a section of a parallel-partitioned continuous-time system, such as that of Figure 4.24, to give a corresponding section $\bar{T}_d(z)$ of the parallel-connected discrete-time system.

A section $T_i(s)$ is expressed in the equivalent forms

$$T_i(s) = \frac{d_1 s + d_0}{(s - p_1)(s - p_1^*)} = \frac{c_1}{s - p_1} + \frac{c_1^*}{s - p_1^*}$$

where the partial fraction coefficients are obtained by Equation 4.72 as

$$c_1 = \left. \frac{d_1 s + d_0}{s - p_1^*} \right|_{s = p_1} = \frac{d_1 p_1 + d_0}{p_1 - p_1^*}$$

and

$$c_1^* = \frac{-d_1 p_1^* - d_0}{p_1 - p_1^*}$$

The time-sampling simulation method maps the two poles p_1 and p_1^* onto the poles $\pi_1 = e^{p_1 T}$ and $\pi_1^* = e^{p_1^* T}$, and the entire section $T_i(s)$ is simulated according to Equation 11.9 as

$$\bar{T}_d(z) = \frac{T c_1 z}{z - \pi_1} + \frac{T c_1^* z}{z - \pi_1^*} = T z \frac{(c_1 + c_1^*)z - (c_1 \pi_1^* + c_1^* \pi_1)}{(z - \pi_1)(z - \pi_1^*)}$$

$$(11.11)$$

This system function has one zero at the origin $\zeta_2 = 0$ and a second zero at

$$\zeta_1 = \frac{c_1 \pi_1^* + c_1^* \pi_1}{c_1 + c_1^*}$$

Expressing the s-plane poles by their real and imaginary parts, $p_1 =$

$\sigma_1 + j\omega_1$ and $p_1^* = \sigma_1 - j\omega_1$, also gives

$$\pi_1 = e^{\sigma_1 T} e^{j\omega_1 T} = r_1 e^{j\omega_1 T} \quad \text{and} \quad \pi_1^* = e^{\sigma_1 T} e^{-j\omega_1 T} = r_1 e^{-j\omega_1 T}$$

$$(11.12)$$

so that

$$\zeta_1 = \frac{[d_1(\sigma_1 + j\omega_1) + d_0]r_1 e^{-j\omega_1 T} - [d_1(\sigma_1 - j\omega_1) + d_0]r_1 e^{j\omega_1 T}}{(c_1 + c_1^*)(p_1 - p_1^*)}$$

But $c_1 + c_1^* = d_1$ and $p_1 - p_1^* = 2j\omega_1$, and rearranging the numerator we have

$$\zeta_1 = \frac{-(d_1\sigma_1 + d_0)r_1(e^{j\omega_1 T} - e^{-j\omega_1 T}) + d_1 j\omega_1 r_1(e^{j\omega_1 T} + e^{-j\omega_1 T})}{2jd_1\omega_1}$$

and, with $z_1 = -d_0/d_1$ representing the zero of $T_i(s)$, this yields

$$\zeta_1 = (z_1 - \sigma_1)\frac{r_1}{\omega_1}\sin \omega_1 T + r_1 \cos \omega_1 T \tag{11.13}$$

which shows that the z-plane location of the second zero is also dependent on the locations of the s-plane poles.

More elaborate systems require several such second-order sections connected in parallel. To find the z-plane locations of the system's zeros involves correspondingly more elaborate algebra.

Example 11.4

We find the z-plane poles and zeros of the time-sampled simulation of a continuous-time system described by the first Laplace transform pair 3.13

$$h(t) = e^{\sigma_1 t} \cos \omega_1 t \, u(t) \quad \overset{\mathcal{L}}{\longleftrightarrow} \quad H(s) = \frac{s - \sigma_1}{(s - \sigma_1)^2 + \omega_1^2}$$

which has two poles $p_1 = \sigma_1 + j\omega_1$ and $p_1^* = \sigma_1 - j\omega_1$ and one real zero at $z_1 = \sigma_1$. The system is represented in the upper half of Figure 11.13, which is fully described in Section 11.3.3.

The poles are mapped as in Equation 11.12, and one system zero goes to the z-plane origin $z = 0$. Since $z_1 = \sigma_1$ the first term of Equation 11.13 vanishes and places the zero ζ_1 at $z = r_1 \cos \omega_1 T$, which also represents the real part of the pole locations, see the z-plane of Figure 11.13.

This z-plane pole-zero configuration corresponds to the z-transform pair 7.30, and leads to the simulated discrete-time system

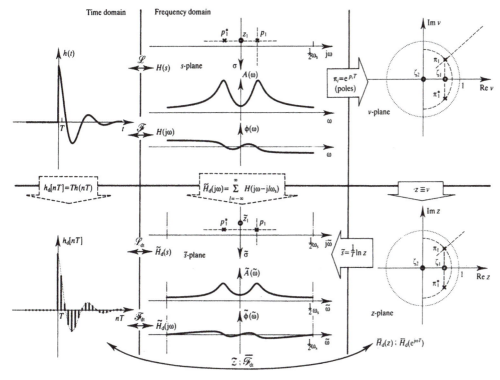

Figure 11.13 Full interpretation of time-sampling method.

$$h_{\mathrm{d}}[nT] = T r_1^n \cos \omega_1 nT \, u[nT]$$

$$\xleftrightarrow{\;\mathscr{Z}\;} \quad \bar{H}_{\mathrm{d}}(z) = T\frac{z^2 - r_1 \cos \omega_1 T \, z}{z^2 - 2r_1 \cos \omega_1 T \, z + r_1^2}$$

whose time domain is a sampled decaying cosine.

11.3.3 Time-sampling simulation process

The full time domain and frequency domain implications of this simulation method are brought together in Figure 11.13, which also sets the pattern for later mapping procedures.

The continuous-time reference system is represented in the upper half of Figure 11.13 by its impulse response $h(t)$, system function $H(s)$, symbolized by the pole-zero locations of the s-plane, and frequency response $H(j\omega)$, represented in terms of amplitude $A(\omega)$ and phase $\phi(\omega)$. These are related by the appropriate Laplace or Fourier transform, as shown.

The conceptual process that gives the method its name is represented in the time domain by the assignment $h_d[nT] = Th(nT)$, while the implementation process is represented by the three-step clockwise sequence of the frequency domain.

The first step consists of mapping the poles and zeros of $H(s)$ onto an auxiliary plane, which we will call the v-plane, according to the results of Section 11.3.2. The poles p_i are mapped to π_i as $\pi_i = e^{p_i T}$, that is, by the law $v = e^{sT}$, which is of the form of the forward $\tilde{s}$-to-z mapping law, while the zeros follow a more elaborate process, as shown earlier.

But **this is not the reversible $\tilde{s}$-plane to z-plane mapping law** of Section 7.2.1, which related the **periodic $\tilde{s}$-plane** associated with the discrete-time Laplace transform to the z-plane associated with the z-transform. In what follows we identify that periodic plane as the $\tilde{s}$-plane, the mapping law as $z = e^{\tilde{s}T}$ and the inversion law as $\tilde{s} = 1/T \ln z$, and these are represented in the lower half of Figure 11.13.

Returning to the implementation process, in the second step we identify the pole and zero locations of the v-plane, however obtained, with points of the z-plane. In other words, we superimpose the z-plane grid onto the v-plane and **interpret the poles and zeros found on the v-plane as belonging to the z-plane.** The full significance of this artifice will be apparent in later sections.

The z-plane poles and zeros thus obtained give sufficient information, within a scaling factor, to determine the system function $\bar{H}_d(z)$ and the frequency response $\bar{H}_d(e^{j\omega T})$, thus completing the simulation part of the process.

The third step, mapping the z-plane to the $\tilde{s}$-plane, provides additional interpretations of the system function $\tilde{H}_d(s)$ and of the frequency response $\tilde{H}_d(j\omega)$, in terms of the discrete-time Laplace transform of $h_d[nT]$, and of the corresponding Fourier subset. Alternatively, the discrete-time impulse response is related by the z-transform to $\bar{H}_d(z)$ and by the corresponding Fourier transform to $\bar{H}_d(e^{j\omega T})$, as shown in the lower half of Figure 11.13.

Finally, the direct link between $H(j\omega)$ and $\tilde{H}_d(j\omega)$, indicated by a dashed arrow, highlights the aliasing problem associated with this method. Time domain sampling with the sampler $s_T(t)$ leads to a periodic function $\tilde{H}_d(j\omega)$ that is a sum of displacements of $H(j\omega)$, as obtained in Equation 6.20,

$$\tilde{H}_d(j\omega) = \sum_{l=-\infty}^{\infty} H(j\omega - jl\omega_s)$$

Consequently this method is unsuitable for simulating systems, such as highpass filters, whose frequency response $H(j\omega)$ does not decay at high frequencies.

The fundamental difference between the various simulation methods lies in the distortions introduced by the s-plane to v-plane mapping. Since the subsequent z-plane to $\tilde{s}$-plane mapping law is common to all methods, the said distortions are fully responsible for differences of the resulting $\tilde{s}$-plane representations.

In the time-sampling method the envelope of $h_d[nT]$ is identical to $h(t)$, for which reason it is called the **impulse invariance method.** Other methods distort the shape of that envelope.

11.3.4   Related methods

The lengthy algebra required to find the system zeros associated with the time sampling method becomes a deterrent. If impulse invariance is not the main consideration, other methods, such as the bilinear mapping of Section 10.6, are preferred. But variants of the time-sampling method are often used, which basically extend the mapping law used for the poles of $H(s)$ to also map the finite zeros of $H(s)$. We illustrate with a typical variant.

Matched z-transform method

Using the imagery of Figure 11.13, this method maps the poles p_i of $H(s)$ onto the v-plane as before, that is, as $\pi_i = e^{p_i T}$. All finite zeros z_i are similarly mapped as $\zeta_i = e^{z_i T}$. Both are one-to-one relationships. Any 'zeros at infinity' of $H(s)$ are arbitrarily placed at $v = -1$ (at $v = 0$, in some versions of this method).

The z-plane grid is superimposed on the v-plane, as was done in the lower half of Figure 11.13, and the poles and zeros of the z-plane are mapped onto the periodic $\tilde{s}$-plane by the reversible law $\tilde{s} = 1/T \ln z$. This has the overall effect of periodically replicating the set of original s-plane poles and zeros as identical sets of the $\tilde{s}$-plane. The additional zeros brought in from the s-plane's 'point at infinity' end up on the imaginary axis of the $\tilde{s}$-plane, at half sampling frequency $\tilde{s} = \frac{1}{2}j\omega_s$, and are periodically repeated.

Example 11.5

We simulate the system of Example 11.4 by the matched z-transform method. The poles are mapped as before, that is, as in Equation 11.12. The zero $z_1 = \sigma_1$ maps to $\zeta_1 = e^{\sigma_1 T} = r_1$, that is, on the same circle as

the poles π_i, while the 'zero at infinity' is effectively placed at $z = -1$.
The simulated system function takes the form

$$\bar{H}_d(z) = \frac{(z + 1)(z - r_1)}{z^2 - 2r_1 \cos \omega_1 T z - r_1^2}$$

11.4 Simulation based on numerical analysis methods

In Section 4.3.4 we showed the equivalence of various time domain and
frequency domain representations of the same Nth-order continuous-
time system, summarizing these in Figure 4.20. We now introduce a
range of seemingly unrelated methods based on simple discrete-time
simulations $D(z)$ and $I(z)$ of the differential and integral operators s
and s^{-1} of the frequency domain representations. In Section 11.5 we
merge these methods into one common mapping method, which we then
illustrate in Section 11.6 by means of the most important case, the
bilinear mapping method.

Numerical analysis offers a host of methods that are suitable for
discrete-time simulation of differentiators and integrators. It also offers
a suitable framework for developing and relating the resulting simulation
methods. In particular it relates certain differentiator simulators $D(z)$ to
appropriate integrator simulators $I(z)$, both of which can be replaced by
a common mapping law.

The simulation objectives are the same as before, namely, to find
a causal and stable discrete-time system that simulates a known causal
and stable continuous-time system. We will show that seemingly minor
differences between approximation criteria can lead to substantially
different discrete-time systems, even to unstable ones.

11.4.1 Discrete-time simulation of differentiator

In Section 10.3.1 we interpreted the ideal continuous-time differentiator
as an elementary system with impulse response $d(t) = \delta'(t)$ and system
response $D(s) = s$, related as the Laplace pair

$$d(t) = \delta'(t) \quad \overset{\mathscr{L}}{\longleftrightarrow} \quad D(s) = s \qquad (11.14)$$

Thus, for an input $f(t)$ we have an output $g(t) = df(t)/dt$, with the
frequency implications shown in the upper left quadrant of Figure 11.14.
In Section 11.1.2 we listed some primitive discrete-time simula-

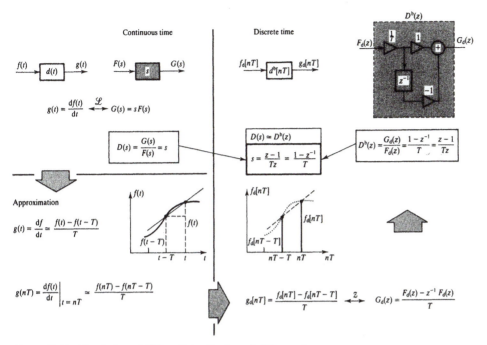

Figure 11.14   Simulation of differentiator (backward difference).

tions of the ideal differentiator. We now formulate the simulation problem using methods from numerical analysis for guidance. The simplest options use two points of the function $f(t)$ separated by an interval T. The straight line these define is interpreted as an approximation to the tangent to the curve $f(t)$, at some point of the interval, see lower left of Figure 11.14.

Backward difference approximation

Interpreting the line thus defined as an **approximation of the tangent at time** t, that is, at the upper limit of the interval T, leads to the **backward difference approximation of the derivative**, expressed as

$$g(t) = \frac{df(t)}{dt} \simeq \frac{f(t) - f(t - T)}{T}$$

Evaluated for a sampling point $t = nT$ this expression subtracts the **preceding sample** $f(nT - T)$ from the **current sample** $f(nT)$ to yield an approximation to the **current derivative**

$$g(nT) = \frac{df(t)}{dt}\bigg|_{t=nT} \simeq \frac{f(nT) - f(nT - T)}{T} \tag{11.15}$$

Conversion to discrete time

This continuous-time relationship is now converted to discrete time by replacing the samples $f(nT)$ and $f(nT - T)$ by discrete-time samples $f_d[nT]$ and $f_d[nT - T]$ of the same value, as indicated in the lower half of Figure 11.14. The resulting difference is assigned to the function $g_d[nT]$, as

$$g_d[nT] = \frac{f_d[nT] - f_d[nT - T]}{T} \tag{11.16}$$

This result represents the difference equation of a discrete-time system. Taking the z-transform of both sides of Equation 11.16 yields

$$G_d(z) = \frac{F_d(z) - z^{-1}F_d(z)}{T}$$

and the discrete-time system function

$$D^b(z) = \frac{G_d(z)}{F_d(z)} = \frac{1 - z^{-1}}{T} = \frac{z - 1}{Tz} \tag{11.17}$$

where the superscript b signifies backward difference. The applicable block diagram is given in the upper right corner of Figure 11.14.

Discrete-time simulator

The discrete-time system $D^b(z)$ is a simulator of the differentiator $D(s)$ in the sense of backward difference approximation. Equating their right sides gives a relationship between the frequency variables s and z of the continuous-time and discrete-time systems, expressed in one of the forms

$$s = \frac{1 - z^{-1}}{T} = \frac{z - 1}{Tz} \quad \text{hence} \quad z = \frac{1}{1 - Ts} \tag{11.18}$$

which represents a simulation or mapping law. We next demonstrate its usage on a simple system, and generalize later in Section 11.5.

11.4.2 System containing one differentiator

We apply the discrete-time simulator $D^b(z)$ to the general first-order system of Section 4.2.4. Our sole objective is to explore various facets of the model, to show that these lead to equivalent representations of the simulated system. We first examine time domain implications, followed by frequency domain simulations of the system function and of the block diagram, and then map the pole-zero locations. These are summarized

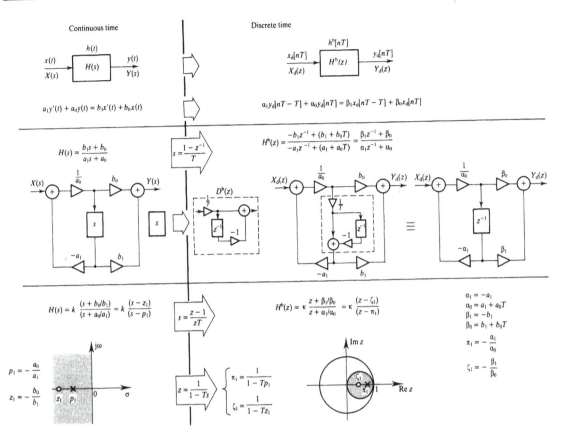

Figure 11.15 Backward difference simulation (first-order system).

in Figure 11.15, where the continuous-time reference system is described in various equivalent forms on the left side and could represent the *RCR* network of Example 4.2.

Simulation of differential equation

In the time domain the system is described by the differential equation

$$a_1 y'(t) + a_0 y(t) = b_1 x'(t) + b_0 x(t) \tag{11.19}$$

Modelling the derivatives $y'(t)$ and $x'(t)$ by the approximation of Equation 11.15, for instance

$$y'(nT) \simeq \frac{y(nT) - y(nT - T)}{T}$$

approximates the **differential equation** 11.19, at an instant $t = nT$, as

$$a_1[y(nT) - y(nT - T)] + a_0 Ty(nT)$$
$$\simeq b_1[x(nT) - x(nT - T)] + b_0 Tx(nT)$$

Converting the samples to discrete time, as before, and rearranging terms, we obtain the **difference equation**

$$-a_1 y_d[nT - T] + (a_1 + a_0 T)y_d[nT]$$
$$= -b_1 x_d[nT - T] + (b_1 + b_0 T)x_d[nT]$$

This represents a general first-order discrete-time system of the form examined in Section 8.2.3,

$$\alpha_1 y_d[nT - T] + \alpha_0 y_d[nT] = \beta_1 x_d[nT - T] + \beta_0 x_d[nT] \quad (11.20)$$

whose coefficients are related to those of Equation 11.19 as

$$\alpha_1 = -a_1 \qquad\qquad \alpha_0 = a_1 + a_0 T$$
$$\beta_1 = -b_1 \qquad\qquad \beta_0 = b_1 + b_0 T$$
$$\text{(11.21)}$$

This time domain result can be applied directly, in the recursive form shown at the end of Section 8.2.2, to give an approximate numerical solution of the differential equation 11.19.

But for system simulation the system function $H^b(z)$ is more relevant. It is obtained from the z-transform of the difference equation 11.20 as

$$H^b(z) = \frac{\beta_1 z^{-1} + \beta_0}{\alpha_1 z^{-1} + \alpha_0} = \frac{\beta_1 + \beta_0 z}{\alpha_1 + \alpha_0 z} \quad (11.22)$$

The objective of this section is to explore alternative means of obtaining this result directly in the frequency domain.

System function substitution

We take a procedural short-cut by expressing the reference system of Equation 11.19 in the Laplace domain as

$$H(s) = \frac{b_1 s + b_0}{a_1 s + a_0} \quad (11.23)$$

and evaluating this function for points $s = (1 - z^{-1})/T$ defined by the simulation law 11.18. This yields

$$H((1 - z^{-1})/T) = \frac{b_1(1 - z^{-1})/T + b_0}{a_1(1 - z^{-1})/T + a_0}$$

which is now a function of the frequency variable z. Written as

$$H^b(z) = \frac{-b_1 z^{-1} + (b_1 + b_0 T)}{-a_1 z^{-1} + (a_1 + a_0 T)} \quad (11.24)$$

and with its coefficients interpreted as in Equations 11.21 it is identical to Equation 11.22.

Block diagram substitution

The block diagram representing the system function $H(s)$ in Figure 11.15 contains one differentiator block s. Replacing this system element by the block derived in Figure 11.14 for the discrete-time simulator $D^b(z)$ yields a discrete-time block digram with redundant components. This diagram can be manipulated (by rules not explained in this book) to the simpler, equivalent form indicated in Figure 11.15.

This representation is consistent with the system function $H^b(z)$ expressed in Equation 11.22, as well as with the difference equation 11.20.

Pole and zero mapping

Finally, we examine the effects of substitution on poles and zeros. We rearrange the reference system function 11.23 as

$$H(s) = \frac{b_1(s + b_0/b_1)}{a_1(s + a_0/a_1)} = k\frac{(s - z_1)}{(s - p_1)}$$

which has an s-plane pole at $p_1 = -a_0/a_1$, a zero at $z_1 = -b_0/b_1$ and a scaling factor $k = b_1/a_1$. Similarly, the second form of the discrete-time system function 11.22 yields

$$H^b(z) = \frac{\beta_0(z + \beta_1/\beta_0)}{\alpha_0(z + \alpha_1/\alpha_0)} = \varkappa\frac{(z - \zeta_1)}{(z - \pi_1)}$$

with a z-plane pole at $\pi_1 = -\alpha_1/\alpha_0$, a zero at $\zeta_1 = -\beta_1/\beta_0$ and a scaling factor $\varkappa = \beta_0/\alpha_0$.

The correspondence (11.21) between coefficients yields the relationships

$$\pi_1 = \frac{a_1}{a_1 + a_0 T} = \frac{1}{1 + Ta_0/a_1} = \frac{1}{1 - Tp_1} \tag{11.25}$$

and

$$\zeta_1 = \frac{1}{1 - Tz_1}$$

We conclude that the s-plane pole p_1 and zero z_1 are mapped to a corresponding pole π_1 and zero ζ_1 of the z-plane by the last form of the simulation law 11.18.

This correspondence is summarized in the lower part of Figure 11.15 and leads to an effective mapping procedure for performing a simulation and for interpreting the results.

11.4.3 Discrete-time simulation of integrator

System representations associated with the system's integral equation, such as those given in the right half of Figure 4.20, can be treated in a formally similar way by approximating the integration operator s^{-1} in some discrete-time form and then replacing each occurrence of this operator according to the resulting mapping law. We derive three such approximations towards the generalization of Section 11.5.

The ideal integrator was interpreted in Section 10.3.1 as an elementary system with input signal $f(t)$ and output signal

$$g(t) = \int_{-\infty}^{t} f(\tau)\,d\tau$$

whose impulse response and system function form the Laplace pair

$$i(t) = u(t) \quad \overset{\mathscr{L}}{\longleftrightarrow} \quad I(s) = s^{-1} \tag{11.26}$$

The left side of Figure 11.16 summarizes the Laplace process. It shows a causal input signal $f(\tau)$, or equivalently a lower integration limit of zero, so that the shaded area under the function $f(\tau)$ represents the integral $g(t)$.

Discrete-time simulations

Many numerical methods are available for approximating an integral. We will consider those involving either the current sample $f[nT]$, or the preceding sample $f[nT - T]$, or both. Our aim is to show how these lead to different mapping laws.

Describing the cumulative area up to the time value $t = nT - T$ by $g_d[nT - T]$, we have three options for computing the integral at $t = nT$. We can simulate the incremental area by holding either the previous value $f_d[nT - T]$, or the current value $f_d[nT]$ or use a linear interpolation between these values. These options are shown in Figure 11.16, where we dropped the subscripts d for brevity.

These three approximations, known as **forward rectangular, backward rectangular** and **trapezoidal** rules, identified by the superscripts f, b and t, are expressed respectively as

$$g^{f}[nT] = g^{f}[nT - T] + Tf_d[nT - T]$$

$$g^{b}[nT] = g^{b}[nT - T] + Tf_d[nT] \tag{11.27}$$

$$g^{t}[nT] = g^{t}[nT - T] + T\frac{f_d[nT] + f_d[nT - T]}{2}$$

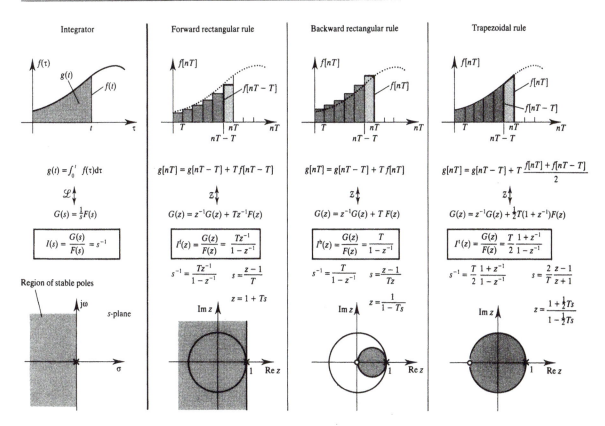

Figure 11.16   Three discrete-time simulators of integrator.

It is clear from the diagrams that the trapezoidal rule gives a closer approximation to the integral than the other two rules, which suggests that it should lead to a better simulation. The two rectangular methods appear to be conceptually similar, and would suggest comparable simulation errors. But we will show that one of these can lead to a simulated system that is unstable.

To obtain the simulator system functions $I^x(z)$, where the superscript x denotes the applicable rule, we take z-transforms of the options 11.27

$$G^f(z) = z^{-1} G^f(z) + T z^{-1} F_d(z)$$

$$G^b(z) = z^{-1} G^b(z) + T F_d(z)$$

$$G^t(z) = z^{-1} G^t(z) + \tfrac{1}{2}T(1 + z^{-1}) F_d(z)$$

which yield three elementary discrete-time systems

$$I^f(z) = \frac{Tz^{-1}}{1 - z^{-1}} = \frac{T}{z - 1}$$

$$I^b(z) = \frac{T}{1 - z^{-1}} = \frac{Tz}{z - 1} \qquad (11.28)$$

$$I^t(z) = \frac{T}{2}\frac{1 + z^{-1}}{1 - z^{-1}} = \frac{T}{2}\frac{z + 1}{z - 1}$$

Each of these is a valid discrete-time simulator of the integrator $I(s)$. Equating the right sides of Equations 11.28 to s^{-1} yields three simulation laws, each of which can be expressed in direct, reciprocal and inverse forms as

$$s^{-1} = \frac{Tz^{-1}}{1 - z^{-1}} \qquad s = \frac{z - 1}{T} \qquad \text{and} \qquad z = 1 + Ts$$

$$(11.29)$$

$$s^{-1} = \frac{T}{1 - z^{-1}} \qquad s = \frac{z - 1}{Tz} \qquad \text{and} \qquad z = \frac{1}{1 - Ts}$$

$$(11.30)$$

$$s^{-1} = \frac{T}{2}\frac{1 + z^{-1}}{1 - z^{-1}} \qquad s = \frac{2}{T}\frac{z - 1}{z + 1} \qquad \text{and} \qquad z = \frac{1 + \frac{1}{2}Ts}{1 - \frac{1}{2}Ts}$$

$$(11.31)$$

The foregoing results are summarized in Figure 11.16, which also gives a guide to the mapping of the imaginary axis of the s-plane, the boundary of the region of stable poles. We elaborate in Section 11.5.3.

11.4.4   System containing one integrator

Consider once again the general first-order system of Section 11.4.2, but now represented by the integral form of the system equation

$$a_1 y(t) + a_0 \int_{-\infty}^{t} y(\tau)\,d\tau = b_1 x(t) + b_0 \int_{-\infty}^{t} x(\tau)\,d\tau \qquad (11.32)$$

Having demonstrated earlier the equivalence of time and frequency domain approaches, we proceed directly in the Laplace domain. Thus, taking the Laplace transform of both sides of Equation 11.32

$$a_1 Y(s) + a_0 s^{-1} Y(s) = b_1 X(s) + b_0 s^{-1} X(s)$$

leads to the integral form of the continuous-time system function

$$H(s) = \frac{Y(s)}{X(s)} = \frac{b_1 + b_0 s^{-1}}{a_1 + a_0 s^{-1}} \qquad (11.33)$$

We can now replace each of the integral operators s^{-1} by one of the discrete-time simulators $I^x(z)$ of Equations 11.28. Selecting the backward rectangular rule $I^b(z)$ for illustration, the form $s^{-1} = T/(1 - z^{-1})$ of the simulation law 11.30 yields

$$H^b(z) = \frac{b_1 + b_0 T/(1 - z^{-1})}{a_1 + a_0 T/(1 - z^{-1})} = \frac{-b_1 z^{-1} + (b_1 + b_0 T)}{-a_1 z^{-1} + (a_1 + a_0 T)} \qquad (11.34)$$

This result is identical to the system function 11.24 obtained by simulating the differentiator $D(s) = s$ according to the backward difference law $D^b(z)$. Simulating other aspects of system representation, such as block diagrams and poles and zeros, would similarly lead to identical results.

11.4.5   Alternative differentiator simulators

Both the backward rectangular simulator $I^b(z)$ of $I(s)$ and the backward difference simulator $D^b(z)$ of $D(s)$ result in the same mapping law 11.30. Applied to equivalent integral or differential system representations $H(s)$, they lead to the same simulation $H^b(z)$.

Selecting alternative integration simulators $I^f(z)$ or $I^t(z)$ would lead to different discrete-time simulations $H^f(z)$ and $H^t(z)$ of the continuous system $H(s)$. We now derive the differentiator counterparts.

In Section 11.4.1 we took the line defined by the two samples $f(t)$ and $f(t - T)$ to represent the derivative of $f(t)$ at the upper limit of the interval T, thus yielding the backward difference approximation 11.15. Interpreting the same line as representing the derivative at the lower limit of the interval T, see Figure 11.14, gives the **forward difference approximation**

$$g(nT - T) = \left.\frac{df(t)}{dt}\right|_{t=nT-T} \simeq \frac{f(nT) - f(nT - T)}{T} \qquad (11.35)$$

Converting to discrete time and taking the z-transform leads to the simulator

$$D^f(z) = \frac{G_d(z)}{F_d(z)} = \frac{1 - z^{-1}}{Tz^{-1}} = \frac{z - 1}{T} \qquad (11.36)$$

The third alternative results from considering the same line to represent the average between the derivatives at $t = nT$ and at $t = nT - T$, that is

$$\frac{g(nT) + g(nT - T)}{2} \simeq \frac{f(nT) - f(nT - T)}{T} \qquad (11.37)$$

and this leads to the simulator

$$D^t(z) = \frac{2}{T} \frac{1 - z^{-1}}{1 + z^{-1}} = \frac{2}{T} \frac{z - 1}{z + 1} \qquad (11.38)$$

These results show a one-to-one correspondence between the differentiator simulators $D^f(z)$, $D^b(z)$ and $D^t(z)$ and the integrator simulators 11.28 of the same designation $I^x(z)$. Each pair is represented by the same set of mapping laws 11.29 to 11.31.

The last of these laws, the set 11.31, which represents both $D^t(z)$ and the trapezoidal rule $I^t(z)$, is known as the **bilinear mapping law** and is the most widely used. It will be examined in detail in Section 11.6.

In terms of mathematical analysis the above rules are consequences of the mean value theorems of the derivative and of the integral of a function. More elaborate simulators, such as those based on Simpson's rule and other integration algorithms, involve more samples and more complex mapping laws.

11.5 Generalization and mapping

Various similarities between the methods and the results of Section 11.4 hint at common properties. We now formalize the equivalence of simulating the differential or the integral representation of a system, interpret the various simulations in terms of a common mapping process involving different mapping laws and examine the distortions these laws introduce.

11.5.1 Equivalence of differentiator and integrator simulation

In Sections 11.4.2 and 11.4.4 a first-order system was simulated using the differential simulator $D^b(z)$ and the integral simulator $I^b(z)$, with identical results. This is no coincidence, such an identity would have been obtained for an arbitrary Nth-order system.

Recall the equivalence of differential and integral representations of a system, first shown in Figure 4.20 and summarized further in the upper half of Figure 11.17, which extends the equivalence to simulation.

The two system functions $H^d(s)$ and $H^i(s)$ are basically identical. Simulating the operator s of the former by $D^b(z)$, and the operator $1/s$ of the latter by $I^b(z) = 1/D^b(z)$ involves the same mapping law and leads to the same simulation result $H^b(z)$. This equivalence extends to all other representation forms.

Example 11.6

We simulate the general second-order system of Figure 11.17 by means

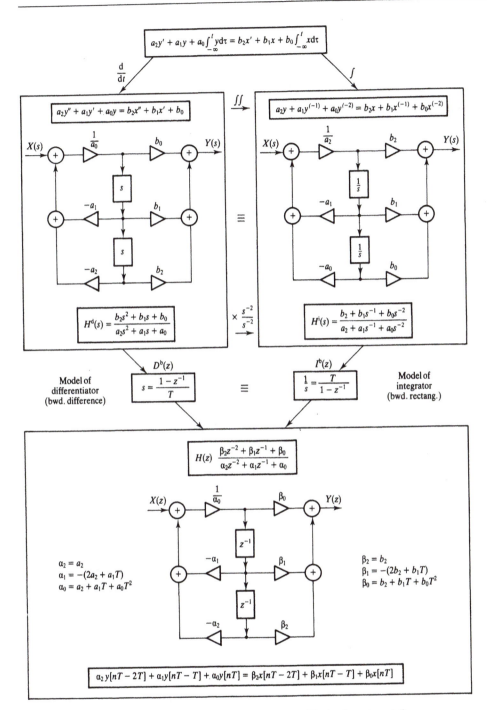

Figure 11.17   Equivalence of two simulation paths (illustrated by backward model).

of the backward simulation law 11.30. The substitution $s = (1 - z^{-1})/T$ in $H^d(s)$ yields

$$H^b(z) = \frac{b_2(1 - z^{-1})^2/T^2 + b_1(1 - z^{-1})/T + b_0}{a_2(1 - z^{-1})^2/T^2 + a_1(1 - z^{-1})/T + a_0}$$

which is reduced to a ratio of polynomials in z^{-1}, of the standard form

$$H^b(z) = \frac{\beta_2 z^{-2} + \beta_1 z^{-1} + \beta_0}{\alpha_2 z^{-2} + \alpha_1 z^{-1} + \alpha_0} \tag{11.39}$$

with coefficients

$$\alpha_0 = a_2 + a_1 T + a_0 T^2 \qquad \beta_0 = b_2 + b_1 T + b_0 T^2$$

$$\alpha_1 = -(2a_2 + a_1 T) \qquad \beta_1 = -(2b_2 + b_1 T)$$

$$\alpha_2 = a_2 \qquad \beta_2 = b_2$$

The block diagram of the lower half of Figure 11.17 completes the simulation. The same result would have been obtained by the substitution $s^{-1} = T/(1 - z^{-1})$ in $H^i(s)$.

To simulate the general Nth-order system it suffices to replace each occurrence of s or s^{-1} by the desired mapping law. In the case of the system function this is expressed as

$$H^d(s) = \frac{\sum_{m=0}^{N} b_m s^m}{\sum_{m=0}^{N} a_m s^m} \xrightarrow{s = D^x(z)} H^x(z) = \frac{\sum_{m=0}^{N} b_m [D^x(z)]^m}{\sum_{m=0}^{N} a_m [D^x(z)]^m}$$

$$\tag{11.40}$$

$$H^i(s) = \frac{\sum_{m=0}^{N} b_m s^{-(N-m)}}{\sum_{m=0}^{N} a_m s^{-(N-m)}} \xrightarrow{s^{-1} = I^x(z)} H^x(z) = \frac{\sum_{m=0}^{N} b_m [I^x(z)]^{-(N-m)}}{\sum_{m=0}^{N} a_m [I^x(z)]^{-(N-m)}}$$

We conclude that the chosen s-to-z mapping law fully characterizes the simulation, regardless of whether the system is described with differentiators or integrators, or even with a mixture of both, as would be the case with an integro-differential equation. In what follows we use the differentiator form $s = D^x(z)$ as a generic description of any particular simulation, that is, the form

$$H^x(z) = H(s)|_{s = D^x(z)}$$

Note that, for the three simulation laws discussed here, the order N of the simulated system is the same as that of the original continuous-time system. This is because each of these simulators has only one delay element z^{-1}, so that the term in the highest power of s in $H(s)$ gives rise to a term in the same power of z^{-1}. More elaborate laws would yield simulated systems of higher order.

11.5.2   Mapping interpretation

So far in this section we interpreted the various simulation laws as mapping points of the continuous s-plane directly to points of the z-plane, and this is the common practice in the available literature. But such interpretation often gives rise to confusion when mapping back to the periodic $\tilde{s}$-plane.

In Section 7.2.1 we defined the forward and inverse forms of the $\tilde{s}$-to-z mapping law, written here with the notation $\tilde{s}$ as

$$z = e^{\tilde{s}T} \qquad \text{hence} \qquad \tilde{s} = \frac{1}{T} \ln z \qquad (11.41)$$

For a given sampling interval T this defines a multi-valued but unique correspondence between the periodic $\tilde{s}$-plane and the z-plane.

In contrast, a simulation law of the form $s = D^{x}(z)$, together with its inversion law, represent a **differently distorted mapping** of features of the two planes. For instance, an s-plane pole located at $s = p_1$, would not be mapped onto the corresponding z-plane point $z = e^{p_1 T}$, but at a point $z = \pi_1$, somewhere in its near (or not so near) vicinity.

Auxiliary v-plane

To make this distortion explicit we make use of the auxiliary v-plane, first introduced with the mapping process of Figure 11.13 for the time-sampling method. We first map the reference system's s-plane features onto the v-plane, according to one of the earlier simulation laws, $v^{x} = D^{x}(s)$, as shown in the upper half of Figure 11.18, to then interpret the results as belonging to the z-plane of the simulated discrete-time system of the lower half of the figure.

We illustrate the backward mapping law $v^{b} = (1 - z^{-1})/T$, which maps the imaginary axis $s = j\omega$ onto the smaller circle located inside the unit circle of the v-plane, see top right of Figure 11.18. Using a third-order Chebyshev lowpass filter for reference, its s-plane poles p_i and zeros z_i are mapped as

$$\pi_i = \frac{1}{1 - Tp_i} \qquad \text{and} \qquad \zeta_i = \frac{1}{1 - Tz_i}$$

The poles are thus located inside that smaller circle and the three s-plane 'zeros at infinity' are mapped onto the origin of the v^{b}-plane $v^{b} = 0$, as shown. We elaborate on grid line distortions in the next section.

These mapped poles and zeros are now interpreted as belonging to the z-plane, that is, we superimpose the polar z-plane grid upon them, as symbolized by the identification of planes, $z \equiv v^{b}$, on the right side of Figure 11.18. The resulting z-plane representation fully describes the simulated discrete-time system, save for an overall scaling factor.

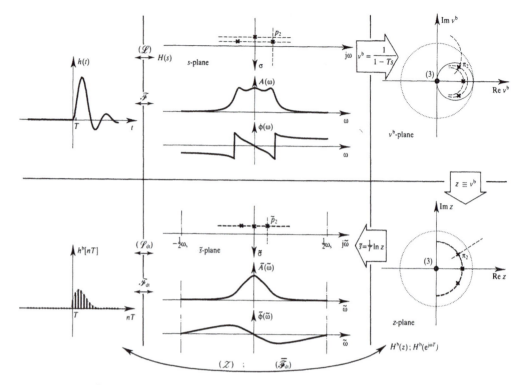

Figure 11.18 Backward mapping law.

All other representations of the simulated system are obtained from the z-plane by the usual transforms, as shown in the lower half of Figure 11.18, by a process identical to that presented in Figure 11.13. The results differ, but this is due entirely to different distortions being introduced by the simulation laws at the s-plane to v-plane mapping stage.

This is clear from the relative location of system poles in the reference s-plane and the simulated $\tilde{s}$-plane, and the consequent distortions of the frequency response and of the impulse response envelope.

11.5.3 Grid distortions under mapping

Earlier, in Figure 11.16, we highlighted the s-plane's region of stable poles and outlined the corresponding mapped regions for three simulations. We now examine the geometrical distortions imposed by these laws on a regular grid and compare them with those of Section 7.2.1 for straight $\tilde{s}$-to-z mapping.

Forward law

Consider the rectangular s-plane grid of Figure 11.19, which highlights the point at the origin (circle), a segment of the imaginary axis (bold line), an arbitrary area from the half-plane of stable poles (shaded) and a possible location of such a pole (cross).

The forward mapping law 11.29, repeated here with v-plane notation,

$$v^{\mathrm{f}} = 1 + Ts$$

is a linear mapping law that involves a mere scaling of the s-plane grid $u = Ts$, followed by a unit translation $v^{\mathrm{f}} = u + 1$, as shown in Figure 11.19 for a scaling factor $T = \frac{1}{2}$. The grid's former origin moves to the v^{f}-plane's reference point $v^{\mathrm{f}} = 1$, but apart from a change of size no grid distortions are involved.

Backward law

The backward mapping law 11.30

$$v^{\mathrm{b}} = \frac{1}{1 - Ts}$$

is interpreted in Figure 11.20 (again for $T = \frac{1}{2}$) in three steps, namely, a negation of the u-plane grid of Figure 11.19 $\bar{u} = -u = -Ts$, which is equivalent to a rotation through an angle π, which is followed by a unit translation $w = \bar{u} + 1$ and by a reciprocal mapping $v^{\mathrm{b}} = 1/w$.

The latter was examined in detail in Section 10.6.2 (see Figure 10.33), where it mapped a grid of orthogonal lines into a grid of orthogonal circles. All the circles pass through the v^{b}-plane origin, which represents the 'point at infinity' of the s-plane, hence also the 'points at $\pm\infty$' of all the s-plane grid lines.

Bilinear law

With v-plane notation the bilinear law 11.31 becomes

$$v = \frac{1 + \frac{1}{2}Ts}{1 - \frac{1}{2}Ts}$$

To aid interpretation we add and subtract unity in the numerator, yielding

$$v = \frac{2}{1 - \frac{1}{2}Ts} - 1$$

Apart from scaling factors the first term is formally identical to the backward mapping law interpreted in Figure 11.20. Indeed, the denomi-

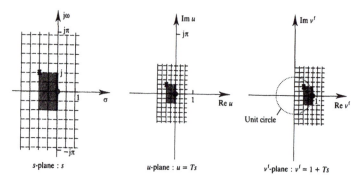

Figure 11.19 Grid distortion, forward law ($T = \frac{1}{2}$).

nator $\bar{w} = 1 - \frac{1}{2}Ts$, shown in the $\bar{w}$-plane of Figure 11.21, is similar to $w = 1 - Ts$ of Figure 11.20, and after reciprocal mapping the grids of $u = 1/\bar{w}$ and $v^b = 1/w$ retain that similarity. The parameter values, $T = 1$ for bilinear mapping and $T = \frac{1}{2}$ for the backward law, used in the figures yields identical grids for the intermediate u-plane of Figure 11.21 and the v^b-plane of Figure 11.20.

The subsequent mapping $v = 2u - 1$ of Figure 11.21 scales and shifts the u-plane grid, such that the small circle, representing the imaginary axis of the s-plane, is superimposed on the unit circle of the v-plane, and the s-plane's 'point at infinity' is shifted to the leftmost

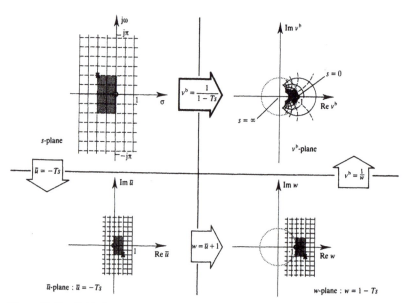

Figure 11.20 Grid distortion, backward law ($T = \frac{1}{2}$).

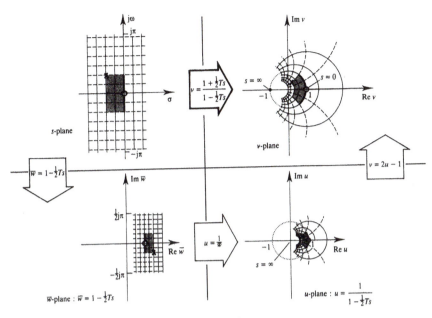

Figure 11.21 Grid distortion, bilinear law $(T = 1)$.

point of the unit circle $v = -1$, as shown. The circular grid shapes, introduced with the reciprocal mapping $u = 1/\overline{w}$, are preserved with scaling.

Comparison

The results of the preceding constructions are repeated in Figure 11.22, using the same value $T = 1$ for comparison. Unlike the mapping law $z = e^{sT}$ illustrated in the upper half of the figure, which defines a multi-valued correspondence between s-plane and z-plane points, each mapping law of this section represents a one-to-one correspondence between s-plane and $v^{\times}$-plane points. In this sense these laws are akin to the frequency mapping laws of Section 10.6, where each point of the prototype $\hat{s}$-plane was mapped to either one point of the s-plane or to a limited number of such points.

Comparing the resulting $v^{\times}$-planes, it is clear that both the bilinear and the backward laws map the stable half of the s-plane inside the unit circle, which represents the region of stable poles of the z-plane. In contrast, the forward law is capable of mapping s-plane poles, especially those located close to the imaginary axis, outside the unit circle, which would lead to an unstable system.

As with $\hat{s}$-to-z mapping, the bilinear law maps the imaginary s-plane axis onto the unit circle. The fundamental difference is that in

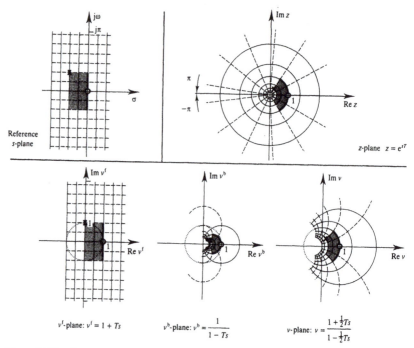

Figure 11.22 Comparative mapping distortions ($T = 1$).

the latter case the entire imaginary axis maps only once onto the unit circle. In the near vicinity of $z = 1$ the two laws lead to comparable frequency responses. In contrast, the other two laws map the imaginary axis of the s-plane onto curves that diverge quickly from the unit circle, making a direct comparison of frequency responses meaningless.

11.6   Bilinear mapping

Several desirable properties combine to make the bilinear mapping law stand out from the rest. Its chief attractions are a close reproduction of the frequency response, particularly in the vicinity of the origin, a reliable conversion of stable continuous-time systems into stable discrete-time systems and a complete absence of aliasing. The main price to be paid is a non-linear distortion of the frequency axis, which is tolerable at low frequencies, but becomes very severe at higher values.

We now examine the bilinear law in some detail. Most of the concepts presented here are also valid for other mapping laws, so that this section may be taken as an illustration of the general mapping process.

11.6.1 Mapping process

The simulation procedure is identical to the mapping process outlined in Section 11.5.2. There we used the backward law, which distorted the s-plane features of Figure 11.18 according to the v^b-plane grid of Figure 11.22. Using the same third-order Chebyshev example, the bilinear law distorts its s-plane features according to the v-plane grid of Figure 11.22, to give the pole and zero locations shown in the v-plane of Figure 11.23. In particular, the three 'zeros at infinity' are mapped onto $v = -1$.

Superimposing the z-plane grid on the v-plane features, as shown on the right side of Figure 11.23, we can now interpret the poles and zeros as belonging to the z-plane and derive suitable expressions for the system function $H_d(z)$ and the frequency response $H_d(e^{j\tilde{\omega} T})$.

The periodic $\tilde{s}$-plane features associated with the discrete-time Laplace transform are obtained from the z-plane by the standard law. Note that the three zeros, originally 'at infinity' but taken to $z = -1$ by the bilinear law, are mapped at 'half sampling frequency' $\frac{1}{2}\tilde{\omega}_s$ and its periodic repetitions.

The resulting frequency response, represented by $\tilde{A}_d(\tilde{\omega})$ and $\tilde{\phi}_d(\tilde{\omega})$ is a frequency-compressed version of the original response. The compression is non-linear in the frequency variable, but does not cause distortions of the actual amplitude or phase values. The two frequency responses are comparable, because both the bilinear mapping law and the standard z-plane law map the corresponding imaginary s-plane axes onto the unit circle of the z-plane, all three lines being the locations of the corresponding frequency responses. This is not true of the forward and backward mapping laws, where frequency responses are mapped onto different lines and are not directly comparable, see Figure 11.18.

Note that the distortions associated with a particular simulation law take place during s-plane to v-plane mapping, compare upper halves of Figures 11.23 and 11.18. The lower halves of the figures simply represent full and self-contained representations of the resulting discrete-time system, linked by the applicable discrete-time transforms.

11.6.2 Effect of sampling interval

In Section 7.2.1 we examined the effect of the sampling interval T on the z-plane representation of a function, see Figure 7.5. The present mapping laws contain the same parameter T, so that similar considerations apply.

We rearrange the diagrams of Figure 11.23 in Figure 11.24, which shows the continuous-time reference system at the top and the discrete-

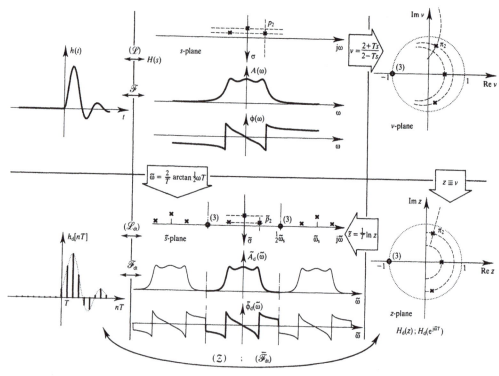

Figure 11.23 Bilinear mapping process.

time results of Figure 11.23 in the middle row of the lower part, identified by $T_0 = T$. The effects of halving and doubling the sampling interval are shown respectively above and below those results.

Halving the sampling interval $T_1 = \frac{1}{2}T$ compresses the v-plane grid towards the v-plane reference point $v = 1$, so that the subsequent z-plane interpretation doubles the sampling frequency $\omega_{s_1} = 2\omega_s$. The overall effect is a lesser compression of the frequency response. Doubling the sampling interval has the converse results, causing very severe compressions of the higher frequency values when fitting them into the available sampling frequency.

A narrowing of the frequency response involves a broadening of the impulse response envelope, as can be observed from the first zero-crossing τ_i of each envelope.

11.6.3 Frequency warping

We saw that the bilinear mapping law fits the entire imaginary axis of the s-plane into one period $\tilde{\omega}_s$ of the imaginary axis of the periodic

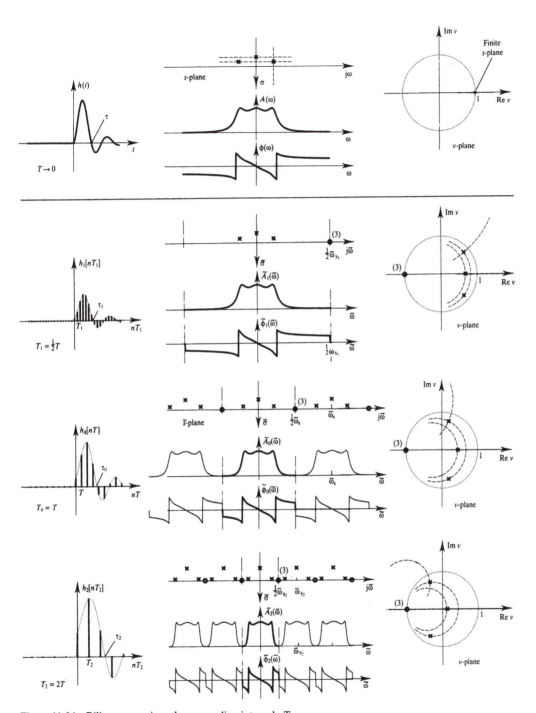

Figure 11.24 Bilinear mapping: three sampling intervals T.

$\tilde{s}$-plane. This causes severe frequency scale distortions called **frequency warping** in the literature.

We now relate points of the two imaginary axes by examining their common location on the unit circle of the z-plane. The standard law $z = e^{\tilde{s}T}$ maps a point $j\tilde{\omega}$ of the $\tilde{s}$-plane's imaginary axis onto a point $z = e^{j\tilde{\omega}T}$ of the unit circle, which in turn is related by the bilinear law 11.31 to some point $s = \sigma + j\omega$ of the s-plane,

$$\sigma + j\omega = \frac{2}{T}\frac{e^{j\tilde{\omega}T} - 1}{e^{j\tilde{\omega}T} + 1}$$

Multiplying numerator and denominator by $e^{-j\tilde{\omega}T/2}$ we have

$$\sigma + j\omega = \frac{2}{T}\frac{e^{j\tilde{\omega}T/2} - e^{-j\tilde{\omega}T/2}}{e^{j\tilde{\omega}T/2} + e^{-j\tilde{\omega}T/2}} = \frac{2}{T}\frac{j\sin\frac{1}{2}\tilde{\omega}T}{\cos\frac{1}{2}\tilde{\omega}T}$$

Equating real and imaginary parts yields $\sigma = 0$ and

$$\omega = \frac{2}{T}\tan\tfrac{1}{2}\tilde{\omega}T \tag{11.42}$$

which confirms that the imaginary axes map onto each other and yields the relationship between continuous and periodic frequency values. Taking the arctangent of both sides yields the inverse relationship

$$\tilde{\omega} = \frac{2}{T}\arctan\tfrac{1}{2}\omega T \tag{11.43}$$

This property of the bilinear method permits a direct mapping of the frequency response $H(j\omega)$ onto the discrete-time frequency response $\tilde{H}(j\tilde{\omega})$ by Equation 11.43, as indicated at the centre of Figure 11.23, bypassing the z-plane.

Graphical interpretation

We interpret the bilinear mapping of the frequency response by means of the graphics of Section 10.6, specifically the form introduced with Figure 10.32. The left side of Figure 11.25 represents the amplitude $A(\omega)$ of the continuous-time frequency response $H(j\omega)$, which is mapped onto the periodic amplitude $\tilde{A}_d(\tilde{\omega})$ on the right. The frequency relationship 11.42 provides the link in the $\omega\tilde{\omega}$-plane of Figure 11.25.

The primary lobe of the tangent function compresses the entire ω-axis into the primary period $(-\pi/T, \pi/T)$ of the $\tilde{\omega}$-axis, thus yielding the primary period of $\tilde{A}_d(\tilde{\omega})$. In the vicinity of the origin the relationship is approximately linear, becoming highly non-linear for increasing frequency, as reflected in the progressive bunching of the hatching lines of Figure 11.25. The shape of the resulting response is similar to that of the original, but compression destroys frequency-related features, such as phase linearity.

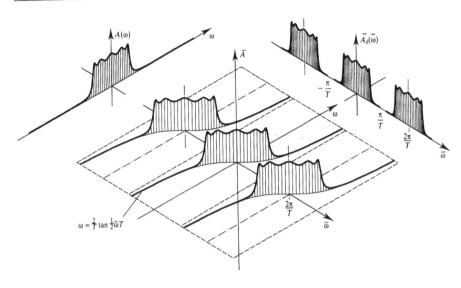

$$\omega = \tfrac{2}{T}\tan\tfrac{1}{2}\tilde{\omega}T$$

Figure 11.25 Bilinear mapping distortions.

Each periodic repetition of the tangent function yields an identical replica on the $\tilde{\omega}$-axis. The great merit of the bilinear law is that each lobe is contained within one sampling period. The infinite lobes do not interact when added, entirely avoiding aliasing.

Comparing the bilinear mapping law with those of Section 10.6, it may be thought of as the limiting case of the progression shown in Figure 10.47, when the number of lobes tends to infinity.

11.6.4 Numerical verification of continuous-time system design

Due to its many desirable properties, the bilinear mapping law makes its way into most computer-aided programs employed for verifying continuous-time system designs. To automate the process the system function $H(s)$ is expressed as a cascaded connection of first-order and second-order modules, as seen in Section 4.4.2. Each module is then simulated by the bilinear mapping law, as indicated in Figure 11.26. The simulated function $\bar{H}(z)$ can then be used to give numerical solutions for arbitrary input signals.

First-order module

Applying the bilinear law 11.31 to the general first-order system

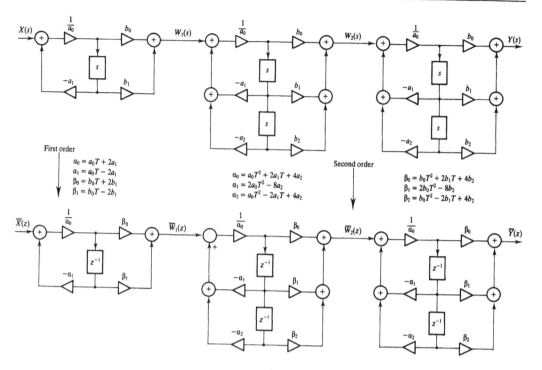

Figure 11.26 Bilinear simulation of basic units.

$$H(s) = \frac{b_1 s + b_0}{a_1 s + a_0}$$

yields

$$\bar{H}(z) = \frac{b_1 2(1 - z^{-1})/T(1 + z^{-1}) + b_0}{a_1 2(1 - z^{-1})/T(1 + z^{-1}) + a_0}$$

which, expressed in the standard form

$$\bar{H}(z) = \frac{\beta_1 z^{-1} + \beta_0}{\alpha_1 z^{-1} + \alpha_0}$$

gives the relationships

$$\alpha_0 = a_0 T + 2a_1 \qquad \beta_0 = b_0 T + 2b_1$$
$$\alpha_1 = a_0 T - 2a_1 \qquad \beta_1 = b_0 T - 2b_1$$

The real numerator coefficients of $H(s)$ determine the real numerator coefficients of $\bar{H}(z)$, and similarly for the denominator coefficients.

Second-order module

The same mapping law applied to the general second-order system

$$H(s) = \frac{b_2 s^2 + b_1 s + b_0}{a_2 s^2 + a_1 s + a_0}$$

gives an expression of the general form of the second-order discrete-time system

$$\bar{H}(z) = \frac{\beta_2 z^{-2} + \beta_1 z^{-1} + \beta_0}{\alpha_2 z^{-2} + \alpha_1 z^{-1} + \alpha_0}$$

and equating coefficients, the relationships

$$\alpha_0 = a_0 T^2 + 2a_1 T + 4a_2$$

$$\alpha_1 = 2a_0 T^2 - 8a_2$$

$$\alpha_2 = a_0 T^2 - 2a_1 T + 4a_2$$

$$\beta_0 = b_0 T^2 + 2b_1 T + 4b_2$$

$$\beta_1 = 2b_0 T^2 - 8b_2$$

$$\beta_2 = b_0 T^2 - 2b_1 T + 4b_2$$

These results are formally similar to those obtained in Example 11.6 for the backward mapping law, but giving different values.

Usage

Having simulated each module, the solution process can be carried out either by successive multiplications of the input function $\bar{X}(z)$ in the frequency domain, or by a sequential application of the recursive time domain process on $x[nT]$ mentioned in Section 8.2.2. In either case the output of each module is taken as the input to the next module.

The process can also be used for designing simple discrete-time systems.

11.7   Formulation of complete design process

The prime objective of this book was to give a systematic treatment of the fundamental principles and methods of continuous- and discrete-time signals and systems, stressing the key role of transforms. Part 3 was structured to exploit formal similarities of the two classes of system to develop related synthesis methods, with an emphasis on frequency mapping. This section brings together and contrasts processes of differ-

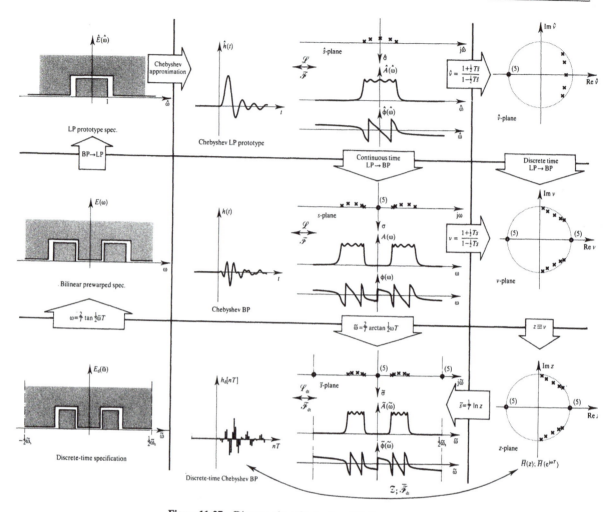

Figure 11.27 Discrete-time Chebyshev BP filter: complete design procedure.

ent types to point out the specific purpose of each and its contribution towards the common design goal.

A typical design problem is formulated in Figure 11.27, which interrelates the necessary design steps. Although the chart looks intricate, most of the processes were covered in two earlier figures. Also, some redundant representations are included for completeness.

We assume the following **design requirements**:

- we wish to design a discrete-time bandpass filter, that fits the specification envelope $E_d(\tilde{\omega})$ of the bottom left corner of Figure 11.27
- the shape of the response shall be of the Chebyshev type

- the sampling interval T is specified
- the bilinear mapping law shall be used for simulation

The basic design process consists essentially of the following steps:

(1) derive the continuous-time lowpass prototype envelope $\hat{E}(\hat{\omega})$ (top row)
(2) fit a prototype Chebyshev approximation into that envelope (top row)
(3) convert the lowpass prototype to a bandpass filter by frequency-mapping (centre)
(4) discrete-time simulate the continuous-time bandpass filter using bilinear mapping (lower right group)

The basic process requires two preliminary steps to convert the specification envelope to the lowpass prototype envelope.

(1) Because the bilinear mapping in step (4) will **warp** the frequency axis, when converting the original specification envelope to continuous time, its frequency axis must be **pre-warped** in anticipation. These distortions are expressed in Figure 11.27 by the complementary mapping laws 11.42 and 11.43,

$$\omega = \frac{2}{T}\tan\tfrac{1}{2}\tilde{\omega}T \qquad \text{and} \qquad \tilde{\omega} = \frac{2}{T}\arctan\tfrac{1}{2}\omega T$$

(2) Because the Chebyshev **lowpass** prototype will be frequency-transformed to a Chebyshev **bandpass** filter, the pre-warped specification must be **bandpass-to-lowpass** transformed by the complementary mapping law, that is, by the pair

$$j\hat{\omega} = f_x(j\omega) \qquad j\omega = f_x^{-1}(j\hat{\omega})$$

This last step, together with the subsequent fitting of the prototype Chebyshev characteristic and mapping back to the bandpass form, was covered in Section 10.6.6. The process was summarized in Figure 10.46 and is now adapted to represent the four cells of the upper left of Figure 11.27.

The last step of the design process, bilinear simulation, was covered in Section 11.6.1. The resulting Figure 11.23 is adapted for the bandpass response, to give the four cells of the lower right of Figure 11.27.

The upper right corner shows an alternative path. Having fitted the required Chebyshev characteristic into the prototype envelope, the bilinear simulation can be performed first, followed by a discrete-time version of the lowpass-to-bandpass frequency mapping. The mapping laws were not given in this text, but are conceptually and formally similar to those of Section 10.6 (Constantinides, 1970).

Exercises

11.1 Draw the block diagram of an Nth-order non-recursive system and show that the multiplier elements b_n, the samples of the impulse response sequence $\{h[nT]\}$ and the polynomial coefficients of the system function $H(z)$ have the same numerical values.

11.2 Draw the block diagram and plot the poles and zeros of a non-recursive system characterized by the impulse response sequence

$$\{h[nT]\} = \ldots, 0, 0, 1, 1, 1, 1, 0, 0, 0, \ldots$$
$$\uparrow$$
$$nT = 0$$

Establish whether it represents a linear-phase system. Plot the frequency response and verify the required z-plane symmetries of the system zeros.

11.3 Describe and sketch the modifications to the time domain and frequency domain processes of Figure 11.6 when the rectangular window $w(t)$ is replaced by a triangular window of the same width $\pm \tau$ and unit magnitude at the origin.

11.4 Use the windowing method described in Figure 11.7, with $\omega_s = 2\omega_c$, to sketch the impulse response $h[nT]$ of a discrete-time FIR linear-phase system resulting from the band-limited Hilbert transformer of Figure 10.14.

11.5 Use the time-sampling method with a sampling interval $T = 0.2$ s to obtain a discrete-time simulation of a system characterized by the impulse response $h(t) = 3e^{-3t} u(t)$. Draw the impulse responses, block diagrams and the s-plane or z-plane representation of both systems.

11.6 Derive a discrete-time simulation of the system function

$$H(s) = \frac{s + 1}{s^2 + 2s + 5}$$

by the time-sampling method, with a sampling frequency $\omega_s = 4\pi$ rad/s. Plot all responses and pole/zero locations to the layout of Figure 11.13 and check that values of $h_d[nT]$ are proportional to corresponding samples of $h(t)$.

11.7 Use the bilinear mapping law (Equations 11.31) for a discrete-time simulation of a continuous-time system described by the differential equation

$$a_1 y'(t) + a_0 y(t) = b_1 x'(t) + b_0 x(t)$$

Arrange the results of equivalent simulation paths according to the layout presented in Figure 11.15 for the backward difference case.

11.8 Contrast the grid distortions gathered in Figure 11.22. Confirm that each of the three mapping laws of Equations 11.29–11.31 represents a one-to-one relationship, by plotting equidistant points of the s-plane's imaginary axis. Show that one of these laws can lead to an unstable discrete-time simulation of a stable continuous-time system.

11.9 With the aid of Figure 11.23 describe the common foundations of the frequency mapping methods. Discuss the process leading from the continuous-time s-plane to the periodic $\tilde{s}$-plane of discrete time and identify the stage at which a specific simulation law is implemented.

11.10 Discuss what effects the sampling interval T of the bilinear mapping law has on the time domain and frequency domain of the simulated system with regard to faithful reproduction of the original system. Does a low sampling frequency lead to aliasing?

11.11 Discuss the meaning of the term 'frequency warping' in the context of the bilinear mapping law and derive the appropriate expression. Explain how its effects are counteracted at the design stage. Give reasons why the term is meaningless in the context of the forward and backward mapping laws.

11.12 Find the relationship between the coefficients of the general second-order continuous-time system and those of the discrete-time system simulated by bilinear mapping. Use these to obtain an alternative simulation of the system function $H(s)$ of Exercise 11.6 for a sampling frequency $\omega_s = 2\pi$ rad/s.

11.13 Obtain a discrete-time simulation of the bandpass filter of Example 10.2 (see end of Chapter 10) using the bilinear mapping law with a sampling interval $T = 1.25 \times 10^{-4}$ s. Interpret the continuous-time system described by Equation 10.58 as a series connection of three second-order bandpass sections with identical numerators and simulate these sections individually. Plot the relevant poles and zeros of the s-plane, z-plane and periodic $\tilde{s}$-plane, as applicable.

11.14 Discuss the overall system design process described in Figure 11.27. Specifically stress the usage of transforms applicable to each signal class appearing in the process and of the frequency mapping laws used to convert systems from one class to another.

Bibliography

The following texts are those which were most influential in shaping this book and they provide additional material. They are not necessarily original sources or latest editions.

Bracewell R. N. (1986). *The Fourier Transform and its Applications* 2nd edn. New York: McGraw-Hill

Brigham E. O. (1988). *The Fast Fourier Transform and its Applications*. Englewood Cliffs NJ: Prentice-Hall

Churchill R. V. and Brown J. W. (1984). *Complex Variables and Applications* 4th edn. New York: McGraw-Hill

Constantinides A. G. (1970). Spectral transformations for digital filters. *Proc IEEE*, **117** (8) pp. 1585–90

Jury E. I. (1964). *Theory and Application of the z-transform Method*. New York: John Wiley

Oberhettinger F. (1957). *Tabellen zur Fourier Transformation*. Berlin: Springer-Verlag

Oppenheim A. V., Willsky A. S. with Young I. T. (1983). *Signals and Systems*. Englewood Cliffs NJ: Prentice-Hall

Papoulis A. (1962). *The Fourier Integral and its Applications*. New York: McGraw-Hill

Papoulis A. (1980). *Circuits and Systems: A Modern Approach*. New York: Holt, Rinehart and Winston

Rabiner L. R. and Gold B. (1975). *Theory and Application of Digital Signal Processing*. Englewood Cliffs NJ: Prentice-Hall

Van Valkenburg M. E. (1982). *Analog Filter Design*, New York: Holt, Rinehart and Winston

Index

Lightning Source UK Ltd.
Milton Keynes UK
UKOW05f1410090216

267963UK00003B/107/P